MEANS MECHANICAL ESTIMATING

Standards and Procedures

MEANS MECHANICAL ESTIMATING

Standards and Procedures

Publisher
E. Norman Peterson, Jr.

Editor In Chief
William D. Mahoney

Senior Editor
John J. Moylan

Contributing Editors
Roger J. Grant
Melville Mossman

Technical Writer
Julia C. Willard

Illustrator
Carl W. Linde

R.S. MEANS COMPANY, INC.
CONSTRUCTION CONSULTANTS & PUBLISHERS
100 Construction Plaza
P.O. Box 800
Kingston, Ma 02364-0800
(617) 747-1270

© 1987

In keeping with the general policy of R.S. Means Company, Inc., its authors, editors, and engineers apply diligence and judgment in locating and using reliable sources for the information published. However, no guarantee or warranty can be given, and all responsibility and liability for loss or damage are hereby disclaimed by the authors, editors, engineers and publisher of this publication with respect to the accuracy, correctness, value and sufficiency of the data, methods and other information contained herein as applied for any particular purpose or use.

No part of this publication may be reproduced, stored in a retrieval system, or transmitted in any form or by any means without prior written permission of R.S. Means Company, Inc.

Printed in the United States of America

10   9   8   7   6   5   4   3

Library of Congress Catalog Card Number 87-404901

ISBN 0-87629-066-7

TABLE OF CONTENTS

FOREWORD	ix
PART I - THE ESTIMATING PROCESS	1
CHAPTER 1: GUIDELINES FOR ESTIMATING	3
CHAPTER 2: TYPES OF ESTIMATES	7
Order of Magnitude Estimates	8
Square Foot and Cubic Foot Estimates	10
Systems (or Assemblies) Estimates	10
Unit Price Estimates	11
CHAPTER 3: BEFORE STARTING THE ESTIMATE	15
CHAPTER 4: THE QUANTITY TAKEOFF	21
CHAPTER 5: PRICING THE ESTIMATE	27
Sources of Cost Information	27
Types of Costs	30
CHAPTER 6: DIRECT COSTS	31
Material	31
Labor	33
Equipment	39
Subcontractors	40
Project Overhead	40
Bonds	41
Sales Tax	44
CHAPTER 7: INDIRECT COSTS	45
Taxes and Insurance	45
Office and Operating Overhead	46
Profit	48
Contingencies	48
CHAPTER 8: THE ESTIMATE SUMMARY	51
CHAPTER 9: PRE-BID SCHEDULING	59
CHAPTER 10: BIDDING STRATEGIES	65
Resource Analysis	65
Market Analysis	65
Bidding Analysis	66
Determining Risk in a New Market Area	66

Analyzing the Bid's Risk	67
Maximizing the Profit-to-Volume Ratio	67
CHAPTER 11: COST CONTROL AND ANALYSIS	**71**
Productivity and Efficiency	79
Overtime Impact	80
Retainage and Cash Flow	82
Life Cycle Costs	86
CHAPTER 12: ESTIMATING WITH A COMPUTER	**87**
Specific Application Programs	88
General Application Programs	100
Applications	102
Summary and Conclusions	108
PART II - COMPONENTS OF MECHANICAL SYSTEMS	**109**
CHAPTER 13: PIPING	**111**
Steel Pipe	112
Screwed Fittings and Flanges for Steel Pipe	117
Steel Weld Fittings and Flanges	120
Grip Type Mechanical Joint Fittings	125
Cast Iron Pipe	126
Cast Iron Soil Pipe Fittings	127
Copper Pipe and Tubing	129
Copper Fittings and Flanges	131
Plastic Pipe and Tubing	134
Fittings for Plastic Pipe and Tubing	135
Preinsulated Pipe or Conduit	137
Valves	138
Pipe Hangers and Supports	141
CHAPTER 14: PUMPS	**149**
CHAPTER 15: PLUMBING SYSTEMS	**151**
Fixtures and Appliances	151
Drains and Carriers	157
CHAPTER 16: FIRE PROTECTION	**163**
Standpipe Systems	163
Sprinkler Systems	165
Chemical, Foam, and Gas Fire Suppression	172
CHAPTER 17: HEATING, VENTILATING, AND AIR CONDITIONING	**177**
Heating Boilers	177
Hydronic Terminal Units	182
Chillers and Hydronic Cooling	185
Air Handling Equipment	191
Fans and Gravity Ventilators	195

CHAPTER 18: DUCTWORK	199
CHAPTER 19: INSULATION	207
CHAPTER 20: AUTOMATIC TEMPERATURE CONTROL	211
CHAPTER 21: SUPPORTS, ANCHORS, AND GUIDES	215
PART III - SAMPLE ESTIMATES	219
CHAPTER 22: USING MEANS MECHANICAL COST DATA	221
Format and Data	221
Section A - Unit Price Costs	223
Section B - Assemblies Cost Tables	235
Section C - Estimating References	237
City Cost Indexes	237
CHAPTER 23: SQUARE FOOT AND SYSTEMS ESTIMATING EXAMPLES	245
Project Description	245
Square Foot Estimating Example	245
Systems Estimating Example	250
The Estimate Summary	285
CHAPTER 24: UNIT PRICE ESTIMATING EXAMPLE	287
Project Description	287
Getting Started	293
Plumbing	293
Fire Protection	303
Heating, Ventilating, and Air Conditioning	315
Alternate Pricing Method	317
Summarizing the Unit Price Estimate	346
APPENDIX	353
INDEX	371

ACKNOWLEDGEMENTS

The editors of *Means Mechanical Estimating* would like to express their appreciation to the following estimators and contractors who contributed to the preparation of this book by sharing their experience and knowledge: Martin E. Ahlstrom, Insulation; Gene E. Corrigan, Sheet Metal; John B. Gwynn, Temperature Control; Fred Thornley, Mechanical Estimator and George H. Twomey, Master Plumber.

FOREWORD

For over 46 years, R.S. Means Co., Inc., has researched and published cost data for the building construction industry. *Means Mechanical Estimating* combines that valuable experience with the senior editor's 42 year involvement with mechanical contracting. This book was written for construction professionals and beginners who desire an understanding of the estimating processes used in the mechanical industry, and is the latest in a series of estimating reference books authored and published by R.S. Means Co., Inc.

Means Mechanical Estimating contains estimating standards and procedures for piping, plumbing, heating, cooling, ventilating, and fire protection. These are dynamic systems, which are constantly in motion serving the structure and its occupants.

Placing boilers, insulating pipes, and other functions accomplished by the mechanical contractor may involve the services of other trades. For this reason, the "total mechanical contractor", much like a general contractor, may employ or subcontract all of the individual mechanical trades, as well as riggers, temperature control workers, electricians, balancing contractors, or excavators. Although there may be thousands of total mechanical contractors, these numbers are far outweighed by the numbers of individual plumbing shops, HVAC firms, sprinkler contractors, and sheet metal contractors accomplishing the bulk of mechanical work today. This book has been designed both to be read in full by the total mechanical contractor, and to be used as a reference tool by the individual trades.

This book is presented in three parts. The first part, "The Estimating Process", defines the types and components of estimates and outlines the progression of each type as a building project advances from an idea to a set of contract plans and specifications. Bidding, purchasing, scheduling, and project management techniques round off this section.

The second part of this book, "Components of Mechanical Systems", describes the individual components used in some or all mechanical systems. This part begins with Chapter 13, "Piping", since pipe is the common material used in all mechanical systems. A description of each component is followed by labor and material pricing and takeoff procedures. Cost adjustment factors are given when appropriate, as well as cost saving "field expedients".

The last part, "Sample Estimates", utilizes a reduced scale set of mechanical building plans as a model for the takeoff and estimating procedures previously discussed, using Means forms and methods to complete square foot, systems, and unit price mechanical estimates. Subcontractor quotes, pricing materials and equipment, and an alternate method of unit pricing are included in "Sample Estimating".

An Appendix of reference information, tables, graphic symbols, and abbreviations concludes *Means Mechancial Estimating*.

Part I
THE ESTIMATING PROCESS

Chapter 1
GUIDELINES FOR ESTIMATING

The term "estimating accuracy" is a relative concept. What is the correct or accurate cost of a given construction project? Is it the total price paid to the contractor? Could another reputable contractor perform the same work for a different cost, whether higher or lower? There is no *one* correct estimated cost for a given project. There are too many variables in construction. At best, the estimator can determine a very close approximation of what his final costs will be. The resulting accuracy of this approximation is directly affected by the amount of detail provided and the amount of time spent on the estimate.

Every cost estimate requires three basic components. The first is the establishment of standard *units of measure*. The second component of an estimate is the determination of the *quantity* of units for each component, an actual counting process: how many linear feet of pipe, how many water closets, pounds of sheet metal, etc. The third component, and perhaps the most difficult to obtain, is the determination of a reasonable *cost* for each unit to purchase and install.

The first element, the designation of measurement units, is the step which determines and defines the level of detail, and thus the degree of accuracy of a cost estimate. In mechanical construction, such units could be as all-encompassing as the HVAC (heating, ventilation, and air conditioning) system per square foot of floor area, or as detailed as a pipe fitting.

Depending on the estimator's intended use, the designation of the "unit" may describe a complete system, or it may imply only an isolated entity. The choice and detail of units also determines the time required to complete an estimate.

The second component of every estimate, the determination of quantity, is more than the counting of units. In construction, this process is called the "material takeoff". In order to perform this function successfully, the estimator should have a working knowledge of the materials, methods, and codes used in mechanical construction. An understanding of the design specification intent is particularly important. This knowledge helps to assure that each quantity is correctly tabulated and that essential items are not forgotten or omitted. The estimator with a thorough knowledge of construction is also more likely to account for all requirements in the estimate whether or not they are called for on the plans or in the specifications. Many of the items to be taken off may not involve any material but, rather, entail labor costs only. Testing is an example of a labor-only item. Experience is, therefore, invaluable to ensure a complete estimate.

The third component is the determination of a reasonable cost for each unit. This aspect of the estimate is significantly responsible for variations in estimating. Rarely do two estimators arrive at exactly the same material cost for a project. However, even if total material costs for an installation are similar for competing contractors, the labor cost estimate for installing that material will account for a variation of bids. Labor costs may vary, due to productivity as well as the pay scales in different geographical areas. The use of specialized equipment can decrease installation time and, therefore, cost. Finally, material prices do fluctuate within the market. These cost differences occur from city to city and even from supplier to supplier in the same town. It is the experienced and well prepared estimator who can keep track of these variations and fluctuations and use them to the best advantage when preparing accurate estimates. Preferential and quantity discounts play a significant part in determining material costs.

This third component of the estimate, the determination of costs, can be defined in three different ways by the estimator. With one approach, the estimator uses a predetermined cost which includes all the elements (i.e., material, installation, overhead, and profit) as one number expressed in dollars per unit. A variation of this approach is to use a unit cost which combines total material and installation only, adding a percent mark-up for overhead and profit to the "bottom line".

A second method is to use unit costs — in dollars — for material and for installation. This is done separately for each item, without mark-ups. These are called "bare costs". Different profit and overhead mark-ups are applied to each before the material and installation prices are added; the result is the total "selling" price.

A third method of unit pricing uses unit costs for materials, and man-hours, man-days or crew-days as the units of labor. Again, these figures are totalled separately; one represents the value for materials (in dollars), the other shows the total man days for installation. The average cost per day of craft labor is determined by allowing for the expected ratios of foremen, journeymen, and apprentices. This is called the "composite labor rate". This rate is multiplied by the total man-days or crew-days to get the total labor cost of installation. Different overhead and profit mark-ups can then be applied to each, material and labor, and the results added to get the total selling price.

The word "*unit*" is used in many ways, as can be seen in the above definitions. Keeping the concepts of units clearly defined is vital to achieving an accurate, professional estimate. For the purposes of this book, the following references to different types of units are used:

- **Unit of Measure:** The standard by which the quantities are counted, such as *linear feet* of pipe, *number* of fixtures, *pounds* of ductwork, *square feet* of blanket insulation.
- **Material Unit Cost:** The cost to purchase each unit of measure: this cost includes any ancillary items that will not be taken off and priced separately, such as pipe nipples, test caps, sheet metal screws, etc.
- **Labor Unit:** The man-hours required to install or fabricate and install a unit of measure. (Note: Labor units multiplied by the labor rate equals the installation unit cost in dollars.)

Chapter 2
TYPES OF ESTIMATES

Building construction estimators use four basic types of estimates. These types may be referred to by different names and may not be recognized by all as definitive. Most estimators, however, will agree that each type has its place in the construction estimating process. The four types of estimates are described below.

Order of Magnitude Estimates: Order of magnitude costs are defined in relation to the usable units that have been designed for a facility. If, for example, a hospital administrator is planning to enlarge a hospital, he needs to know the projected cost per bed. If an estimator knows the quantity of beds in a proposed hospital (or the number of apartments in an apartment building, or the tons of air conditioning) the cost of the project can be estimated. Accuracy may be plus or minus 20%.

Square Foot and Cubic Foot Estimates: This type is most often useful when only the proposed size and use of a planned building is known. This method can be completed within an hour or two. Depending on the source of cost information an accuracy of plus or minus 15% can be expected.

Systems (or Assemblies) Estimates: A Systems Estimate is best used as a budgetary tool in the planning stages of a project when some parameters have been decided (e.g., size of the space, owner's requirements, etc.). This type of estimate could require as much as one day to complete. Because more specific information is known about the project, a plus or minus 10% accuracy can be attained from this estimating method.

Unit Price Estimates: Working drawings and full specifications are required to complete a unit price estimate. It is the most accurate of the four types but is also the most time consuming. Used primarily for bidding purposes, the accuracy of a unit price estimate can be plus or minus 5%.

Figure 2.1 graphically demonstrates the relative relationship of required time versus resultant accuracy of a complete building estimate for each of these four basic estimate types. It should be recognized that, as an estimator *and* his company gain repetitive experience on similar or identical projects, the accuracy of all four types of estimates should improve dramatically. In fact, given enough experience, Square Foot, and Systems estimates may closely approach the accuracy of Unit Price Estimates.

Order of Magnitude Estimates

The Order of Magnitude Estimate, also known as a Conceptual Estimate, can be completed with only a minimum of information. The "units", as described in the first chapter of this book, can be very general for this type and need not be well defined. For example: "The mechanical work for the office building of a small service company in a suburban industrial park will cost about $20,000". This type of statement (or estimate) can be made after a few minutes of thought used to draw upon experience and to make comparisons with similar projects from the past. While this rough figure might be appropriate for a project in one region of the country, substantial adjustments may be required for a change of geographic location due to climate and for cost changes over time, due to, for example, changes of materials, inflation, or code changes.

Figure 2.2, from *Means Mechanical Cost Data*, 1987, includes data for a refined approach to the Order of Magnitude Estimate. This format is based on unit of use. Please note at the bottom of the category "Nursing homes", for example, that costs are given "per bed" or "per person". The proposed use and magnitude of the planned structure — such as the number of beds for a hospital building or the number of apartments in a complex — may be the only parameters known at the time the Order of Magnitude Estimate is done. The data given in Figure 2.2 does not require that details of the proposed project be known in order to determine rough costs; the only required information is the intended use and capacity of the building. What is lacking in accuracy (plus or minus 20%) is more than compensated by the minimal time required to complete the Order of Magnitude Estimate — a matter of minutes. Some mechanical units which are frequently used are "tons of air conditioning", "number of radiators", "number of plumbing fixtures", "CFM of air" to be handled, "number of sprinkler heads", etc.

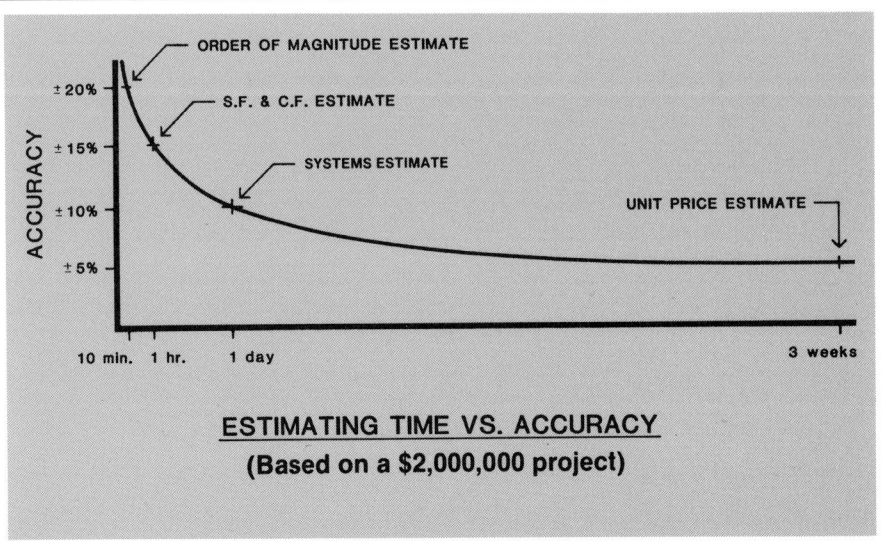

Figure 2.1

17.1 S.F., C.F. and % of Total Costs

				UNIT COSTS			% OF TOTAL			
			UNIT	¼	MEDIAN	¾	¼	MEDIAN	¾	
500	0010	HOUSING Public (low-rise)	S.F.	33.80	47.05	64.20				500
	0020	Total project costs	C.F.	2.85	3.70	4.68				
	2720	Plumbing	S.F.	2.46	3.40	4.35	7.10%	9%	11.50%	
	2730	Heating, ventilating, air conditioning		1.31	2.46	2.79	4.40%	6%	6.40%	
	2900	Electrical		2.15	3.11	4.36	4.90%	6.50%	8.20%	
	3100	Total: Mechanical & Electrical		6.40	9.20	12.40	15.60%	19.20%	22.10%	
	9000	Per apartment, total cost	Apt.	37,600	42,600	53,100				
	9500	Total: Mechanical & Electrical	"	6,300	8,650	10,800				
510	0010	ICE SKATING RINKS	S.F.	32.05	44.75	73.45				510
	0020	Total project costs	C.F.	1.82	2.27	2.68				
	2720	Plumbing	S.F.	.96	1.42	2.17	3.10%	3.20%	4.60%	
	2900	Electrical		2.53	3.32	4.60	5.70%	7%	10.10%	
	3100	Total: Mechanical & Electrical		4.58	6.50	9.80	12.40%	16.40%	25.90%	
520	0010	JAILS	S.F.	103	117	135				520
	0020	Total project costs	C.F.	7.60	9.45	11.85				
	2720	Plumbing	S.F.	6	10.35	12.45	7%	8.30%	12%	
	2770	Heating, ventilating, air conditioning		6.05	10.90	15.80	6.30%	9.40%	12.10%	
	2900	Electrical		8.80	11.70	14.65	7.80%	10.10%	12.40%	
	3100	Total: Mechanical & Electrical		22.60	33.30	42.10	23.20%	29.90%	35.30%	
530	0010	LIBRARIES		57.95	71	88.35				530
	0020	Total project costs	C.F.	4.02	4.87	6.10				
	2720	Plumbing	S.F.	2.40	3.30	4.40	3.60%	4.50%	5.80%	
	2770	Heating, ventilating, air conditioning		5.85	8.30	10.85	8.70%	11%	13.20%	
	2900	Electrical		5.90	7.25	9.65	8.40%	10.90%	12.10%	
	3100	Total: Mechanical & Electrical		12.45	17.30	24.75	19.40%	25.50%	29.40%	
550	0010	MEDICAL CLINICS		55.80	68.30	84.85				550
	0020	Total project costs	C.F.	4.11	5.55	7.20				
	2720	Plumbing	S.F.	3.84	5.30	7.25	6.10%	8.40%	10.20%	
	2770	Heating, ventilating, air conditioning		4.65	6.05	8.80	6.70%	9%	11.70%	
	2900	Electrical		5.25	6.70	8.80	8.10%	9.90%	12%	
	3100	Total: Mechanical & Electrical		12.10	15.55	21.70	19%	24.30%	30.10%	
570	0010	MEDICAL OFFICES		52	64.55	78.05				570
	0020	Total project costs	C.F.	3.94	5.20	7				
	2720	Plumbing	S.F.	3.14	4.66	6.45	5.70%	6.90%	9.40%	
	2770	Heating, ventilating, air conditioning		3.72	5.50	7.05	6.50%	8%	10.40%	
	2900	Electrical		4.43	6.25	8.30	7.60%	9.80%	11.70%	
	3100	Total: Mechanical & Electrical		10.15	14.40	18.85	17.30%	22.40%	27.20%	
590	0010	MOTELS		48.30	50.35	74.70				590
	0020	Total project costs	C.F.	2.96	5.65	6.85				
	2720	Plumbing	S.F.	3.37	4.21	5.15	9.40%	10.60%	12.50%	
	2770	Heating, ventilating, air conditioning		1.78	3.06	5.50	4.90%	5.60%	10%	
	2900	Electrical		3.03	3.80	4.80	7.40%	9.10%	10.40%	
	3100	Total: Mechanical & Electrical		7.40	8.80	12.50	18.60%	23.70%	26.20%	
	9000	Per rental unit, total cost	Unit	13,700	25,500	35,500				
	9500	Total: Mechanical & Electrical	"	3,825	5,125	5,625				
600	0010	NURSING HOMES	S.F.	52.60	68.15	84.05				600
	0020	Total project costs	C.F.	4.17	5.40	7.15				
	2720	Plumbing	S.F.	4.57	5.60	8.30	8.30%	10.30%	14.10%	
	2770	Heating, ventilating, air conditioning		4.89	7.05	8.30	10.60%	11.70%	11.80%	
	2900	Electrical		5.25	6.65	8.20	9.70%	11%	12.50%	
	3100	Total: Mechanical & Electrical		12	16.35	23.60	22%	28.10%	33.20%	
	9000	Per bed or person, total cost	Bed	20,500	26,500	32,900				
610	0010	OFFICES Low-Rise (1 to 4 story)	S.F.	42.15	54.15	71.30				610
	0020	Total project costs	C.F.	3.10	4.34	5.80				
	2720	Plumbing	S.F.	1.61	2.43	3.48	3.60%	4.50%	6.10%	
	2770	Heating, ventilating, air conditioning		3.47	4.79	7	7.20%	10.40%	11.90%	
	2900	Electrical		3.53	4.90	6.76	7.40%	9.60%	11%	
	3100	Total: Mechanical & Electrical		7.25	10.80	15.85	14.50%	20.50%	26.90%	

For expanded coverage of these items see *Means Square Foot Cost Data 1987*

Figure 2.2

Square Foot and Cubic Foot Estimates

The use of Square Foot and Cubic Foot estimates is most appropriate prior to the preparation of plans or preliminary drawings, when budgetary parameters are being analyzed and established. Please refer again to Figure 2.2 and note that costs for each type of project are presented first as "Total project costs" by square foot and by cubic foot. These costs are broken down into different components, and then into the relationship of each component to the project as a whole, in terms of costs per square foot. This breakdown enables the designer, planner, or estimator to adjust certain components according to the unique requirements of the proposed project. The costs on this and other pages of *Means Mechanical Cost Data* were derived from more than 10,500 projects contained in the Means data bank of construction costs. These costs include the contractor's overhead and profit but do not include architectural fees or land costs.

Historical data for square foot costs of new construction are plentiful (see *Means Mechanical Cost Data*, Division 17.1). However, the best source of square foot costs is the estimator's own cost records for similar projects, adjusted to the parameters of the project in question. While helpful for preparing preliminary budgets, Square Foot and Cubic Foot estimates can also be useful as checks against other, more detailed estimates. While slightly more time is required than with Order of Magnitude Estimates, a greater accuracy (plus or minus 15%) is achieved due to a more specific definition of the project.

Systems (or Assemblies) Estimates

One of the primary advantages of Systems (or Assemblies) Estimating is to enable alternate construction techniques to be readily compared for budgetary purposes. Rapidly rising design and construction costs in the past have made budgeting and cost effectiveness studies increasingly important in the early stages of building projects. Never before has the estimating process had such a crucial role in the initial planning process. Unit Price Estimating, because of the time and detailed information required, is not possible as a budgetary or planning tool. A faster and more cost-effective method is needed for the planning phase of a building project; this is the Systems, or Assemblies Estimate.

The Systems method is a logical, sequential approach that reflects how a building is constructed. Twelve "Uniformat" divisions organize building construction into major components that can be used in Systems Estimates. These Uniformat divisions are listed below:

Systems Estimating Divisions:
Division 1 - Foundations
Division 2 - Substructures
Division 3 - Superstructure
Division 4 - Exterior Closure
Division 5 - Roofing
Division 6 - Interior Construction
Division 7 - Conveying
Division 8 - Mechanical
Division 9 - Electrical
Division 10 - General Conditions
Division 11 - Special
Division 12 - Site Work

Each division is further broken down into systems. Division 8, which covers mechanical construction, is comprised of the following major groups of systems: Plumbing, Fire Protection, Heating and Air Conditioning, and Special Systems.

Each system incorporates several different components into an assemblage that is commonly used in construction. Figure 2.3 is an example of a typical system, in this case "Plumbing — Three Fixture Bathroom" (from *Means Mechanical Cost Data*, 1987).

A great advantage of the Systems Estimate is that the estimator/designer is able to substitute one system for another during design development and can quickly determine the relative cost differential. The owner can then anticipate budget requirements before the final details and dimensions are established.

The Systems method does not require the degree of final design detail needed for a Unit Price Estimate, but estimators who use this approach must have a solid background knowledge of construction materials and methods, code requirements, design options, and budget considerations.

The Systems Estimate should not be used as a substitute for the Unit Price Estimate. While the Systems approach can be an invaluable tool in the planning stages of a project, it should be supported by Unit Price Estimating when greater accuracy is required.

Unit Price Estimates

The Unit Price Estimate is the most accurate and detailed of the four estimate types and therefore takes the most time to complete. Detailed working drawings and specifications must be available to the unit price estimator. All decisions regarding the project's design, materials, and methods must have been made in order to complete this type of estimate. Because there are fewer variables, the estimate can be more accurate. Working drawings and specifications are used to determine the quantities of materials, equipment, and labor. Current and accurate unit costs for these items are a necessity. These costs can come from different sources: actual quotations for the particular job from normal sources of supply; manufacturers' price sheets; or prices from an up-to-date industry source book, such as *Means Mechanical Cost Data*.

Because of the detail involved and the need for accuracy, completion of a Unit Price Estimate entails a great deal of time and expense. Unit Price Estimating is best suited for construction bidding. It can also be an effective method for determining certain detailed costs in a conceptual budget or during design development.

Most construction specification manuals and cost reference books, such as *Means Mechanical Cost Data*, compile and present unit price information into the 16 divisions of the "Masterformat" of the Construction Specifications Institute, Inc.

CSI Masterformat Divisions:
Division 1 - General Requirements
Division 2 - Site Work
Division 3 - Concrete
Division 4 - Masonry
Division 5 - Metals

PLUMBING — B8.1-630 — Three Fixture Bathrooms

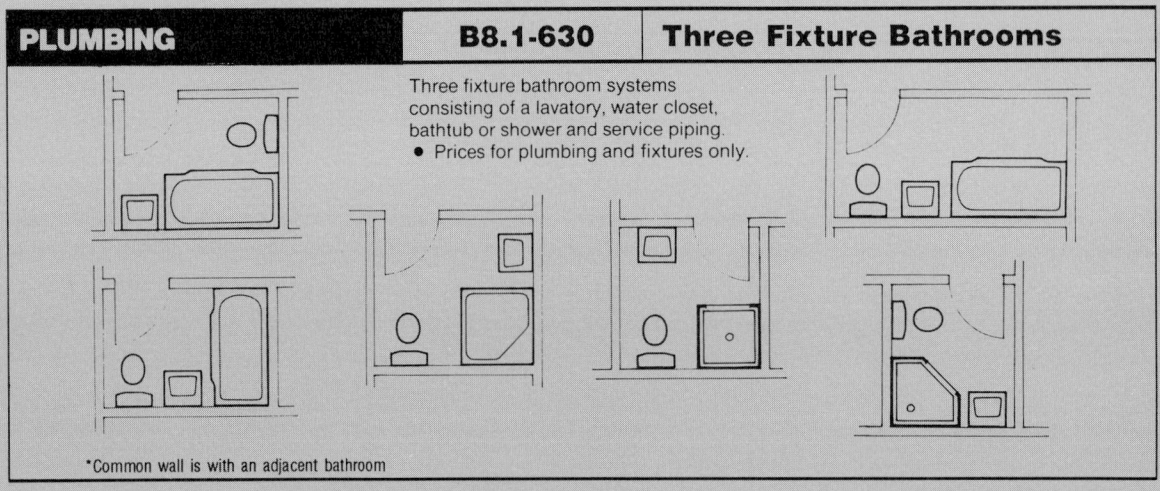

Three fixture bathroom systems consisting of a lavatory, water closet, bathtub or shower and service piping.
- Prices for plumbing and fixtures only.

*Common wall is with an adjacent bathroom

System Components	QUANTITY	UNIT	MAT.	INST.	TOTAL
SYSTEM 08.1-630-1170					
BATHROOM, LAVATORY, WATER CLOSET & BATHTUB					
ONE WALL PLUMBING, STAND ALONE					
Wtr closet, 2 pc close cpld vit china flr mntd w/seat supply & stop	1.000	Ea.	126.50	93.50	220
Water closet, rough-in waste & vent	1.000	Set	90.97	249.03	340
Lavatory w/ftngs, wall hung, white, PE on CI, 20" x 18"	1.000	Ea.	137.50	62.50	200
Lavatory, rough-in waste & vent	1.000	Set	122.21	292.79	415
Bathtub, white PE on CI, w/ftgs, mat bottom, recessed, 5' long	1.000	Ea.	280.50	109.50	390
Baths, rough-in waste and vent	1.000	Set	69.16	250.34	319.50
TOTAL			826.84	1,057.66	1,884.50

8.1-630	Three Fixture Bathroom, One Wall Plumbing	MAT.	INST.	TOTAL
1150	Bathroom, three fixture, one wall plumbing			
1160	Lavatory, water closet & bathtub			
1170	Stand alone	825	1,050	1,875
1180	Share common plumbing wall *	720	800	1,520

8.1-630	Three Fixture Bathroom, Two Wall Plumbing	MAT.	INST.	TOTAL
2130	Bathroom, three fixture, two wall plumbing			
2140	Lavatory, water closet & bathtub			
2160	Stand alone	830	1,075	1,905
2180	Long plumbing wall common *	750	880	1,630
3610	Lavatory, bathtub & water closet			
3620	Stand alone	885	1,225	2,110
3640	Long plumbing wall common *	840	1,100	1,940
4660	Water closet, corner bathtub & lavatory			
4680	Stand alone	1,500	1,075	2,575
4700	Long plumbing wall common *	1,425	800	2,225
6100	Water closet, stall shower & lavatory			
6120	Stand alone	895	1,375	2,270
6140	Long plumbing wall common *	855	1,275	2,130
7060	Lavatory, corner stall shower & water closet			
7080	Stand alone	1,050	1,250	2,300
7100	Short plumbing wall common *	980	890	1,870

Figure 2.3

Division 6 - Wood & Plastics
Division 7 - Moisture-Thermal Control
Division 8 - Doors, Windows & Glass
Division 9 - Finishes
Division 10 - Specialties
Division 11 - Equipment
Division 12 - Furnishings
Division 13 - Special Construction
Division 14 - Conveying Systems
Division 15 - Mechanical
Division 16 - Electrical

Division 15, Mechanical, is further divided into the following subdivisions in *Means Mechanical Cost Data*:

15.1 Pipe and Fittings
15.2 Plumbing Fixtures
15.3 Plumbing Appliances
15.4 Fire Extinguishing Systems
15.5 Heating
15.6 HVAC Piping Specialties
15.7 Air Conditioning and Ventilating

This method of organizing the various components provides a standard of uniformity that is widely used by construction industry professionals: contractors, material suppliers, engineers, and architects. A sample unit price page from the 1987 edition of *Means Mechanical Cost Data* is shown in Figure 2.4. This page lists various types of elements. (Please note that the heading "15.1 Pipe and Fittings" denotes the subdivision classification for these items in Division 15.) Each page contains a wealth of information useful in Unit Price Estimating. The type of work to be performed is described in detail: typical crew make-ups, daily outputs, units of measure, and separate costs for material and installation. Total costs are extended to include the installing contractor's overhead and profit.

Pipe and fittings are installed by plumbers, pipefitters, sprinkler fitters, and gas fitters. Rather than duplicating subdivision 15.1 for each trade, only one trade is shown in the crew makeup in question. Many of the materials in 15.1 are used in each of the mechanical trades.

A fifth type of estimate warrants mention, a refinement of the unit price method. This is the "Scheduling Estimate", which involves the application of realistic manpower allocations. A complete Unit Price Estimate is a prerequisite for the preparation of a Scheduling Estimate. The purpose of the Scheduling Estimate is to determine costs based upon actual working conditions. This is done using data obtained from the Unit Price Estimate. A "human factor" can be applied. For example, if a task requires 7.5 hours to complete, based on unit price data, a tradesman will most likely take 8 hours to complete the work, and will, in any event, be paid for 8 hours work. Costs can be adjusted accordingly. A thorough discussion of scheduling and scheduling estimating is beyond the scope of this book, but a brief overview is included in Chapter Nine; this subject is covered in more detail in *Means Scheduling Manual*, 2nd edition, by F. William Horsley.

13.1	Special Construction	CREW	DAILY OUTPUT	MAN-HOURS	UNIT	BARE COSTS MAT.	LABOR	EQUIP.	TOTAL	TOTAL INCL O&P	
770	7000	Vinyl coated fabric pillow tanks, freestanding, 5000 gallons	4 Clab	4	8	Ea.	1,700	130		1,830	2,050	770
	7100	Supporting embankment not included, 25,000 gallons	6 Clab	2	24		6,000	385		6,385	7,175	
	7200	50,000 gallons	8 Clab	1.50	42.670		8,700	685		9,385	10,600	
	7300	100,000 gallons	9 Clab	.90	80		13,500	1,300		14,800	16,700	
	7400	150,000 gallons		.50	144		17,050	2,325		19,375	22,200	
	7500	200,000 gallons		.40	180		20,500	2,900		23,400	26,800	
	7600	250,000 gallons	↓	.30	240	↓	23,900	3,875		27,775	32,000	

15.1	Pipe & Fittings	CREW	DAILY OUTPUT	MAN-HOURS	UNIT	BARE COSTS MAT.	LABOR	EQUIP.	TOTAL	TOTAL INCL O&P	
010	0010	**AVERAGE** Square foot and percent of total										010
	0100	job cost for plumbing, see division 17.1										
040	0010	**BACKFLOW PREVENTER** Includes gate valves,										040
	0020	and four test cocks, corrosion resistant, automatic operation										
	1000	Double check principle										
	1080	Threaded										
	1100	¾" pipe size	1 Plum	16	.500	Ea.	115	11.50		126.50	145	
	1120	1" pipe size		14	.571		130	13.15		143.15	160	
	1140	1-½" pipe size		10	.800		245	18.40		263.40	295	
	1160	2" pipe size		7	1.140	↓	325	26		351	395	
	1300	Flanged										
	1380	3" pipe size	Q-1	4.50	3.560	Ea.	875	74		949	1,075	
	1400	4" pipe size	"	3	5.330		1,310	110		1,420	1,600	
	1420	6" pipe size	Q-2	3	8	↓	1,990	170		2,160	2,450	
	4000	Reduced pressure principle										
	4100	Threaded										
	4120	¾" pipe size	1 Plum	16	.500	Ea.	155	11.50		166.50	185	
	4140	1" pipe size		14	.571		190	13.15		203.15	230	
	4150	1-¼" pipe size		12	.667		310	15.35		325.35	365	
	4160	1-½" pipe size		10	.800		340	18.40		358.40	400	
	4180	2" pipe size	↓	7	1.140	↓	420	26		446	500	
	5000	Flanged, bronze										
	5060	2-½" pipe size	Q-1	5	3.200	Ea.	1,260	66		1,326	1,475	
	5080	3" pipe size		4.50	3.560		1,600	74		1,674	1,875	
	5100	4" pipe size	↓	3	5.330		2,230	110		2,340	2,625	
	5120	6" pipe size	Q-2	3	8		4,700	170		4,870	5,425	
	5600	Flanged, iron										
	5660	2-½" pipe size	Q-1	5	3.200	Ea.	1,150	66		1,216	1,350	
	5680	3" pipe size		4.50	3.560		1,310	74		1,384	1,550	
	5700	4" pipe size	↓	3	5.330		1,840	110		1,950	2,175	
	5720	6" pipe size	Q-2	3	8		2,810	170		2,980	3,350	
	5740	8" pipe size		2	12		6,140	260		6,400	7,125	
	5760	10" pipe size	↓	1	24	↓	8,140	515		8,655	9,700	
070	0010	**CLEANOUTS**										070
	0060	Floor type										
	0080	Round or square, scoriated nickel bronze top										
	0100	2" pipe size	1 Plum	10	.800	Ea.	34.50	18.40		52.90	65	
	0120	3" pipe size		8	1		36.75	23		59.75	74	
	0140	4" pipe size		6	1.330		50.25	31		81.25	100	
	0160	5" pipe size	↓	4	2		84.75	46		130.75	160	
	0180	6" pipe size	Q-1	6	2.670		84.75	55		139.75	175	
	0200	8" pipe size	"	4	4	↓	97.25	83		180.25	230	
	0340	Recessed for tile, same price										

42

Figure 2.4

Chapter 3
BEFORE STARTING THE ESTIMATE

Selecting a project to bid on can be done in numerous ways. A mechanical contractor may be invited to bid by an architect, general contractor, or an owner. The contractor may hear of a project from a construction publication or an advertisement for bids from a local newspaper. Whichever source is used, the process may involve weeks of hard work with only a chance of bidding success. It can also mean the opportunity to obtain a contract for a successful and lucrative project. It is not uncommon for the contractor to bid ten or more jobs in order to win just one. The successful bid depends heavily upon estimating accuracy and thus, the preparation, organization, and care that go into the estimating process. (Conversely, if a job is won due to omissions in the estimate, it is likely *not* to be a "successful" bid.)

The first step before starting the estimate is to obtain copies of the plans and specifications in *sufficient quantities*. Most estimators mark up plans with colored pencils and make numerous notes and references. For one estimator to work from plans that have been used by another is difficult at best and may easily lead to errors. Most often, only one complete set is provided to the mechanical contractor. Additional sets must be purchased, if needed. When more than one estimator works on a particular project, especially if they are working from the same plans, careful coordination is required to prevent omissions as well as duplications. Color coding of the various piping circuits, when neatly done, can be used to mark the takeoff progress and will allow for another estimator to finish a takeoff, if necessary.

The estimator should be aware of and note any instructions to bidders, which may be included in the specifications. To avoid future confusion, the bid's due date, time, and place should be clearly stated and understood upon receipt of the construction documents (plans and specifications). The due date should be marked on a calendar to avoid conflicting bid dates and closing times. Completion of the takeoff should be made as soon as possible, not two days prior to the bid deadline.

Utility companies should be contacted to obtain the service charges, and the local municipal building department contacted for inspection charges, permit fees, and building service charges.

If bid security, or a bid bond, is required, it should be arranged for at this time, especially if the bonding capability or capacity of a contractor is limited or has not previously been established. If a bond or bid security is to be based on a percentage of the bid, then a preliminary square foot or systems estimate must be prepared to figure the amount of the bid security. This allows the bonding company to analyze the project up front, and thus prevents the contractor from estimating and bidding on a job that cannot be bonded.

The estimator should attend any pre-bid meetings with the owner or architect, preferably *after* reviewing the plans and specifications. Important points are often brought up at such meetings and details clarified. Attendance is important, not only to show the owner your interest in the project, but also to assure equal and competitive bidding. For many projects, attendance is required or the bid will not be accepted. It is to the estimator's advantage to examine and review the plans and specifications before any such meetings and before the initial site visit. It is important to become familiar with the project as soon as possible.

All contract documents should be read thoroughly. They exist to protect all parties involved in the construction process. The contract documents are written so that the contractors will be bidding equally and competitively, and to ensure that all items in a project are included. The contract documents protect the designer (the architect or engineer) by ensuring that all work is supplied and installed as specified. The owner also benefits from thorough and complete construction documents by being assured of a quality job and a complete, functional project. Finally, the contractor benefits from good contract documents because the scope of work is well defined, eliminating the gray areas of what is implied but not stated. "Extras" are more readily avoided. Change orders, if required, are accepted with less argument if the original contract documents are complete, well stated, and most importantly, read by all concerned parties. The Appendix of this book contains a reproduction of the two mechanical pages from SPEC-AID, an R. S. Means Co., Inc. publication from *Means Forms for Building Construction Professionals*. This SPEC-AID was created not only to assist designers and planners when developing project specifications, but also as an aid to the contractor's estimator. The mechanical section of the SPEC-AID lists some some typical mechanical components and variables that are normally included in building construction. The estimator can use this form as a means of outlining a project's requirements, and as a checklist to be sure that all items have been included. A preprinted summary sheet (shown in Figure 3.1) serves both as the SPEC-AID and as a pricing estimate sheet.

During the first review of the specifications, any items to be priced should be identified and noted. All work to be sub-contracted should be examined for "related work" required from other trades or contractors. Such work is usually referenced and described in a thorough project specification. "Work by others" or "Not in Contract" should be clearly defined both on the drawings and in the specifications. Certain materials are sometimes specified by the designer and purchased by the owner to be installed (labor only) by the contractor. These items should be noted and the responsibilities of each party clearly understood. An example of such an item might involve allocating responsibility for receiving, temporary storage, and protection of restaurant fixtures furnished by others but installed by plumbers.

The "General Conditions", "Supplemental Conditions", and "Special Conditions" sections of the specifications should be examined carefully by

NAME OF JOB:	DUE DATE:
LOCATION:	BID SENT TO:
ARCHITECT:	ENGINEER:

		1	2
	Labor w/Payroll Taxes, Welfare		
	Travel		
	Fixtures		
	Drains, Carriers, etc.		
	Sanitary - Interior		
	Storm - Interior		
	Water - Interior		
	Valves - Interior		
	Water Exterior - Street Connection		
	Water Exterior - Extension to Building		
	Sanitary & Storm Exterior - Street Connection		
	Sanitary & Storm Exterior - Extension to Building		
	Water Meter		
	Gas Service		
	Gas Interior		
	Accessories		
	Flashings for: Vents, Floor & Roof Drains, Pans		
	Hot Water Tanks: Automatic w/trim		
	Hot Water Tanks: Storage, w/stand and trim		
	Hot Water Circ. Pumps, Aquastats, Mag. Starters		
	Sump Pumps and/or Ejectors; Controls, Mag. Starters		
	Flue Piping		
	Rigging, Painting, Excavation and Backfill		
	Therm. Gauges, P.R.V. Controls, Specialties, etc.		
	Record Dwgs., Tags and Charts		
	Fire Extinguishers, Equipment, and/or Standpipes		
	Permits		
	Sleeves, Inserts and Hangers		
	Staging		
	Insulation		
	Acid Waste System		
	Sales Tax		
	Trucking and Cartage		

Figure 3.1

all parties involved in the project. These sections describe the items that have a direct bearing on the proposed project, but may not be part of the actual, physical installation. Temporary heat and water are examples of these kinds of items. Also included in these sections is information regarding completion dates, payment schedules (e.g. retainage), submittal requirements, allowances, alternates, and other important project requirements. Each of these conditions can have a significant bearing on the ultimate cost of the project. They must be read and understood prior to performing the estimate. The most popular General Conditions will soon be memorized by an estimator, who will then need only to skim through them looking for any deletions or changes. *Study* the supplemental or special conditions, as these are prepared for each particular job.

While analyzing the plans and specifications, the estimator should evaluate the different portions of the project to determine which areas warrant the most attention. The estimator should focus first on those items which represent the largest cost centers of the project or which entail the greatest risk. These cost centers are not always the portions of the job that require the most time to estimate but are those items that will have the most significant impact on the estimate.

Other sections of the specifications should be read because items such as electric motor controls may be duplicated under the mechanical and electrical specifications.

Figure 3.2 from Division 17 of *Means Mechanical Cost Data*, 1987, is a chart showing typical percentages of mechanical and electrical work relative to projects as a whole. These charts have been developed to represent the overall average percentages for new construction. Commonly used building types are included.

When the overall scope of the work has been identified, the drawings should be examined to confirm the information in the specifications. This is the time to clarify details while reviewing the general content. The estimator should note which sections, elevations, and detail drawings are for which plans. At this point and throughout the whole estimating process, the estimator should note and list any discrepancies between the plans and specifications, as well as any possible omissions. It is often stated in bid documents that bidders are obliged to notify the owner or architect/engineer of any such discrepancies. When so notified, the designer will most often issue an addendum to the contract documents in order to properly notify all parties concerned and to assure equal and competitive bidding. Competition can be fair only if the same information is provided for all bidders. Notifying services such as F.W. Dodge of intent to bid will increase receipt of sub-bids and suppliers' material quotations.

Once familiar with the contract documents, the estimator should solicit bids by notifying appropriate subcontractors, manufacturers, and vendors. Those whose work may be affected by the site conditions should accompany the estimator on a job site visit (especially in cases of renovation and remodeling, where existing conditions can have a significant effect on the cost of a project). This notification is usually done by post card or telephone call. It may also be beneficial to inform these vendors of other locations where the bid documents can be reviewed, to avoid tie-up of your plans or plan room.

17.1 S.F., C.F. and % of Total Costs

			UNIT	UNIT COSTS ¼	UNIT COSTS MEDIAN	UNIT COSTS ¾	% OF TOTAL ¼	% OF TOTAL MEDIAN	% OF TOTAL ¾	
500	0010	HOUSING Public (low-rise)	S.F.	33.80	47.05	64.20				500
	0020	Total project costs	C.F.	2.85	3.70	4.68				
	2720	Plumbing	S.F.	2.46	3.40	4.35	7.10%	9%	11.50%	
	2730	Heating, ventilating, air conditioning		1.31	2.46	2.79	4.40%	6%	6.40%	
	2900	Electrical		2.15	3.11	4.36	4.90%	6.50%	8.20%	
	3100	Total: Mechanical & Electrical		6.40	9.20	12.40	15.60%	19.20%	22.10%	
	9000	Per apartment, total cost	Apt.	37,600	42,600	53,100				
	9500	Total: Mechanical & Electrical	"	6,300	8,650	10,800				
510	0010	ICE SKATING RINKS	S.F.	32.05	44.75	73.45				510
	0020	Total project costs	C.F.	1.82	2.27	2.68				
	2720	Plumbing	S.F.	.96	1.42	2.17	3.10%	3.20%	4.60%	
	2900	Electrical		2.53	3.32	4.60	5.70%	7%	10.10%	
	3100	Total: Mechanical & Electrical		4.58	6.50	9.80	12.40%	16.40%	25.90%	
520	0010	JAILS	S.F.	103	117	135				520
	0020	Total project costs	C.F.	7.60	9.45	11.85				
	2720	Plumbing	S.F.	6	10.35	12.45	7%	8.30%	12%	
	2770	Heating, ventilating, air conditioning		6.05	10.90	15.80	6.30%	9.40%	12.10%	
	2900	Electrical		8.80	11.70	14.65	7.80%	10.10%	12.40%	
	3100	Total: Mechanical & Electrical		22.60	33.30	42.10	23.20%	29.90%	35.30%	
530	0010	LIBRARIES		57.95	71	88.35				530
	0020	Total project costs	C.F.	4.02	4.87	6.10				
	2720	Plumbing	S.F.	2.40	3.30	4.40	3.60%	4.50%	5.80%	
	2770	Heating, ventilating, air conditioning		5.85	8.30	10.85	8.70%	11%	13.20%	
	2900	Electrical		5.90	7.25	9.65	8.40%	10.90%	12.10%	
	3100	Total: Mechanical & Electrical		12.45	17.30	24.75	19.40%	25.50%	29.40%	
550	0010	MEDICAL CLINICS		55.80	68.30	84.85				550
	0020	Total project costs	C.F.	4.11	5.55	7.20				
	2720	Plumbing	S.F.	3.84	5.30	7.25	6.10%	8.40%	10.20%	
	2770	Heating, ventilating, air conditioning		4.65	6.05	8.80	6.70%	9%	11.70%	
	2900	Electrical		5.25	6.70	8.80	8.10%	9.90%	12%	
	3100	Total: Mechanical & Electrical		12.10	15.55	21.70	19%	24.30%	30.10%	
570	0010	MEDICAL OFFICES		52	64.55	78.05				570
	0020	Total project costs	C.F.	3.94	5.20	7				
	2720	Plumbing	S.F.	3.14	4.66	6.45	5.70%	6.90%	9.40%	
	2770	Heating, ventilating, air conditioning		3.72	5.50	7.05	6.50%	8%	10.40%	
	2900	Electrical		4.43	6.25	8.30	7.60%	9.80%	11.70%	
	3100	Total: Mechanical & Electrical		10.15	14.40	18.85	17.30%	22.40%	27.20%	
590	0010	MOTELS		48.30	50.35	74.70				590
	0020	Total project costs	C.F.	2.96	5.65	6.85				
	2720	Plumbing	S.F.	3.37	4.21	5.15	9.40%	10.60%	12.50%	
	2770	Heating, ventilating, air conditioning		1.78	3.06	5.50	4.90%	5.60%	10%	
	2900	Electrical		3.03	3.80	4.80	7.40%	9.10%	10.40%	
	3100	Total: Mechanical & Electrical		7.40	8.80	12.50	18.60%	23.70%	26.20%	
	9000	Per rental unit, total cost	Unit	13,700	25,500	35,500				
	9500	Total: Mechanical & Electrical	"	3,825	5,125	5,625				
600	0010	NURSING HOMES	S.F.	52.60	68.15	84.05				600
	0020	Total project costs	C.F.	4.17	5.40	7.15				
	2720	Plumbing	S.F.	4.57	5.60	8.30	8.30%	10.30%	14.10%	
	2770	Heating, ventilating, air conditioning		4.89	7.05	8.30	10.60%	11.70%	11.80%	
	2900	Electrical		5.25	6.65	8.20	9.70%	11%	12.50%	
	3100	Total: Mechanical & Electrical		12	16.35	23.60	22%	28.10%	33.20%	
	9000	Per bed or person, total cost	Bed	20,500	26,500	32,900				
610	0010	OFFICES Low-Rise (1 to 4 story)	S.F.	42.15	54.15	71.30				610
	0020	Total project costs	C.F.	3.10	4.34	5.80				
	2720	Plumbing	S.F.	1.61	2.43	3.48	3.60%	4.50%	6.10%	
	2770	Heating, ventilating, air conditioning		3.47	4.79	7	7.20%	10.40%	11.90%	
	2900	Electrical		3.53	4.90	6.76	7.40%	9.60%	11%	
	3100	Total: Mechanical & Electrical		7.25	10.80	15.85	14.50%	20.50%	26.90%	

For expanded coverage of these items see *Means Square Foot Cost Data 1987*

Figure 3.2

During a site visit, the estimator should take notes, and possibly photographs, of any unusual situations pertinent to the construction and, thus, to the project estimate. If unusual site conditions exist, or if questions arise during the takeoff, a second site visit is recommended.

In some areas, questions are likely to arise that cannot be answered clearly by the plans and specifications. It is crucial that the owner or responsible party be notified quickly, preferably in writing, so that these questions may be resolved before unnecessary problems arise. Often such items involve more than one contractor and can only be resolved by the owner or architect/engineer. A proper estimate cannot be completed until all such questions are answered.

Chapter 4
THE QUANTITY TAKEOFF

The quantity takeoff is the cornerstone of construction bidding, therefore, it should be organized so that the information gathered can be used to future advantage. Purchasing and scheduling can be made easier if items are taken off and listed by system, material type, construction phase, or floor. Material purchasing will similarly benefit.

Much of the methods and materials utilized by the mechanical trades are "understood" or standardized, and not usually indicated on the drawings. Thus, in the quantity takeoff, the mechanical estimator must count not only the items shown on the drawings but must envision the completed design and include the proper number of fittings, hangers, shields, inserts, bolts, nuts, and gaskets, to name but a few. The estimator should also be familiar with or have access to all codes and regulations. To effectively cover all aspects of the project, certain steps should be followed.

Units for each item should be used consistently throughout the whole project — from takeoff to cost control. If the units in a takeoff are consistent, the original estimate can be equitably compared to progress reports and final cost reports (See Chapter 11, "Cost Control"). Part Two of this book is devoted to descriptions of over 80 mechanical components. In that section, a takeoff procedure is suggested for the components. Typical material and labor units are also given for component installation. Each material and labor unit consists of a list of items that are generally included in the cost of the component. Also given are the units by which the component is measured or counted.

Quantities should be taken off by one person if the project is not too large and if time allows. For larger projects, the plans are often split and the work assigned to two or more estimators. In this case, a project leader ought to be assigned to coordinate and assemble the estimate.

Traditionally, quantities are taken off from the drawings in the same sequence as they are erected or installed. Recently, however, the sequence of takeoff and estimating is more often based on the systems or type of work. In this case, preference in the estimating process is given to certain items or components based on the relative costs.

When working with the plans during the quantity takeoff, consistency is a very important consideration. If each job is approached in the same manner, a pattern will develop, such as moving from the lower floors to the top, clockwise or counterclockwise. The choice of method is not important, but consistency is. The purpose of being consistent is to avoid duplications as well as omissions and errors. Pre-printed forms provide an excellent means for developing consistent patterns. Figures 4.1 and 4.2 are two examples of such forms. These Quantity Sheets are designed purely for quantity accumulation. Parts Two and Three of this book contain examples of the use of quantity sheets. Figure 4.3, a Consolidated Estimate sheet, is designed to be used for both quantity takeoff and pricing on one form. There are many other variations.

Every contractor could benefit from designing custom company forms. If employees of a company use the same types of forms, communications and coordination of the estimating process will proceed more smoothly. One estimator will be able to more easily understand the work of another. R.S. Means has published a book completely devoted to forms and their use, entitled *Means Forms for Building Construction Professionals*. Scores of forms, examples, and instructions for use are included.

Appropriate and easy-to-use forms are the first, and most important, of the "tools of the trade" for estimators. Other tools useful to the estimator include tapes, scales, rotometers, mechanical counters, and colored pencils.

A number of other short cuts can be used for the quantity takeoff. If approached logically and systematically, these techniques help to save time without sacrificing accuracy. Consistent use of accepted abbreviations saves the time of writing things out. An abbreviations list similar to those that appear in the Appendix of this book might be posted in a conspicuous place for each estimator to provide a consistent pattern of definitions for use within an office.

All dimensions — whether printed, measured, or calculated — that can be used for determining quantities of more than one item should be listed on a separate sheet and posted for easy reference. Posted gross dimensions can also be used to quickly check for order of magnitude errors.

Rounding off, or decreasing the number of significant digits, should be done only when it will not statistically affect the resulting product. The estimator must use good judgement to determine instances when rounding is appropriate. An overall two or three percent variation in a competitive market can often be the difference between getting or losing a job, or between profit or no profit. The estimator should establish rules for rounding to achieve a consistent level of precision. In general, it is best not to round numbers until the final summary of quantities. Pennies should be rounded off, and, when appropriate, dollars can be rounded off to the closest five or ten.

The final summary is also the time to convert units of measure into standards for practical use (linear feet of copper tube to twenty foot divisions, for example). This is done to keep the numerical value of the unit equitable to what will be purchased and handled.

Be sure to quantify (count) and include "labor only" items that are not shown on the plans. Such items may or may not be indicated in the specifications and might include cleanup, special labor for handling materials, testing, code inspectors, etc.

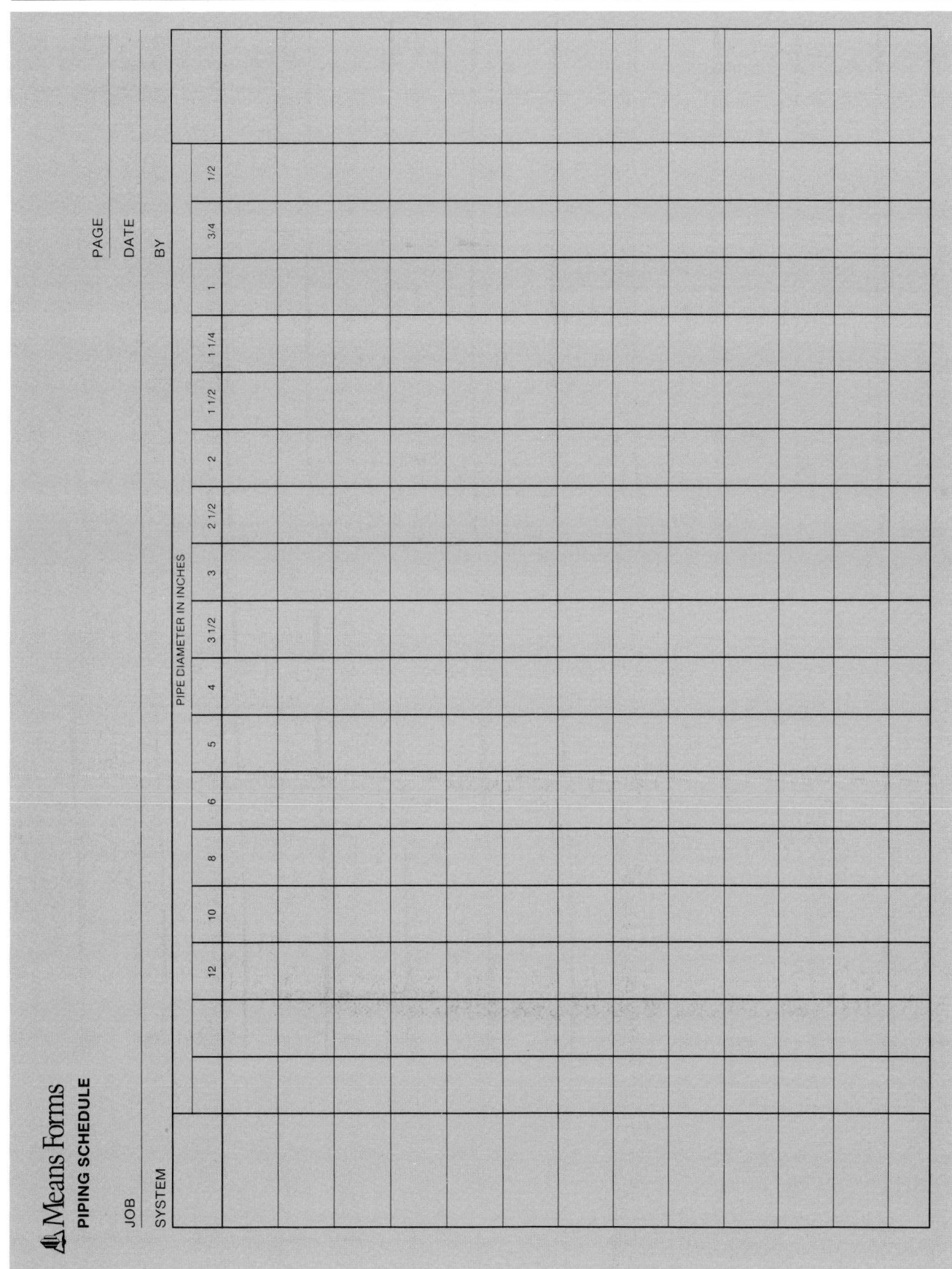

Figure 4.1

Means Forms

DUCTWORK SCHEDULE

JOB _____ PAGE _____ OF _____
SYSTEM _____ DATE _____
MATERIAL _____ BY _____
                              DRAWING NO. _____

| DUCT SIZE | | LINING | INSUL. | GAUGE | LENGTH | TOTAL LENGTH | LBS./FOOT | TOTAL POUNDS |
HEIGHT	WIDTH							

Figure 4.2

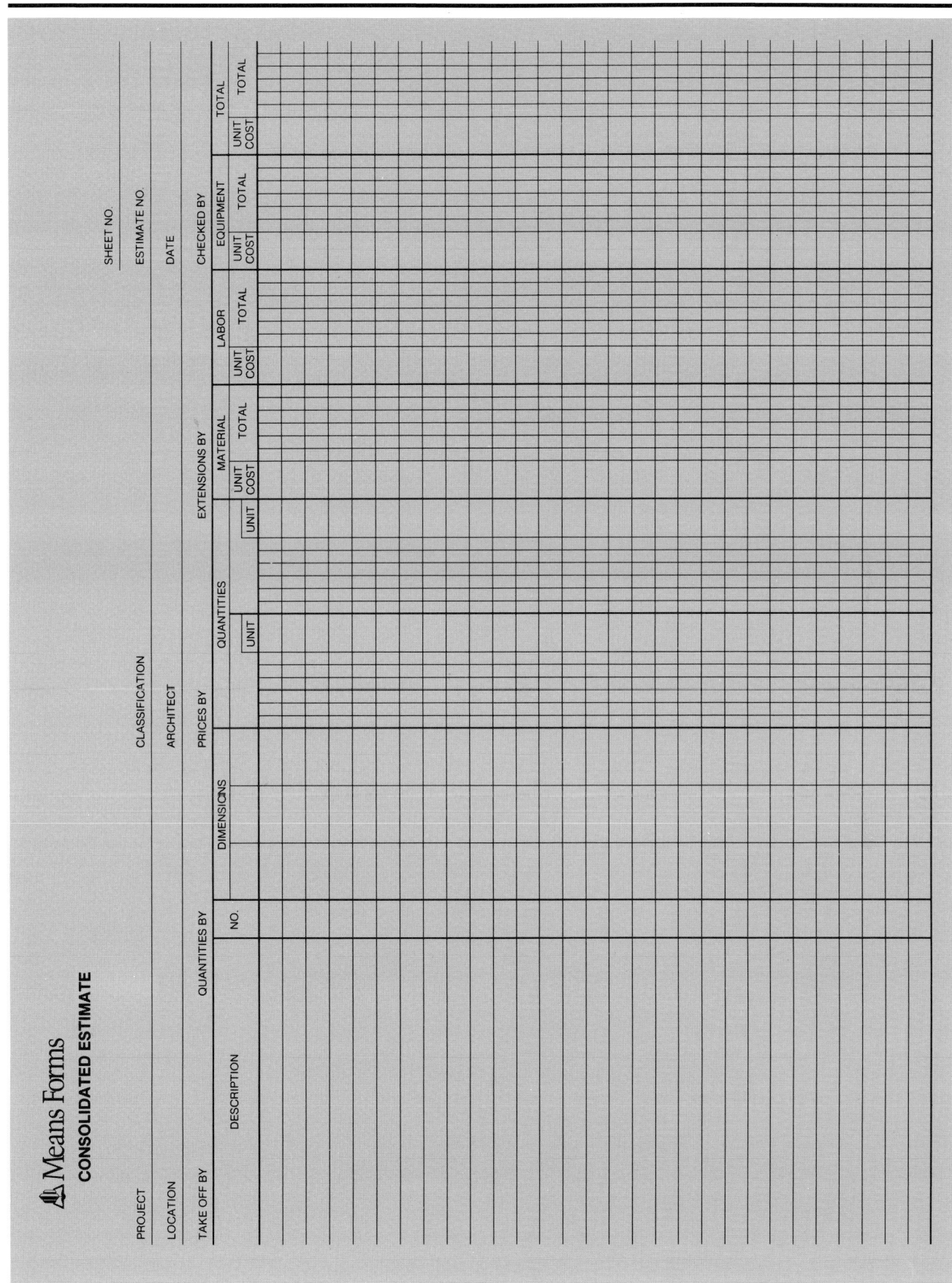

Figure 4.3

The following list summarizes the aforementioned suggestions plus a few more guidelines which will be helpful during the quantity take off:

- Use preprinted forms.
- Use only the front side of each piece of paper.
- Transfer carefully when copying numbers from one sheet to the next.
- List dimensions (width, length) in a consistent order.
- Verify the scale of drawings before using them as a basis for measurement, a good check is to scale off a bath tub, (usually five feet), a light fixture (two or four feet), or a doorway (three feet.)
- Mark drawings neatly and consistently as quantities are counted.
- Be alert for changes in scale, or notes such as "N.T.S." (not to scale). Sometimes these drawings have been photographically reduced.
- Include required items which may not appear in the plans and specs.
- Be alert for discrepancies between the plans and the specifications.

And perhaps the four most important points:

- Print legibly.
- Be organized.
- Use common sense.
- Be consistent.

Chapter 5
PRICING THE ESTIMATE

When the quantities have been counted, values, in the form of unit costs, must be applied and project burdens (overhead and profit) added in order to determine the total selling price (the quote). Depending upon the chosen estimating method (based on the degree of accuracy required) and the level of detail, these unit costs may be direct or "bare", or may include overhead, profit, or contingencies. In Unit Price Estimating, the unit costs most commonly used are "bare", or "unburdened". Most extend labor quantities as man-hours or man-days and price material quantities in dollars. When the total man-days are determined, it is necessary to multiply these figures by the appropriate daily rate, before adding mark-ups. Items such as overhead and profit are usually added to the total direct costs at the estimate summary.

Sources of Cost Information

One of the most difficult aspects of the estimator's job is determining accurate and reliable bare cost data. Sources for such data are varied, but can be categorized in terms of their relative reliability. The most reliable of any cost information is direct quotations from usual sources of supply for this particular job, and accurate, up-to-date, well-kept records of completed work by the estimator's own company. There is no better cost for a particular construction item than the *actual* cost to the contractor of that item from another recent job, modified (if necessary) to meet the requirements of the project being estimated.

Bids from responsible subcontractors are the next most reliable source of cost data. Any estimating inaccuracies are essentially absorbed by the subcontractor. A subcontract bid is a known, fixed cost, prior to the project. Whether the price is "right" or "wrong" does not matter (as long as it is a responsible competitive bid with no apparent errors). The bid is what the appropriate portion of work will cost. No estimating is required of the prime contractor, except for possible verification of the quote and comparison with other subcontractors' quotes.

Quotations by vendors for material costs are, for the same reasons, as reliable as subcontract bids. In this case, however, the estimator must apply estimated labor costs. Thus the "installed" price for a particular item may be more variable.

Whenever possible, all price quotations from vendors or subcontractors should be confirmed in writing. Qualifications and exclusions should be clearly stated. The items quoted should be checked to be sure that they are complete and as specified. One way to assure these requirements is to prepare a form on which all subcontractors and vendors must submit their quotations. This form, generally called a "request for quote", can suggest all of the appropriate questions. It also provides a format to organize the information needed by the estimator and, later, the purchasing agent. This technique can be especially useful to the estimator in a smaller organization, since he must often act as purchasing agent as well.

The above procedures are ideal, but in the realistic haste of estimating and bidding, quotations are often received verbally, either in person or by telephone. The importance of gathering all pertinent information is heightened because omissions are more likely. A preprinted form, such as the one shown in Figure 5.1, is essential to assure that all required information and qualifications are obtained and understood. How often has the subcontractor or vendor said, "I didn't know that I was supposed to include that"? With the help of such forms, the appropriate questions may be covered.

If the estimator has no cost records for a particular item and is unable to obtain a quotation, then another reliable source of price information is current unit price cost books such as *Means Mechanical Cost Data*. R. S. Means presents all such data in the form of national averages; these figures can be adjusted to local conditions, a procedure that will be explained in Part Three of this book. In addition to being a source of primary costs, unit price books are useful as a reference or cross-check for verifying costs obtained from unfamiliar sources.

Lacking cost information from any of the above-mentioned sources, the estimator may have to rely on experience and personal knowledge of the field to develop costs.

No matter which source of cost information is used, the system and sequence of pricing should be the same as that used for the quantity takeoff. This consistent approach should continue through both accounting and cost control during work on the project.

Figure 5.1

Types of Costs

All costs included in a Unit Price Estimate can be divided into two types: direct and indirect. Direct costs are those dedicated solely to the physical construction of a specific project. Material, labor, equipment, and subcontract costs, as well as project overhead costs are all direct.

Types of Costs in a Construction Estimate:

Direct Costs	Indirect Costs
Material	Taxes
Labor	Insurance
Equipment	Office Overhead
Subcontractors	Profit
Project Overhead	Contingencies
Sales tax	
Bonds	

Indirect costs are usually added to the estimate at the summary stage and are most often calculated as a percentage of the direct costs. They include such items as taxes, insurance, overhead, profit, and contingencies. The indirect costs account for great variation in estimates among different bidders.

A clear understanding of direct and indirect cost factors is a fundamental part of pricing the estimate. Chapters 6 and 7 address the components of direct and indirect costs in detail.

Chapter 6
DIRECT COSTS

Direct costs can be defined as those necessary for the completion of the project, in other words, the hard costs. Material, labor, and equipment are among the more obvious items in this category. While subcontract costs include the overhead and profit (indirect costs) of the subcontractor, they are considered to be direct costs to the prime mechanical contractor. Also included are certain project overhead costs for items that are necessary for construction. Examples are a storage trailer, telephone, tools, and possibly temporary power and lighting. Sales tax and bonds are additional direct costs, since they are essential for the performance of the project. On most jobs, items such as temporary lighting and temporary sanitary facilities are stated in the Supplementary General Conditions as being the responsibility of the general or prime contractor.

Material

When quantities have been carefully taken off, estimates of material cost can be very accurate. For a high level of accuracy, the material unit prices must be reliable and current. The most reliable source of material costs is a quotation from a vendor for the particular job in question. Ideally, the vendor should have access to the plans and specifications for verification of quantities and specified products. Providing material quotes (and submittals for approval) is a service which most suppliers will perform.

Material pricing appears relatively simple and straightforward. There are, however, certain considerations that the estimator must address when analyzing material quotations. The reputation of the vendor is a significant factor. Can the vendor "deliver", both figuratively and literally? Estimators may choose not to rely on a "competitive" lower price from an unknown vendor, but will instead use a slightly higher price from a known, reliable vendor. Experience is the best judge for such decisions.

There are many other questions that the estimator should ask. How long is the price guaranteed? When does the guarantee begin? Is there an escalation clause? Does the price include delivery charges or sales tax, if required? Where is the point of FOB? This can be an extremely important factor. Are guarantees and warranties in compliance with the specification requirements? Will there be adequate and appropriate storage space available? If not, can staggered shipments be made? Note that most of these questions can be addressed on the form shown in Chapter 5, Figure 5.1. More information should be obtained, however, to assure that a quoted price is accurate and competitive.

The estimator must be sure that the quotation or obtained price is for the materials as per plans and specifications. Architects and engineers may write into the specifications that: a) a particular type or brand of product must be used with no substitution, b) a particular type or brand of product is specified, but alternate brands of equal quality and performance may be accepted *upon approval,* or c) no particular type or brand is specified. Depending upon the options, the estimator may be able to find an acceptable, less expensive alternative. In some cases, these substitutions can substantially lower the cost of a project. Note also that most specifications will require that "catalogue cuts" or "shop drawings" be submitted for certain materials as part of the contract conditions. In this case, there is pressure on the estimator to obtain the lowest possible price on materials that he believes will meet the specified criteria and gain the architect's approval. A typical submittal or transmittal cover sheet is shown in Figure 6.1.

When the estimator has received material quotations, there are still other considerations which should have a bearing on the final choice of a vendor. Lead time — the amount of time between order and delivery — must be determined and considered. It does not matter how competitive or low a quote is if the material cannot be delivered to the job site on time to support the schedule. If a delivery date is promised, is there a guarantee, or a penalty clause for late delivery?

Delivery and lead time problems can be substantially eliminated by having the major equipment encased and delivered immediately after approval. This equipment is then shipped to a staging area for the job. The staging area may be the contractor's premises, the job site, supply house, etc. The most efficent method for decreasing delivery and lead time, particularly for large equipment, is to ship it to the yard of the rigger who is going to put this equipment in position. Payment for this equipment can be arranged through the architect, providing insurance and ownership have been established.

The estimator should also determine if there are any unusual payment requirements. Cash flow for a company can be severely affected if a large material purchase, thought to be payable in 30 days, is delivered C.O.D. or "site draft". Truck drivers may not allow unloading until payment has been received. Such requirements must be determined during the estimating stage so that the cost of borrowing money, if necessary, can be included.

If unable to obtain the quotation of a vendor from whom the material would be purchased, the estimator has other sources for obtaining material prices. These include, in order of reliability:

1. Current price lists from manufacturers' catalogues. Be sure to check that the list indicates "contractor discounts".
2. Cost records from previous jobs. Historical costs must be updated for present market conditions.
3. Reputable and current annual unit price cost books, such as *Means Mechanical Cost Data,* 1987. Such books usually represent national averages and must be factored to local markets.

No matter which price source is used, the estimator must be sure to include an allowance for any burdens, such as delivery, taxes or finance charges, over the actual cost of the material. (The same kinds of concerns that apply to vendor quotations should be taken into consideration when using these other price sources.)

Labor

In order to determine the installation cost for each item of construction, the estimator must know two pieces of information: first, the labor rate (hourly wage or salary) of the worker, and second, how much time a worker will need to complete a given unit or task of the installation — in other words, the productivity or output. Wage rates are usually known going into a project, but productivity may be very difficult to determine.

To estimators working for contractors, the construction labor rates that the contractor pays will be known, well-documented, and constantly updated. Projected labor rates for construction jobs of long duration should also be contended with. Estimators for owners, architects, or engineers must determine labor rates from outside sources. Unit price data books, such as *Means Mechanical Cost Data*, provide national average labor wage rates, with location factors for 162 major U.S. and Canadian cities. The unit costs for labor are based on these averages. Figure 6.2 shows national average *union* rates for the construction industry based on January 1, 1987. Figure 6.3 lists national average *open shop* rates, again based on January 1, 1987.

Note the column entitled "Worker's Compensation". Plumbers, pipe fitters, sprinkler installers, and sheet metal workers have among the lowest rates. These are average numbers which vary from state to state and between trades (see Figure 6.4). Worker's Compensation insurance is often considered a direct cost when applied to field labor.

If more accurate union labor rates are required, the estimator has different options. Union locals can provide rates (as well as negotiated increases) for a particular location. Employer bargaining groups can usually provide labor cost data as well. R. S. Means Co., Inc. publishes *Labor Rates for the Construction Industry* on an annual basis. This book lists the union labor rates by trade for over 300 U.S. and Canadian cities.

Determination of non-union, or "open shop" rates is much more difficult. In larger cities, employer organizations often exist to represent non-union contractors. These organizations may have records of local pay scales, but ultimately, wage rates are determined by individual contractors.

Labor units (or man-hours) are the least predictable of all factors for building projects. It is important to determine as accurately as possible prevailing productivity. The best source of labor productivity or labor units (and therefore labor costs) is the estimator's well-kept records from previous projects. If there are no company records for productivity, cost data books, such as *Means Mechanical Cost Data*, and productivity reference books, such as *Means Man-Hour Standards*, can be invaluable. Included with the listing for each individual construction item is the designation of a suggested crew make-up (usually 2-4 workers). The crew is the minimum grouping of workers which can be expected to accomplish the task efficiently. Figure 6.5, a typical page from *Means Mechanical Cost Data*, 1987, includes this data and indicates the man-hours as well as the average daily output of the designated crew for each construction item. When more than one or two tradespersons are designated, refer to the crew listings in the foreword of *Means Mechanical Cost Data*.

Means Forms
LETTER OF TRANSMITTAL

FROM: DMV Mechanical Contractors
Braintree, Mass. 02184

TO: Raymond Supply Co.
Foxboro, Mass.

DATE: Nov. 8, 1987
PROJECT: Office Building
LOCATION: Kingston, Mass.
ATTENTION: P. V. Howard
RE:

Gentlemen:
WE ARE SENDING YOU ☑ HEREWITH ☐ DELIVERED BY HAND ☐ UNDER SEPARATE COVER
VIA _____ THE FOLLOWING ITEMS:
☐ PLANS ☐ PRINTS ☑ SHOP DRAWINGS ☐ SAMPLES ☐ SPECIFICATIONS
☐ ESTIMATES ☐ COPY OF LETTER ☐ _____

COPIES	DATE OR NO.	DESCRIPTION
2	354	H. B. Jones Cast Iron Boiler
2	VF9	Maco Pump Submittals

THESE ARE TRANSMITTED AS INDICATED BELOW
☐ FOR YOUR USE ☐ APPROVED AS NOTED ☐ RETURN ___ CORRECTED PRINTS
☐ FOR APPROVAL ☑ APPROVED FOR CONSTRUCTION ☐ SUBMIT ___ COPIES FOR ___
☐ AS REQUESTED ☐ RETURNED FOR CORRECTIONS ☐ RESUBMIT ___ COPIES FOR ___
☐ FOR REVIEW AND COMMENT ☐ RETURNED AFTER LOAN TO US ☐ FOR BIDS DUE ___
☐ _____

REMARKS: Approved as submitted — Release for ASAP shipment to riggers yard as per our P.O. # 0895

IF ENCLOSURES ARE NOT AS INDICATED, PLEASE NOTIFY US AT ONCE.

SIGNED: John J. Moylan

Figure 6.1

Installing Contractor's Overhead & Profit

Below are the **average** installing contractor's percentage mark-ups applied to base labor rates to arrive at typical billing rates.

Column A: Labor rates are based on union wages averaged for 30 major U.S. cities. Base rates including fringe benefits are listed hourly and daily. These figures are the sum of the wage rate, employer-paid fringe benefits such as vacation pay, employer-paid health and welfare costs, pension costs, plus appropriate training and industry advancement funds costs.

Column B: Workers' Compensation rates are the national average of state rates established for each trade.

Column C: Column C lists average fixed overhead figures for all trades. Included are Federal and State Unemployment costs set at 5.5%; Social Security Taxes (FICA) set at 7.15%; Builder's Risk Insurance costs set at 1.14%; and Public Liability costs set at 0.82%. All the percentages except those for Social Security Taxes vary from state to state as well as from company to company.

Column D and E: Percentages in Columns D and E are based on the presumption that the installing contractor has annual billing of $500,000 and up. Overhead percentages may increase with smaller annual billing. The overhead percentages for any given contractor may vary greatly and depend on a number of factors, such as the contractor's annual volume, engineering and logistical support costs, and staff requirements. The figures for overhead and profit will also vary depending on the type of job, the job location, and the prevailing economic conditions. All factors should be examined very carefully for each job.

Column F: Column F lists the total of columns B, C, D, and E.

Column G: Column G is Column A (hourly base labor rate) multiplied by the percentage in Column F (O&P percentage).

Column H: Column H is the total of Column A (hourly base labor rate) plus Column G (Total O&P).

Column I: Column I is Column H multiplied by eight hours.

Abbr.	Trade	A Base Rate Incl. Fringes Hourly	A Daily	B Workers' Comp. Ins.	C Average Fixed Overhead	D Overhead	E Profit	F Total Overhead & Profit %	G Amount	H Rate with O & P Hourly	I Daily
Skwk	Skilled Workers Average (35 trades)	$20.80	$166.40	10.4%	14.6%	12.6%	10%	47.6%	$ 9.90	$30.70	$245.60
	Helpers Average (5 trades)	15.90	127.20	11.1		12.7		48.4	7.70	23.60	188.80
	Foremen Average, Inside (50¢ over trade)	21.30	170.40	10.4		12.6		47.6	10.15	31.45	251.60
	Foremen Average, Outside ($2.00 over trade)	22.80	182.40	10.4		12.6		47.6	10.85	33.65	269.20
Clab	Common Building Laborers	16.10	128.80	11.4		10.8		46.8	7.55	23.65	189.20
Asbe	Asbestos Workers	23.00	184.00	8.8		15.7		49.1	11.30	34.30	274.40
Boil	Boilermakers	23.00	184.00	7.2		16.0		47.8	11.00	34.00	272.00
Bric	Bricklayers	20.55	164.40	8.5		10.7		43.8	9.00	29.55	236.40
Brhe	Bricklayer Helpers	16.15	128.40	8.5		10.7		43.8	7.05	23.20	185.60
Carp	Carpenters	20.55	164.40	11.4		10.8		46.8	9.60	30.15	241.20
Cefi	Cement Finishers	19.70	157.60	6.6		10.8		42.0	8.30	27.95	223.60
Elec	Electricians	22.65	181.20	4.5		15.9		45.0	10.20	32.85	262.80
Elev	Elevator Constructors	23.05	184.40	6.0		15.9		46.5	10.70	33.75	270.00
Eqhv	Equipment Operators, Crane or Shovel	21.20	169.60	7.8		13.8		46.2	9.80	31.00	248.00
Eqmd	Equipment Operators, Medium Equipment	20.75	166.00	7.8		13.8		46.2	9.60	30.35	242.80
Eqlt	Equipment Operators, Light Equipment	19.60	156.80	7.8		13.8		46.2	9.05	28.65	229.20
Eqol	Equipment Operators, Oilers	17.55	140.40	7.8		13.8		46.2	8.10	25.65	205.20
Eqmm	Equipment Operators, Master Mechanics	22.00	176.00	7.8		13.8		46.2	10.15	32.15	257.20
Glaz	Glaziers	20.75	166.00	8.7		10.8		44.1	9.15	29.90	239.20
Loth	Lathers	20.50	164.00	7.2		10.8		42.6	8.75	29.25	234.00
Marb	Marble Setters	20.25	162.00	8.5		10.7		43.8	8.85	29.10	232.80
Mill	Millwrights	21.25	170.00	7.2		10.8		42.6	9.05	30.30	242.40
Mstz	Mosaic and Terrazzo Workers	20.05	160.40	6.0		10.7		41.3	8.30	28.35	226.80
Pord	Painters, Ordinary	19.55	156.40	8.9		10.8		44.3	8.65	28.20	225.60
Psst	Painters, Structural Steel	20.20	161.60	29.4		10.1		64.1	12.95	33.15	265.20
Pape	Paper Hangers	19.75	158.00	8.9		10.8		44.3	8.75	28.50	228.00
Pile	Pile Drivers	20.35	162.80	18.3		15.7		58.6	11.95	32.30	258.40
Plas	Plasterers	20.25	162.00	8.7		10.9		44.2	8.95	29.20	233.60
Plah	Plasterer Helpers	16.65	133.20	8.7		10.9		44.2	7.35	24.00	192.00
Plum	Plumbers	23.00	184.00	5.3		15.9		45.8	10.55	33.55	268.40
Rodm	Rodmen (Reinforcing)	22.10	176.80	18.6		13.3		56.5	12.50	34.60	276.80
Rofc	Roofers, Composition	19.15	153.20	20.8		10.4		55.8	10.70	29.85	238.80
Rots	Roofers, Tile & Slate	19.25	154.00	20.8		10.4		55.8	10.75	30.00	240.00
Rohe	Roofer Helpers (Composition)	14.20	113.60	20.8		10.4		55.8	7.90	22.10	176.80
Shee	Sheet Metal Workers	23.10	184.80	7.0		15.8		47.4	10.95	34.05	272.40
Spri	Sprinkler Installers	23.85	190.80	6.1		15.9		46.6	11.10	34.95	279.60
Stpi	Steamfitters or Pipefitters	23.30	186.40	5.3		15.9		45.8	10.70	33.95	271.60
Ston	Stone Masons	20.60	164.80	8.5		10.7		43.8	9.00	29.60	236.80
Sswk	Structural Steel Workers	22.10	176.80	21.5		13.7		59.8	13.20	35.30	282.40
Tilf	Tile Layers (Floor)	20.00	160.00	6.0		10.7		41.3	8.25	28.25	226.00
Tilh	Tile Layer Helpers	16.10	128.80	6.0		10.7		41.3	6.65	22.75	182.00
Trlt	Truck Drivers, Light	16.70	133.60	9.8		10.7		45.1	7.55	24.25	194.00
Trhv	Truck Drivers, Heavy	16.95	135.60	9.8		10.7		45.1	7.65	24.60	196.80
Sswl	Welders, Structural Steel	22.10	176.80	21.5		13.7		59.8	13.20	35.30	282.40
Wrck	*Wrecking	16.10	128.80	23.1	▼	10.4	▼	58.1	9.35	25.45	203.60

*Not included in Averages.

Figure 6.2

Installing Contractor's Overhead & Profit

Below are the **average** installing contractor's percentage markups applied to base labor rates to arrive at typical billing rates.

Column A: Labor rates are based on average wages for 7 major U.S. regions. Base rates including fringe benefits are listed hourly and daily. These figures are the sum of the wage rate and employer-paid fringe benefits such as vacation pay and employer-paid health costs.

Column B: Workers' Compensation rates are the national average of state rates established for each trade.

Column C: Column C lists average fixed overhead figures for all trades. Included are Federal and State Unemployment costs set at 5.5%; Social Security Taxes (FICA) set at 7.15%; Builder's Risk Insurance costs set at 1.14%; and Public Liability costs set at 0.82%. All the percentages except those for Social Security Taxes vary from state to state as well as from company to company.

Column D and E: Percentages in Columns D and E are based on the presumption that the installing contractor has annual billing of $500,000 and up. Overhead percentages may increase with smaller annual billing. The overhead percentages for any given contractor may vary greatly and depend on a number of factors, such as the contractor's annual volume, engineering and logistical support costs, and staff requirements. The figures for overhead and profit will also vary depending on the type of job, the job location, and the prevailing economic conditions. All factors should be examined very carefully for each job.

Column F: Column F lists the total of columns B, C, D, and E.

Column G: Column G is Column A (hourly base labor rate) multiplied by the percentage in Column F (O&P percentage).

Column H: Column H is the total of Column A (hourly base labor rate) plus Column G (Total O&P).

Column I: Column I is Column H multiplied by eight hours.

Abbr.	Trade	A Base Rate Incl. Fringes Hourly	A Daily	B Workers' Comp. Ins.	C Average Fixed Overhead	D Subs Overhead	E Subs Profit	F Subs Total Overhead & Profit %	G Amount	H Rate with Subs O & P Hourly	I Daily
Skwk	Skilled Workers Average	$12.30	$ 98.40	10.4%	14.6%	22.8%	10%	57.8%	$ 7.10	$19.40	$155.20
	Helpers Average ($2.00 under trade)	10.30	82.40	11.1		23.0		58.7	6.05	16.35	130.80
	Foremen Average, ($2.00 over trade)	14.30	114.40	10.4		22.8		57.8	8.25	22.55	180.40
Clab	Laborers	9.70	77.60	11.4		21.0		57.0	5.55	15.25	122.00
Asbe	Pipe or Duct Insulators	13.55	108.40	8.8		26.0		59.4	8.05	21.60	172.80
Boil	Boilermakers	13.55	108.40	7.2		26.0		57.8	7.85	21.35	170.80
Bric	Brick or Block Masons	12.10	96.80	8.5		21.0		54.1	6.55	18.65	149.20
Carp	Carpenters	12.20	97.60	11.4		21.0		57.0	6.95	19.15	153.20
Cefi	Cement Finishers	11.70	93.60	6.6		21.0		52.2	6.10	17.80	142.40
Elec	Electricians	13.30	106.40	4.5		26.0		55.1	7.35	20.65	165.20
Elev	Elevator Constructors	13.50	108.00	6.0		26.0		56.6	7.65	21.15	169.20
Eqhv	Equipment Operators, Crane	12.50	100.00	7.8		24.0		56.4	7.05	19.55	156.40
Eqmd	Equipment Operators	12.25	98.00	7.8		24.0		56.4	6.90	19.15	153.20
Eqmm	Equipment Mechanics	13.00	104.00	7.8		24.0		56.4	7.35	20.35	162.80
Glaz	Glaziers	12.30	98.40	8.7		21.0		54.3	6.70	19.00	152.00
Lath	Lathers	12.20	97.60	7.2		21.0		52.8	6.45	18.65	149.20
Mill	Millwrights	12.50	100.00	7.2		21.0		52.8	6.60	19.10	152.80
Pord	Painters	11.60	92.80	8.9		21.0		54.5	6.30	17.90	143.20
Pile	Pile Drivers	12.05	96.40	18.3		26.0		68.9	8.30	20.35	162.80
Plas	Plasterers	11.95	95.60	8.7		21.0		54.3	6.50	18.45	147.60
Plum	Plumbers	13.45	107.60	5.3		26.0		55.9	7.50	20.95	167.60
Rodm	Rodmen (Reinforcing)	13.00	104.00	18.6		24.0		67.2	8.75	21.75	174.00
Rofc	Roofers	11.40	91.20	20.8		21.0		66.4	7.55	18.95	151.60
Shee	Sheet Metal Workers	13.50	108.00	7.0		26.0		57.6	7.75	21.25	170.00
Spri	Sprinkler Installers	13.85	110.80	6.1		26.0		56.7	7.85	21.70	173.60
Stpi	Pipefitters	13.65	109.20	5.3		26.0		55.9	7.65	21.30	170.40
Ston	Stone Masons	12.20	97.60	8.5		21.0		54.1	6.60	18.80	150.40
Sswk	Structural Steel Erectors	13.00	104.00	21.5		24.0		70.1	9.10	22.10	176.80
Tilf	Flooring Installers	11.90	95.20	6.0		21.0		51.6	6.15	18.05	144.40
Trhv	Truck Drivers	10.20	81.60	9.8		21.0		55.4	5.65	15.85	126.80
Wrck	Wreckers	9.75	78.00	23.1		21.0		68.7	6.70	16.45	131.60

Figure 6.3

GENERAL REQUIREMENTS | **C10.2-200** | **Workers' Compensation**

Table 10.2-203 Worker's Compensation by Trade and State

STATE	CARPENTRY – 3 stories or less	CARPENTRY – interior cab. work	CARPENTRY – general	CONCRETE WORK – NOC	CONCRETE WORK – flat (flr. sdwk.)	ELECTRICAL WIRING – inside	EXCAVATION – earth NOC	EXCAVATION – rock	GLAZIERS	INSULATION WORK	LATHING	MASONRY	PAINTING & DECORATING	PILE DRIVING	PLASTERING	PLUMBING	ROOFING	SHEET METAL WORK (HVAC)	STEEL ERECTION – door & sash	STEEL ERECTION – inter. ornam.	STEEL ERECTION – structure	TILE WORK – NOC	WATERPROOFING	WRECKING
	5651	5437	5403	5213	5221	5190	6217	6217	5462	5479	5443	5022	5474	6003	5480	5183	5551	5538	5102	5102	5040	5057	5348	9014	5701
AL	6.59	4.62	7.11	8.39	3.67	3.36	6.09	6.09	6.26	6.30	4.11	5.55	8.39	17.55	6.17	3.29	11.36	7.38	4.42	4.42	6.86	12.57	4.33	3.60	12.57
AK	13.79	7.51	11.65	11.16	8.41	8.02	7.64	7.64	10.39	11.49	9.24	14.69	10.81	21.63	11.78	8.09	21.58	8.08	13.60	13.60	37.87	21.74	6.65	3.93	37.87
AZ	11.41	8.34	21.37	9.41	8.51	5.88	7.83	7.83	13.92	11.47	8.97	11.52	7.43	18.77	14.45	3.86	20.41	6.35	10.19	10.19	12.04	11.17	6.56	5.77	5.77
AR	6.41	4.13	7.24	6.44	4.07	2.62	4.50	4.50	5.47	6.26	5.18	4.71	6.14	12.81	5.29	3.06	10.13	5.55	4.18	4.18	27.09	7.41	3.14	2.99	27.09
CA	13.91	6.16	11.12	6.50	6.50	4.38	6.19	6.19	12.72	16.37	8.64	8.36	10.29	22.00	13.12	6.10	26.87	7.02	8.45	8.45	21.04	13.65	7.05	10.29	19.23
CO	11.11	7.99	11.23	11.07	8.99	3.64	9.09	9.09	12.31	12.65	7.56	13.71	9.51	28.75	17.72	5.57	33.24	7.55	9.77	9.77	21.97	25.35	5.16	4.13	25.35
CT	16.97	11.02	18.26	16.67	12.22	4.35	8.78	8.78	12.55	13.17	11.12	19.00	12.61	23.22	9.39	7.20	37.02	9.90	10.02	10.02	49.69	31.60	8.96	3.92	49.69
DE	11.24	11.24	11.24	8.72	5.95	4.10	8.68	8.68	10.03	11.24	8.15	6.73	10.54	11.26	8.15	4.19	25.26	8.27	9.50	9.50	22.07	9.50	6.56	6.73	24.10
DC	8.82	8.92	14.48	23.54	11.00	11.04	15.40	15.40	12.06	12.32	9.16	21.96	10.23	31.92	10.53	13.94	28.12	11.50	19.55	19.55	40.45	39.66	23.31	5.16	40.45
FL	13.22	6.96	15.72	16.71	10.82	6.83	12.15	12.15	11.97	11.10	10.27	12.02	10.97	22.18	11.80	8.57	30.77	10.12	9.37	9.37	22.84	21.69	6.35	5.66	22.84
GA	8.19	4.34	10.39	6.54	3.95	4.12	6.64	6.64	6.86	5.26	4.68	4.59	5.99	12.41	6.32	4.30	13.17	5.56	3.38	3.38	12.48	15.89	3.12	3.26	15.89
HI	16.08	16.55	56.46	28.24	14.36	13.69	28.94	28.94	34.71	16.88	16.05	19.48	17.31	46.20	23.91	12.07	46.26	11.92	11.95	11.95	38.71	43.93	14.50	12.12	38.71
ID	8.54	5.71	10.38	7.29	4.45	3.66	7.17	7.17	8.61	8.43	5.87	8.43	7.26	13.95	7.64	3.89	15.82	7.02	5.80	5.80	13.15	13.30	4.75	4.55	13.15
IL	11.32	7.55	11.46	16.65	7.87	5.55	8.28	8.28	11.43	9.08	8.29	11.26	10.05	22.50	7.01	7.94	30.22	9.32	9.84	9.84	35.06	49.40	8.65	4.01	49.40
IN	3.32	2.00	3.56	2.75	1.78	1.11	2.45	2.45	3.11	2.45	1.81	2.24	3.00	4.92	3.32	1.46	4.34	2.43	1.63	1.63	3.00	6.86	1.29	1.63	6.86
IA	3.71	3.63	7.29	7.47	3.75	3.88	4.50	4.50	4.09	5.06	3.95	4.80	7.22	9.63	4.35	5.10	9.61	4.87	5.66	5.66	15.43	17.86	2.76	2.62	15.43
KS	5.67	4.64	6.59	4.03	4.48	2.62	3.34	3.34	4.96	10.72	5.65	4.73	8.07	16.13	4.78	3.15	13.04	4.23	4.13	4.13	8.77	19.46	3.33	2.52	19.46
KY	7.24	3.68	9.04	4.36	4.32	3.34	4.47	4.47	5.14	5.67	4.85	4.59	7.79	11.30	4.82	3.00	10.27	4.00	4.85	4.85	15.03	7.75	3.56	3.31	15.03
LA	9.70	6.09	10.64	6.31	4.98	3.45	6.41	6.41	7.33	5.97	4.30	4.69	8.49	22.62	6.84	4.64	14.99	6.15	6.10	6.10	15.35	10.38	5.11	4.57	15.35
ME	6.06	7.03	18.31	13.41	7.09	5.27	10.66	10.66	10.29	9.53	6.94	9.24	10.42	27.89	9.49	6.80	25.02	8.66	8.34	8.34	33.58	27.46	6.66	4.56	33.58
MD	10.09	8.83	10.31	22.62	7.82	7.25	10.20	10.20	13.13	12.45	8.24	9.36	21.04	26.39	9.00	9.48	26.49	11.38	11.42	11.42	37.95	28.06	8.37	4.24	37.95
MA	8.83	4.77	19.00	13.33	5.85	3.64	5.54	5.54	8.74	6.42	6.33	10.94	11.47	14.78	7.52	5.46	55.36	7.63	8.29	8.29	32.06	30.28	5.88	5.22	27.54
MI	7.54	6.27	10.24	14.50	8.88	3.52	9.97	9.97	11.79	9.40	9.08	9.60	9.71	17.68	10.09	5.01	19.17	7.52	6.74	6.74	17.29	19.78	10.12	NA	17.29
MN	13.84	13.84	28.60	19.32	10.79	5.28	11.77	11.27	12.77	12.43	12.93	14.09	14.11	33.20	12.93	10.38	39.85	11.16	14.74	14.74	39.07	38.04	9.58	7.82	39.07
MS	6.12	4.13	7.60	4.96	3.50	3.27	4.86	4.86	5.06	4.20	4.61	3.72	3.95	15.63	6.40	2.00	9.25	4.71	4.48	4.48	13.30	7.95	4.01	2.79	13.30
MO	4.58	4.24	4.78	5.98	3.49	2.18	4.19	4.19	5.33	6.83	4.30	5.11	6.41	16.95	4.59	3.02	10.27	4.36	5.00	5.00	10.23	11.92	2.86	3.81	11.92
MT	13.17	9.18	18.37	11.36	10.80	7.72	13.30	13.30	11.32	13.71	11.41	12.89	13.71	32.45	15.24	8.04	40.58	11.80	10.48	10.48	38.92	25.62	7.83	9.87	38.92
NE	5.80	3.40	5.74	5.30	3.98	2.54	5.12	5.12	5.54	4.54	4.67	4.84	4.15	12.41	6.35	3.16	12.29	3.89	3.27	3.27	8.65	13.99	3.03	4.59	13.99
NV	10.38	10.38	10.38	8.69	8.69	5.46	8.69	8.69	7.38	10.38	10.38	10.74	7.62	8.69	10.38	6.37	16.00	6.37	10.38	10.38	19.45	19.45	7.38	8.69	19.45
NH	11.45	5.85	11.18	15.59	7.44	5.03	15.17	15.17	8.77	10.49	9.66	8.86	10.76	26.95	12.67	6.85	36.18	6.82	8.09	8.09	24.11	19.47	6.43	4.28	24.11
NJ	5.62	4.73	5.62	5.91	4.87	2.55	5.31	5.31	3.98	6.76	7.03	7.40	8.37	6.84	7.03	2.79	13.28	3.87	6.03	6.03	15.45	7.62	2.68	2.96	21.05
NM	11.33	10.26	12.73	17.13	9.18	4.79	7.69	7.69	12.36	13.76	8.22	11.01	8.17	47.64	13.16	8.25	28.26	9.27	12.17	12.17	19.90	21.01	8.75	5.28	19.90
NY	8.76	5.01	8.95	11.68	7.79	4.09	8.04	8.04	11.71	6.52	7.53	9.70	8.41	17.58	12.17	5.85	NA	6.99	4.88	4.88	16.62	15.67	7.05	3.40	24.31
NC	4.94	3.50	7.13	4.29	2.50	3.71	3.25	3.25	4.17	4.59	3.31	3.60	3.86	14.84	4.29	3.01	9.42	4.20	3.02	3.02	16.55	5.05	2.92	1.69	16.55
ND	5.93	5.93	5.93	5.25	5.25	2.20	5.42	5.42	8.16	3.95	3.08	3.10	4.92	13.65	3.08	6.22	8.20	6.22	5.93	5.93	13.65	13.65	2.90	8.20	NA
OH	5.13	5.13	5.13	5.44	5.44	1.98	5.44	5.44	5.62	5.08	5.08	5.31	5.62	5.44	5.08	3.51	11.85	7.65	NA	NA	21.25	21.25	2.12	11.85	NA
OK	10.62	6.70	10.23	8.96	6.10	4.25	8.46	8.46	8.80	9.50	6.70	6.18	6.50	22.67	10.16	6.23	28.07	7.85	8.03	8.03	32.71	18.31	5.08	5.24	32.71
OR	23.29	11.36	20.02	20.24	12.97	6.02	14.97	14.97	13.68	22.40	15.14	16.70	20.00	39.32	16.21	8.92	47.47	14.54	15.61	15.61	38.51	39.13	17.39	13.94	38.51
PA	10.38	9.71	10.38	14.98	6.69	4.47	10.30	10.30	9.46	10.38	9.66	9.57	13.98	11.00	9.66	5.45	22.77	8.03	9.79	9.79	22.54	9.79	6.64	9.57	53.22
RI	10.44	7.03	12.15	13.12	11.09	6.73	10.00	10.00	12.14	10.24	7.82	10.96	12.02	31.97	10.87	4.89	31.59	8.13	9.60	9.60	52.45	47.35	6.24	5.38	52.45
SC	8.04	5.35	8.10	5.13	3.82	6.19	5.21	5.21	4.96	7.19	4.38	6.63	8.68	12.04	5.85	2.55	15.79	7.24	3.77	3.77	15.49	13.35	3.89	2.45	15.49
SD	6.08	4.38	8.58	8.15	4.95	4.10	6.18	6.18	5.56	5.81	4.67	4.62	6.92	12.57	5.29	4.88	17.20	4.31	5.02	5.02	12.32	11.48	3.69	2.47	12.32
TN	5.57	4.08	7.09	5.25	3.71	2.41	4.51	4.51	5.24	6.27	4.02	5.76	6.00	10.63	3.94	2.92	10.48	4.82	4.03	4.03	11.40	7.31	2.72	2.43	11.40
TX	7.64	5.29	7.64	8.00	5.55	3.83	4.90	4.90	4.41	8.64	3.44	5.68	5.45	11.94	6.84	4.16	16.97	7.47	4.34	4.34	15.69	8.36	3.28	4.40	16.77
UT	NA	NA	6.55	7.87	3.69	2.05	5.72	5.72	5.36	7.51	6.76	6.34	7.83	10.82	6.24	3.42	15.44	3.95	4.10	4.10	NA	11.15	3.08	3.46	11.15
VT	5.29	3.91	6.53	8.84	4.15	3.44	6.69	6.69	6.30	6.46	4.86	6.77	14.64	13.27	5.99	3.10	12.07	5.63	5.70	5.70	15.11	13.56	3.70	2.94	15.11
VA	6.90	6.07	9.06	8.24	4.58	4.11	5.27	5.27	6.39	10.12	9.43	6.95	7.44	10.22	4.88	4.14	19.11	7.89	4.66	4.66	13.72	19.25	3.74	2.60	13.72
WA	7.75	7.75	7.75	6.51	5.36	2.65	7.21	7.21	8.15	6.25	7.75	9.05	7.54	15.11	8.08	2.81	7.42	3.23	6.92	6.92	9.69	9.69	6.06	8.53	9.29
WV	8.34	8.34	8.34	11.46	11.46	3.08	6.52	6.52	3.18	3.18	12.16	5.41	12.16	7.62	12.16	2.71	9.93	3.18	7.64	7.64	6.12	7.64	5.41	2.10	6.12
WI	5.37	3.91	9.45	6.74	4.93	3.45	4.39	4.39	6.99	6.92	5.33	7.08	5.68	12.31	7.60	4.47	15.40	6.36	5.78	5.78	27.84	20.35	4.42	4.38	24.94
WY	5.00	5.00	5.00	5.00	5.00	5.00	5.00	5.00	5.00	5.00	5.00	5.00	5.00	5.00	5.00	5.00	5.00	5.00	5.00	5.00	5.00	5.00	5.00	5.00	5.00
AVG	8.94	6.64	11.41	10.30	6.61	4.46	7.81	7.81	8.73	8.80	7.21	8.53	8.91	18.33	8.73	5.31	20.77	7.00	7.51	7.51	21.51	18.57	5.96	5.09	23.08

Figure 6.4

15.6 HVAC Piping Specialties

			CREW	DAILY OUTPUT	MAN-HOURS	UNIT	MAT.	LABOR	EQUIP.	TOTAL	TOTAL INCL O&P	
040	0010	**AUTOMATIC AIR VENT**										040
	0020	Cast iron body, stainless steel internals										
	0060	½" NPT inlet, 300 psi	1 Stpi	12	.667	Ea.	42	15.55		57.55	69	
	0140	¾" NPT inlet, 300 psi		12	.667		42	15.55		57.55	69	
	0180	½" NPT inlet, 250 psi		10	.800		136	18.65		154.65	175	
	0220	¾" NPT inlet, 250 psi		10	.800		136	18.65		154.65	175	
	0260	1" NPT inlet, 250 psi		10	.800		204	18.65		222.65	250	
	0340	1-½" NPT inlet, 250 psi	Q-5	12	1.330		415	28		443	495	
	0380	2" NPT inlet, 250 psi	"	12	1.330		415	28		443	495	
	0600	Forged steel body, stainless steel internals										
	0640	½" NPT inlet, 750 psi	1 Stpi	12	.667	Ea.	415	15.55		430.55	480	
	0680	¾" NPT inlet, 750 psi		12	.667		415	15.55		430.55	480	
	0760	¾" NPT inlet, 1000 psi		10	.800		625	18.65		643.65	715	
	0800	1" NPT inlet, 1000 psi	Q-5	12	1.330		625	28		653	730	
	0880	1-½" NPT inlet, 1000 psi		10	1.600		1,750	34		1,784	1,975	
	0920	2" NPT inlet, 1000 psi		10	1.600		1,750	34		1,784	1,975	
	1100	Formed steel body, non corrosive										
	1110	⅛" NPT inlet 35 psi	1 Stpi	32	.250	Ea.	6.20	5.85		12.05	15.30	
	1120	¼" NPT inlet 150 psi		32	.250		15.65	5.85		21.50	26	
	1130	¾" NPT inlet 150 psi		32	.250		15.65	5.85		21.50	26	
	1300	Chrome plated brass, automatic/manual, for radiators										
	1310	⅛" NPT inlet, nickel plated brass	1 Stpi	32	.250	Ea.	1.82	5.85		7.67	10.50	
080	0010	**AIR CONTROL** With strainer										080
	0040	2" diameter	Q-5	6	2.670	Ea.	290	56		346	400	
	0080	2-½" diameter		5	3.200		320	67		387	450	
	0100	3" diameter		4	4		495	84		579	665	
	0120	4" diameter		3	5.330		710	110		820	945	
	0140	6" diameter	Q-6	3.40	7.060		1,075	155		1,230	1,400	
	0160	8" diameter		3	8		1,625	175		1,800	2,050	
	0180	10" diameter		2.20	10.910		2,475	235		2,710	3,075	
	0200	12" diameter		1.70	14.120		3,575	305		3,880	4,375	
120	0010	**CIRCUIT SENSOR** Flow meter,										120
	0060	2-½" pipe size	Q-5	12	1.330	Ea.	70	28		98	120	
	0100	3" pipe size		11	1.450		80.50	31		111.50	135	
	0140	4" pipe size		8	2		94.50	42		136.50	165	
	0180	5" pipe size		7.30	2.190		116	46		162	195	
	0220	6" pipe size		6.40	2.500		133	52		185	225	
	0260	8" pipe size	Q-6	5.30	4.530		189	98		287	350	
	0280	10" pipe size		4.60	5.220		224	115		339	410	
	0360	12" pipe size		4.20	5.710		350	125		475	565	
	0010	**CIRCUIT SETTER** Balance valve										160
	0020	¾" pipe size	1 Stpi	20	.400	Ea.	31.50	9.30		40.80	48	
	0040	1" pipe size		18	.444		40.60	10.35		50.95	60	
	0060	1-½" pipe size		12	.667		65.10	15.55		80.65	94	
	0080	2" pipe size		10	.800		95.90	18.65		114.55	135	
	0100	2-½" pipe size	Q-5	15	1.070		184	22		206	235	
	0120	3" pipe size		10	1.600		276	34		310	355	
	0140	4" pipe size		3	5.330		390	110		500	590	
	0200	For differential meter, add					455			455	500	
200	0010	**EXPANSION JOINTS** Bellows type, neoprene cover, flanged spool										200
	0100	6" face to face, ½" diameter	1 Stpi	14	.571	Ea.	171	13.30		184.30	210	
	0110	¾" diameter		14	.571		171	13.30		184.30	210	
	0120	1" diameter		13	.615		171	14.35		185.35	210	
	0140	1-¼" diameter		11	.727		171	16.95		187.95	215	
	0160	1-½" diameter		10.60	.755		180	17.60		197.60	225	
	0180	2" diameter	Q-5	13.30	1.200		183	25		208	240	
	0200	3" diameter	"	11.40	1.400		204	29		233	265	

Figure 6.5

The estimator who has neither company records nor the sources described above must put together the appropriate crews and determine the expected output or productivity. This type of estimating should only be attempted based upon strong experience and considerable exposure to construction methods and practices. There are rare occasions when this approach is necessary to estimate a particular item or a new technique. Even then, the new labor units are often extrapolated from existing figures for similar work, rather than being created from scratch.

Equipment

In recent years, construction equipment has become more important, not only because of the incentive to reduce labor costs, but also as a response to new, high-technology construction methods and materials. As a result, these costs represent an increasing percentage of total project costs in construction. Estimators must carefully address the issue of equipment and related expenses. Equipment costs can be divided into the two following categories:

- **Rental, lease or ownership costs**: These costs may be determined based on hourly, daily, weekly, monthly or annual increments. These fees or payments only buy the "right" to use the equipment (i.e., exclusive of operating costs).
- **Operating costs**: Once the "right" of use is obtained, costs are incurred for actual use or operation. These costs may include fuel, lubrication, maintenance, and parts.

Equipment costs, as described above, do not include the labor expense of operators. However, some cost books and suppliers may include the operator in the quoted price for equipment as an "operated" rental cost. In other words, the equipment is priced as if it were a subcontract cost. In mechanical construction, equipment is often procured in this way, via an arrangement with the general contractor for equipment already on site. The advantage of this approach is that it can be used on an "as needed" basis and not carried as a weekly or monthly cost.

Equipment ownership costs apply to both leased and owned equipment. The operating costs of equipment, whether rented, leased or owned, are available from the following sources (listed in order of reliability):

1. The company's own records
2. Annual cost books containing equipment operating costs, such as *Means Mechanical Cost Data*
3. Manufacturers' estimates
4. Text books dealing with equipment operating costs

These operating costs consist of fuel, lubrication, expendable parts replacement, minor maintenance, transportation, and mobilizing costs. For estimating purposes, the equipment ownership and operating costs should be listed separately. In this way, the decision to rent, subcontract or purchase can be decided project by project.

There are two commonly used methods for including equipment costs in a construction estimate. The first is to include the equipment as a part of the construction task for which it is used. In this case, costs are included in each line item as a separate unit price. The advantage of this method is that costs are allocated to the division or task that actually incurs the expense. As a result, more accurate records can be kept for each installed component. A disadvantage of this method occurs in the pricing of equipment that may be used for many different tasks. Duplication of costs can occur in this instance. Another disadvantage is that the budget may be left short for the following reason: the estimate may only reflect two hours for a crane truck, when the minimum cost of the crane is usually a daily (8-hour) rental charge.

The second method for including equipment costs in the estimate is to keep all such costs separate and to include them in Division 1 as a part of Project Overhead. The advantage of this method is that all equipment costs are grouped together, and that machines used for several tasks are included (without duplication). One disadvantage is that for future estimating purposes, equipment costs will be known only on a job basis and not per installed unit.

Whichever method is used, the estimator must be consistent, and must be sure that all equipment costs are included but not duplicated. The estimating method should be the same as that chosen for cost monitoring and accounting. In this way, the data will be available both for monitoring the project's costs and for bidding future projects.

A final word of caution about equipment is to consider its age and reliability. If an older item, such as a pick-up truck, needs frequent repair, it may cost far more in lost man-hours to the project than is reflected in its calculated cost rate.

Subcontractors

Subcontractors often account for a large percentage of the mechanical estimator's bid. When subcontractors are used, quotations should be solicited and analyzed in the same way as material quotes. A primary concern is that the bid covers the work as per plans and specifications, and that all appropriate work, alternates, and allowances, if any, are included. Any exclusions should be clearly stated and explained. If the bid is received verbally, a form such as that shown in Chapter 5 (Figure 5.1) will help to assure that it is documented accurately. Any unique scheduling or payment requirements must be noted and evaluated prior to submission of your bid. Such requirements could affect (restrict or enhance) the normal progress of the project, and should therefore be known in advance.

The estimator should note how long the subcontract bid will be honored. This time period usually varies from 30 to 90 days and is usually included as a condition in complete bids.

The estimator should know or verify the bonding capability and capacity of unfamiliar subcontractors. Taking such action may be necessary when bidding in a new location. Other than word of mouth, these inquiries may be the only way to confirm subcontractor reliability.

Project Overhead

Project Overhead represents those construction costs that are usually included in Division One — General Requirements. Site management is covered in this section. Typical items are supervisory personnel, job engineers, cleanup, and temporary heat and power. While these items may not be directly part of the physical structure, they are a part of the project. Project Overhead, like all other direct costs, can be separated into material, labor, and equipment components. Figures 6.6 and 6.7 are examples of forms that can help ensure that all appropriate costs are included.

Some may not agree that certain items (such as equipment or scaffolding) should be included in Project Overhead, and might prefer to list such items in another division. Ultimately, it is not important *where* each item is incorporated into the estimate but that *every item is included somewhere.*

Project Overhead often includes time-related items; equipment rental, supervisory labor, and temporary utilities are examples. The cost for these items depends upon the duration of the project. A preliminary schedule should, therefore, be developed *prior* to completion of the estimate so that time-related items can be properly counted. This will be further discussed in Chapter Nine, "Pre-Bid Scheduling".

Bonds

Although bonds are really a type of "direct cost", they are priced and based upon the total "bid" or "selling price". For this reason, they are generally figured after indirect costs have been added. Bonding requirements for a project will be specified in Division 1 — "General Requirements", and will be included in the construction contract. Various types of bonds may be required. Listed below are a few common types:

- **Bid Bond**: A form of bid security executed by the bidder or principle and by a surety (bonding company) to guarantee that the bidder will enter into a contract within a specified time and furnish any required Performance or Labor and Material Payment bonds.

- **Completion Bond**: Also known as "Construction" or "Contract" bond. The guarantee by a surety that the construction contract will be completed and that it will be clear of all liens and encumbrances.

- **Labor and Material Payment Bond**: The guarantee by a surety to the owner that the contractor will pay for all labor and materials used in the performance of the contract as per the construction documents. The claimants under the bond are those having direct contracts with the contractor or any subcontractor.

- **Performance Bond**: (1) A guarantee that a contractor will perform a job according to the terms of the contracts. (2) A bond of the contractor in which a surety guarantees to the owner that the work will be performed in accordance with the contract documents. Except where prohibited by statute, the performance bond is frequently combined with the labor and material payment bond.

- **Surety Bond**: A legal instrument under which one party agrees to answer to another party for the debt, default or failure to perform of a third party.

Means Forms

PROJECT OVERHEAD SUMMARY

PROJECT
LOCATION ARCHITECT
QUANTITIES BY: PRICES BY: EXTENSIONS BY: CHECKED BY:

SHEET NO.
ESTIMATE NO.
DATE

DESCRIPTION	QUANTITY	UNIT	MATERIAL/EQUIPMENT UNIT	MATERIAL/EQUIPMENT TOTAL	LABOR UNIT	LABOR TOTAL	TOTAL COST UNIT	TOTAL COST TOTAL
Job Organization: Superintendent								
Project Manager								
Timekeeper & Material Clerk								
Clerical								
Safety, Watchman & First Aid								
Travel Expense: Superintendent								
Project Manager								
Engineering: Layout								
Inspection/Quantities								
Drawings								
CPM Schedule								
Testing: Soil								
Materials								
Structural								
Equipment: Cranes								
Concrete Pump, Conveyor, Etc.								
Elevators, Hoists								
Freight & Hauling								
Loading, Unloading, Erecting, Etc.								
Maintenance								
Pumping								
Scaffolding								
Small Power Equipment/Tools								
Field Offices: Job Office								
Architect/Owner's Office								
Temporary Telephones								
Utilities								
Temporary Toilets								
Storage Areas & Sheds								
Temporary Utilities: Heat								
Light & Power								
Water								
PAGE TOTALS								

Page 1 of 2

Figure 6.6

Means Forms

DESCRIPTION	QUANTITY	UNIT	MATERIAL/EQUIPMENT UNIT	MATERIAL/EQUIPMENT TOTAL	LABOR UNIT	LABOR TOTAL	TOTAL COST UNIT	TOTAL COST TOTAL
Totals Brought Forward								
Winter Protection: Temp. Heat/Protection								
Snow Plowing								
Thawing Materials								
Temporary Roads								
Signs & Barricades: Site Sign								
Temporary Fences								
Temporary Stairs, Ladders & Floors								
Photographs								
Clean Up								
Dumpster								
Final Clean Up								
Punch List								
Permits: Building								
Misc.								
Insurance: Builders Risk								
Owner's Protective Liability								
Umbrella								
Unemployment Ins. & Social Security								
Taxes								
City Sales Tax								
State Sales Tax								
Bonds								
Performance								
Material & Equipment								
Main Office Expense								
Special Items								
TOTALS:								

Figure 6.7

Sales Tax

Sales tax varies from state to state and often from city to city within a state (see Figure 6.8). Larger cities may have a sales tax in addition to the state sales tax. Some localities also impose separate sales taxes on labor and equipment.

When bidding takes place in unfamiliar locations, the estimator should check with local agencies regarding the amount and the method of payment of sales tax. Local authorities may require owners to withhold payments to out-of-state contractors until payment of all required sales tax has been verified. Sales tax is often taken for granted or even omitted and, as can be seen in Figure 6.8, can be as much as 7.5% of material costs. Indeed, this can represent a significant portion of the project's total cost. Conversely, some clients and/or their projects may be tax exempt. If this fact is unknown to the estimator, a large dollar amount for sales tax might be needlessly included in a bid.

Sales Tax Percentages on Materials by State

State	Tax	State	Tax	State	Tax	State	Tax
Alabama	4%	Illinois	5%	Montana	0%	Rhode Island	6%
Alaska	0	Indiana	5	Nebraska	3.5	South Carolina	5
Arizona	5	Iowa	4	Nevada	5.75	South Dakota	4
Arkansas	4	Kansas	3	New Hampshire	0	Tennessee	5.5
California	6	Kentucky	5	New Jersey	6	Texas	4
Colorado	3	Louisiana	4	New Mexico	3.75	Utah	5.5
Connecticut	7.5	Maine	5	New York	4	Vermont	4
Delaware	0	Maryland	5	North Carolina	3	Virginia	4
District of Columbia	6	Massachusetts	5	North Dakota	4	Washington	6.5
Florida	5	Michigan	4	Ohio	5.5	West Virginia	5
Georgia	3	Minnesota	6	Oklahoma	3	Wisconsin	5
Hawaii	4	Mississippi	6	Oregon	0	Wyoming	3
Idaho	4	Missouri	6.225	Pennsylvania	6	Average	4.25%

Figure 6.8

Chapter 7
INDIRECT COSTS

Indirect costs are those "costs of doing business" that are incurred by the general staff. These expenses are sometimes referred to as a "burden" to the project. Indirect costs may include certain fixed, or known, expenses and percentages, as well as costs which can be variable and subjectively determined. Government authorities require payment of certain taxes and insurance, usually based upon labor costs and determined by trade. These are a type of fixed indirect cost. Office overhead, if well understood and established, can also be considered as a relatively fixed percentage. Profit and contingencies, however, are more variable and subjective. These figures are often determined based on the judgement and discretion of the person responsible for the company's growth and success.

If the direct costs for the same project have been carefully determined, they should not vary significantly from one estimator to another. It is the indirect costs that are often responsible for variations between bids.

The direct costs of a project must be itemized, tabulated, and totalled before the indirect costs can be applied to the estimate. Indirect costs include:
- Taxes and Insurance
- Office or Operating Overhead (vs. Project Overhead)
- Profit
- Contingencies

Taxes and Insurance

The taxes and insurance included as indirect costs are most often related to the costs of labor and/or the type of work. This category may include Worker's Compensation, Builder's Risk, and Public Liability insurance, as well as employer-paid Social Security tax and Unemployment Insurance. By law, the employer must pay these expenses. Rates are based on the type and salary of the employees, as well as the location and/or type of business.

Office or Operating Overhead

Office overhead, or the cost of doing business, is perhaps one of the main reasons why so many contractors are unable to realize a profit, or even to stay in business. This is manifested in two ways. Either a company does not know its true overhead cost and, therefore, fails to mark up its costs enough to recover them; or management does not restrain or control overhead costs effectively and fails to remain competitive.

If a contractor does not know the costs of operating the business, then, more than likely, these costs will not be recovered. Many companies survive, and even turn a profit, by simply adding an arbitrary percentage for overhead to each job, without knowing how the percentage is derived or what is included. When annual volume changes significantly, whether by increase or decrease, the previously used percentage for overhead may no longer be valid. When such a volume change occurs, the owner often finds that the company is not doing as well as before and cannot determine the reasons. Chances are, overhead costs are not being fully recovered. As an example, Figure 7.1 lists annual office costs and expenses for a "typical" mechanical contractor. It is assumed that the anticipated annual volume of the company is $1,500,000. Each of the items is described briefly below.

Owner: This includes only a reasonable base salary and does not include profits. An owner's salary is *not* a company's profit.

Engineer/Estimator: Since the owner is primarily on the road getting business, this is the person who runs the daily operation of the company and is responsible for estimating. In some operations, the estimator who successfully wins a bid, then becomes the "project manager" and is responsible to the owner for its profitability.

Secretary/Receptionist: This person manages office operations and handles paperwork. A talented individual in this position can be a tremendous asset.

Office Worker Insurance & Taxes: These costs are for main office personnel only and, for this example, are calculated as 37% of the total salaries based on the following breakdown:

Worker's Compensation	6%
FICA	7%
Unemployment	4%
Medical & other insurance	10%
Profit sharing, pension, etc.	10%
	37%

Physical Plant Expenses: Whether the office, warehouse, and yard are rented or owned, roughly the same costs are incurred. Telephone and utility costs will vary depending on the size of the building and the type of business. Office equipment includes items such as the rental of a copy machine and typewriters.

Professional Services: Accountant fees are primarily for quarterly audits. Legal fees go towards collecting and contract disputes. In addition, a prudent contractor will have *every* contract read by his lawyer prior to signing.

Miscellaneous: There are many expenses that could be placed in this category. Included in the example are just a few of the possibilities. Advertising includes the Yellow Pages, promotional materials, etc.

Uncollected Receivables: This amount can vary greatly, and is often affected by the overall economic climate. Depending upon the timing of "uncollectables", cash-flow can be severely restricted and can cause serious financial problems, even for large companies. Sound cash planning and anticipation of such possibilities can help to prevent severe repercussions. While the office example used here is feasible within the industry, keep in mind that it is hypothetical and that conditions and costs vary widely from company to company.

Annual Main Office Expenses

Assume: $1,500,000 Annual Volume in the Field
        30% Material, 70% Labor

Office/Operating Expenses:

Owner	$ 60,000
Engineer/Estimator	44,000
Secretary/Receptionist	18,000
Personnel Insurance & Taxes	44,030
Office Rent	10,000
Utilities	1,800
Telephone	6,000
Vehicles (2)	13,000
Office Equipment	2,400
Legal/Accounting Services	5,000
Miscellaneous:	
Advertising	1,750
Seminars	3,000
Travel & Entertainment	8,000
Uncollected Receivables	18,000
Total	$234,980

$$\frac{\text{Expenses}}{\text{Labor Volume}} = \frac{\$234,980}{\$1,050,000} = 22.4\%$$

To support this overhead, job site staffing would have to average approximately 23 workers throughout the year.

Figure 7.1

In order for this example company to stay in business without losses (profit is not yet a factor), not only must all direct construction costs be paid, but an additional $234,980 must be recovered during the year (as a percentage of volume) in order to operate the office. The percentage may be calculated and applied in two ways:

- Office overhead applied as a percentage of labor costs only. This method requires that labor and material costs be estimated separately.
- Office overhead applied as a percentage of total project costs. This can be used whether or not material and labor costs are estimated separately.

Remember that the anticipated volume is $1,500,000 for the year, 70% of which is expected to be labor. Office overhead costs, therefore, will be approximately 22.4% of the labor cost, or 15.7% of annual volume *for this example*. The most common method for recovering these costs is to apply this percentage to each job over the course of the year.

The estimator must also remember that, if volume changes significantly, then the percentage for office overhead should be recalculated for current conditions. The same is true if there are changes in office staff. Salaries are the major portion of office overhead costs. It should be noted that a percentage is commonly applied to material costs, for handling, regardless of the method of recovering office overhead costs. This percentage is more easily calculated if material costs are estimated and listed separately.

Profit

Determining a fair and reasonable percentage to be included for profit is not an easy task. This responsibility is usually left to the owner or chief estimator. Experience is crucial in anticipating what profit the market will bear. The economic climate, competition, knowledge of the project, and familiarity with the architect, engineer, or owner all affect the way in which profit is determined. Chapter 10 includes a method to mathematically determine the profit margin based on historical bidding information. As with all facets of estimating, experience is the key to success.

Contingencies

Like profit, contingencies can also be difficult to quantify. Especially appropriate in preliminary budgets, the addition of a contingency is meant to protect the contractor as well as to give the owner a realistic estimate of potential project costs.

A contingency percentage should be based on the number of "unknowns" in a project, or the level of risk involved. This percentage should be inversely proportional to the amount of planning detail that has been done for the project. If complete plans and specifications are supplied, the estimate is thorough and precise, and the market is stable, then there is little need for a contingency. Figure 7.2, from *Means Mechanical Cost Data, 1987*, lists suggested contingency percentages that may be added to an estimate based on the stage of planning and development.

1.1	Overhead		CREW	DAILY OUTPUT	MAN-HOURS	UNIT	BARE COSTS				TOTAL INCL O&P	
							MAT.	LABOR	EQUIP.	TOTAL		
040	0011	CLEANING UP After job completion, minimum										040
	0040	Maximum				Project					1%	
	0050	Cleanup of floor area, continuous, per day	A-5	12	1.500	M.S.F.	1.50	24	1.20	26.70	39	
	0100	Final	"	11.50	1.570	"	1.60	25	1.25	27.85	40	
	1000	Mechanical demolition, see division 02.2										
060	0011	CONSTRUCTION COST INDEX For 162 major U.S. and										060
	0020	Canadian cities, total cost, min. (Greensboro, NC)				%					80.40%	
	0050	Average	C13.1-100								100%	
	0100	Maximum (Anchorage, AK)									132.60%	
110	0010	CONTINGENCIES Allowance to add at conceptual stage				Project					15%	110
	0050	Schematic stage									10%	
	0100	Preliminary working drawing stage									7%	
	0150	Final working drawing stage									2%	
120	0010	CONTRACTOR EQUIPMENT See division 1.5	C10.3-300									120
140	0010	CREWS For building construction, see foreword										140
150	0010	ENGINEERING FEES Educational planning consultant, minimum				Project					.50%	150
	0100	Maximum				"					2.50%	
	0200	Electrical, minimum	C10.1-103			Contrct					4.10%	
	0300	Maximum									10.10%	
	0400	Elevator & conveying systems, minimum									2.50%	
	0500	Maximum									5%	
	0600	Food service & kitchen equipment, minimum									8%	
	0700	Maximum									12%	
	1000	Mechanical (plumbing & HVAC), minimum									4.10%	
	1100	Maximum									10.10%	
160	0011	HISTORICAL COST INDEXES Back to 1943	C13.1-100									160
180	0010	INSURANCE Builders risk, standard, minimum				Job					.19%	180
	0050	Maximum									1.14%	
	0200	All-risk type, minimum	C10.1-301								.20%	
	0250	Maximum									1.16%	
	0400	Contractor's equipment floater, minimum				Value					.50%	
	0450	Maximum				"					2.50%	
	0600	Public liability, average				Job					.82%	
	0610											
	0800	Workers' compensation & employer's liability, average	C10.2-200									
	0850	by trade, carpentry, general				Payroll		11.41%				
	1000	Electrical						4.46%				
	1150	Insulation						8.80%				
	1450	Plumbing						5.31%				
	1550	Sheet metal work (HVAC)						7%				
190	0010	JOB CONDITIONS Modifications to total										190
	0020	project cost summaries										
	0100	Economic conditions, favorable, deduct				Project					2%	
	0200	Unfavorable, add									5%	
	0300	Hoisting conditions, favorable, deduct									1%	
	0400	Unfavorable, add									5%	
	0700	Labor availability, surplus, deduct									1%	
	0800	Shortage, add									10%	
	0900	Material storage area, available, deduct									1%	
	1000	Not available, add									2%	
	1100	Subcontractor availability, surplus, deduct									5%	
	1200	Shortage, add									12%	
	1300	Work space, available, deduct									1%	
	1400	Not available, add									4%	
200	0011	LABOR INDEX For 162 major U.S. and Canadian cities										200
	0020	Minimum (Charleston, SC)				%		66.40%				

For expanded coverage of these items see *Means Building Construction Cost Data 1987*

Figure 7.2

If an estimate is priced and each individual item is rounded upward, or "padded", this is, in essence, adding a contingency. This method can cause problems, however, because the estimator can never be quite sure of what is the actual cost and what is the "padding", or safety margin, for each item. At the summary, the estimator cannot determine exactly how much has been included as a contingency factor for the project as a whole. A much more accurate and controllable approach is to price the estimate precisely and then add one contingency amount at the bottom line.

Chapter 8
THE ESTIMATE SUMMARY

At the pricing stage of the estimate, there is typically a large amount of paperwork that must be assembled, analyzed and tabulated. Generally, the information contained in this paperwork could be recorded on any or all of the following major categories:

- Quantity Takeoff Sheets for all mechanical work items (Figure 8.1)
- Material supplier's written quotations (see note below)
- Equipment or material supplier's or subcontractor's Telephone Quotations (Figure 8.2)
- Subcontractor's written quotations
- Equipment supplier's quotations
- Cost Analysis or Consolidated Estimate Sheets (Pricing Sheets) (Figures 8.3 and 8.4)
- Recap Summary Sheets or Estimate Summary Sheets

   Note: Additional forms, such as Request for Quote postal cards, are often prepared by the individual contractor.

In the "real world" of estimating, many quotations, especially for large equipment and for subcontracts, are not received until the last minute before the bidding deadline. Therefore, a system is needed to efficiently handle the paperwork and to ensure that everything will get transferred once (and only once) from the quantity takeoff to the cost analysis sheets. Some general rules for this process are as follows:

- The piping, fixtures, labor, etc., should have been previously priced and entered in the estimate or summary sheet.
- Write on only one side of any document.
- Use Telephone Quotation forms for uniformity in recording prices received by phone.
- Document the source of every quantity and price.
- Keep each type of document in its "pile" (Quantities, Material, Subcontractors, Equipment) piled in order by classifications.
- Keep the entire estimate in one or more compartmentalized folders.
- If you are pricing your own materials, number and code each takeoff sheet and each pricing extension sheet as it is created. At the same time, keep an index list of each sheet by number. If a sheet is to be abandoned, write "VOID" on it, but do not discard it. Keep it until the bid is accepted to be able to account for all pages and sheets.

   Note: A helpful technique for organizing these sheets and forms involves the use of pastel colors to code each type of category of cost sheets. This system makes locating and revising sheets both easier and quicker.

All subcontract costs should be properly noted and listed separately. These costs contain the subcontractor's markups and may be treated differently from other direct costs when the estimator calculates the prime contractor's overhead and profit.

After all the unit prices and allowances have been entered on the pricing sheets, the costs are extended. In making the extensions, ignore the cents column and round all totals to the nearest five or ten dollars. In a column of figures, the cents will average out and will not be of consequence. Finally, each subdivision is added and the results checked, preferably by someone other than the person doing the extensions.

It is important to check the larger items for order of magnitude errors. If the total costs are divided by the building area, the resulting square foot cost figures can be used to quickly check with expected square foot costs. These cost figures should be recorded for comparison to past projects and as a resource for future estimating.

The takeoff and pricing method, as discussed, has been to utilize a Quantity Sheet for the material takeoff (see Figure 8.1), and to transfer the data to an analysis form for pricing the material, labor, and equipment items (see Figure 8.3).

An alternative to this method is a consolidation of the takeoff task and pricing on a single form. This approach works well for smaller bids and for change orders. An example, the Consolidated Estimate Form, is shown in Figure 8.4. The same sequences and recommendations used to complete the Quantity Sheet and Cost Analysis form are to be followed when using the Consolidated Estimate form to price the estimate.

When the pricing of all direct costs is complete, the estimator has two choices: 1) to make all further price changes and adjustments on the Cost Analysis or Consolidated Estimate sheets, *or* 2) to transfer the total costs for each subdivision to an Estimate Summary sheet so that all further price changes, until bid time, will be done on one sheet. Any indirect cost markups and burdens will be figured on this sheet also.

Unless the estimate has a limited number of items, it is recommended that costs be transferred to an Estimate Summary sheet. This step should be double-checked since an error of transposition may easily occur. Pre-printed forms can be useful, although a plain columnar form may suffice. This summary with page numbers from each extension sheet can also serve as an index of the mechanical specifications.

A company that repeatedly uses certain standard listings can save valuable time by having a custom Estimate Summary sheet printed with these items listed. The printed UCI division and subdivision headings may serve as another type of checklist, ensuring that all required costs are included. Appropriate column headings or categories for any estimate summary form could be as follows:

- Material
- Labor
- Equipment
- Subcontractor
- Miscellaneous
- Total

As items are listed in the proper columns, each category is added and appropriate markups applied to the total dollar values. Different percentages may be added to the sum of each column at the estimate summary. These percentages may include the following items, as discussed in Chapter 7:

- Taxes and Insurance
- Overhead
- Profit
- Contingencies

Figure 8.1

Means Forms
TELEPHONE QUOTATION

PROJECT	DATE
FIRM QUOTING	TIME
ADDRESS	PHONE ( )
ITEM QUOTED	BY
	RECEIVED BY

WORK INCLUDED	AMOUNT OF QUOTATION

DELIVERY TIME — **TOTAL BID**

DOES QUOTATION INCLUDE THE FOLLOWING: If ☐ NO is checked, determine the following:

STATE & LOCAL SALES TAXES	☐ YES	☐ NO	MATERIAL VALUE
DELIVERY TO THE JOB SITE	☐ YES	☐ NO	WEIGHT
COMPLETE INSTALLATION	☐ YES	☐ NO	QUANTITY
COMPLETE SECTION AS PER PLANS & SPECIFICATIONS	☐ YES	☐ NO	DESCRIBE BELOW

EXCLUSIONS AND QUALIFICATIONS

ADDENDA ACKNOWLEDGEMENT — **TOTAL ADJUSTMENTS**

ADJUSTED TOTAL BID

ALTERNATES

ALTERNATE NO.
ALTERNATE NO.
ALTERNATE NO.
ALTERNATE NO.
ALTERNATE NO.
ALTERNATE NO.
ALTERNATE NO.

Figure 8.2

Means Forms

COST ANALYSIS

SHEET NO.

PROJECT ESTIMATE NO.

ARCHITECT DATE

TAKE OFF BY: QUANTITIES BY: PRICES BY: EXTENSIONS BY: CHECKED BY:

DESCRIPTION	SOURCE/DIMENSIONS			QUANTITY	UNIT	MATERIAL		LABOR		EQ./TOTAL	
						UNIT COST	TOTAL	UNIT COST	TOTAL	UNIT COST	TOTAL

Figure 8.3

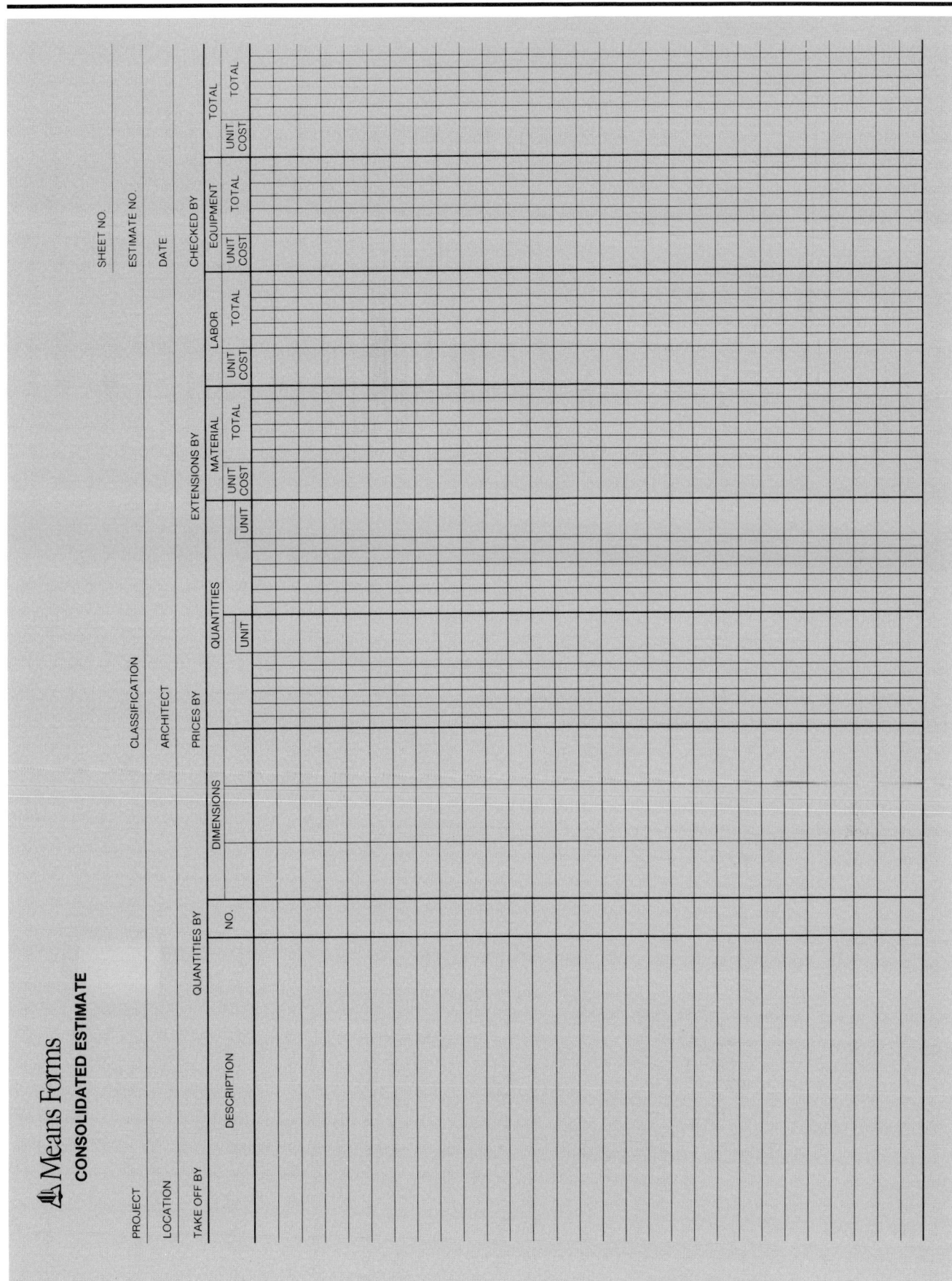

Figure 8.4

Chapter 9
PRE-BID SCHEDULING

The need for planning and scheduling is clear once the contract is signed and work commences on the project. However, some scheduling is also important during the bidding stage for the following reasons:
- To determine if the project can be completed in the allotted or specified time using normal crew sizes.
- To identify potential overtime requirements.
- To determine the time requirements for supervision.
- To anticipate possible temporary heat and power requirements.
- To price certain general requirement items and overhead costs.
- To budget for equipment usage.
- To anticipate and justify material and equipment delivery requirements. (An awareness of these requirements may, for example, justify using a more expensive vendor quote with better delivery terms.)

The schedule produced prior to bidding may be a simple bar chart or network diagram that includes overall quantities, probable delivery times, and available manpower. Network scheduling methods, such as the Critical Path Method (CPM) and the Precedence Chart simplify pre-bid scheduling because they do not require time-scaled line diagrams.

In the CPM Diagram, the activity is represented by an arrow. Nodes indicate start/stop between activities. The Precedence Diagram, on the other hand, shows the activity as a node with arrows used to denote precedence relationships between the activities. The precedence arrows may be used in different configurations to represent the sequential relationships between activities. Examples of CPM and Precedence diagrams are shown in Figures 9.1 and 9.2, respectively. In both systems, duration times are indicated along each path. The sequence (path) of activities requiring the most total time represents the shortest possible time (critical path) in which those activities may be completed.

For example, in both Figure 9.1 and Figure 9.2, activities A, B, and C require 20 successive days for completion before activity G can begin. Activity paths for D and E (15 days), and for F (12 days) are shorter and can easily be completed during the 20-day sequence. Therefore, this 20-day sequence is the shortest possible time (i.e., the "critical path") for the completion of these activities — before activity G can begin.

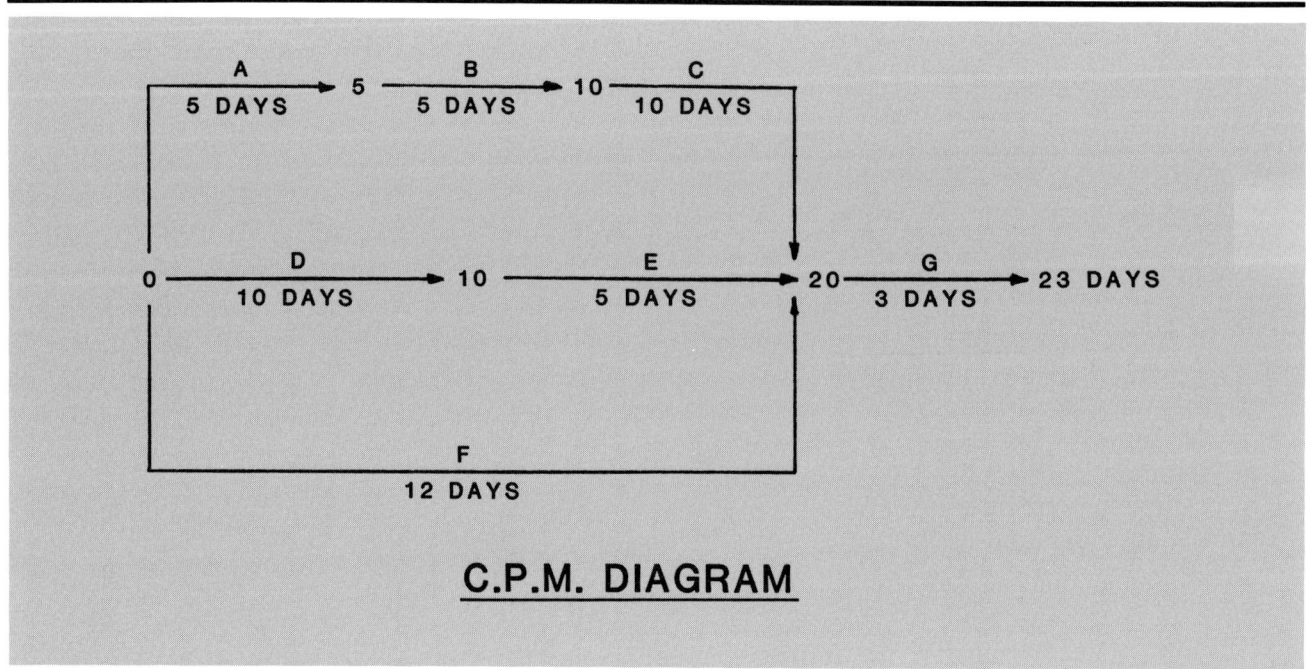

Figure 9.1

Past experience or a prepared rough schedule may suggest that the time specified in the bidding documents is insufficient to complete the required work. In such cases, a more comprehensive schedule should be produced prior to bidding; this schedule will help to determine the added overtime or premium time work costs required to meet the completion date.

A three-story office building project is used for the Sample Estimates in Part Three of this book. A preliminary schedule is needed to determine the supervision and manning requirements of the job. The specifications state that the building must be completed within one year. Normally, the excavation, foundations, and superstructure can be completed in approximately one half of the construction period. Therefore, the major portion of the mechanical work is restricted to the last six months of the one year allotted.

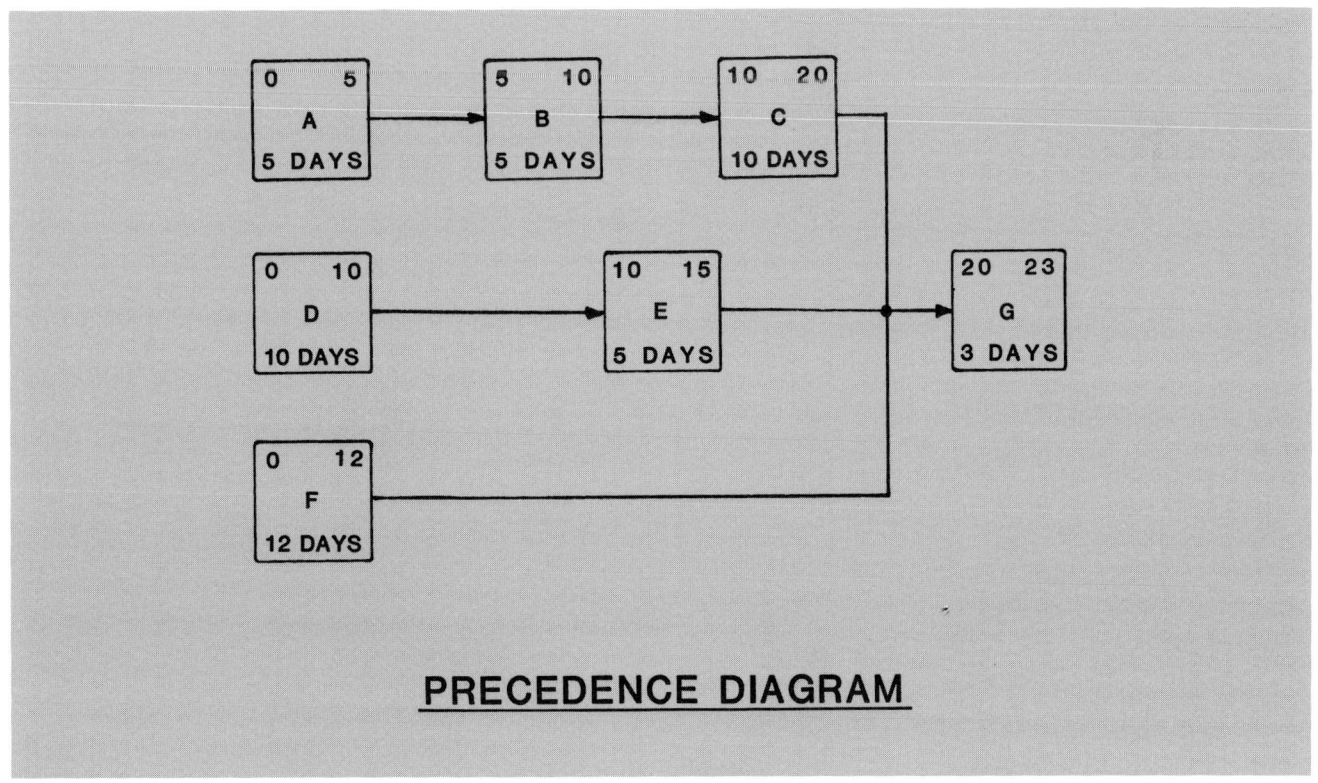

PRECEDENCE DIAGRAM

Figure 9.2

A rough schedule for the mechanical work might be produced as shown in Figure 9.3. The man-days used to develop this schedule are derived from output figures determined in the estimate. Output can be determined based on the figures in *Means Mechanical Cost Data*. Man-days can also be figured by dividing the total labor cost shown on the estimate by the cost per man-day for each appropriate tradesperson.

As shown, the preliminary schedule can be used to determine supervision requirements, to develop appropriate crew sizes, and as a basis for ordering materials. All of these factors must be considered at this preliminary stage in order to determine how to meet the required one-year completion date.

A pre-bid schedule can provide much more information than simple job duration. It can be used to refine the estimate by introducing realistic manpower projections. The schedule may also help the contractor to adjust the structure and size of the company based on projected requirements for months, even years ahead. A schedule can also become an effective tool for negotiating contracts.

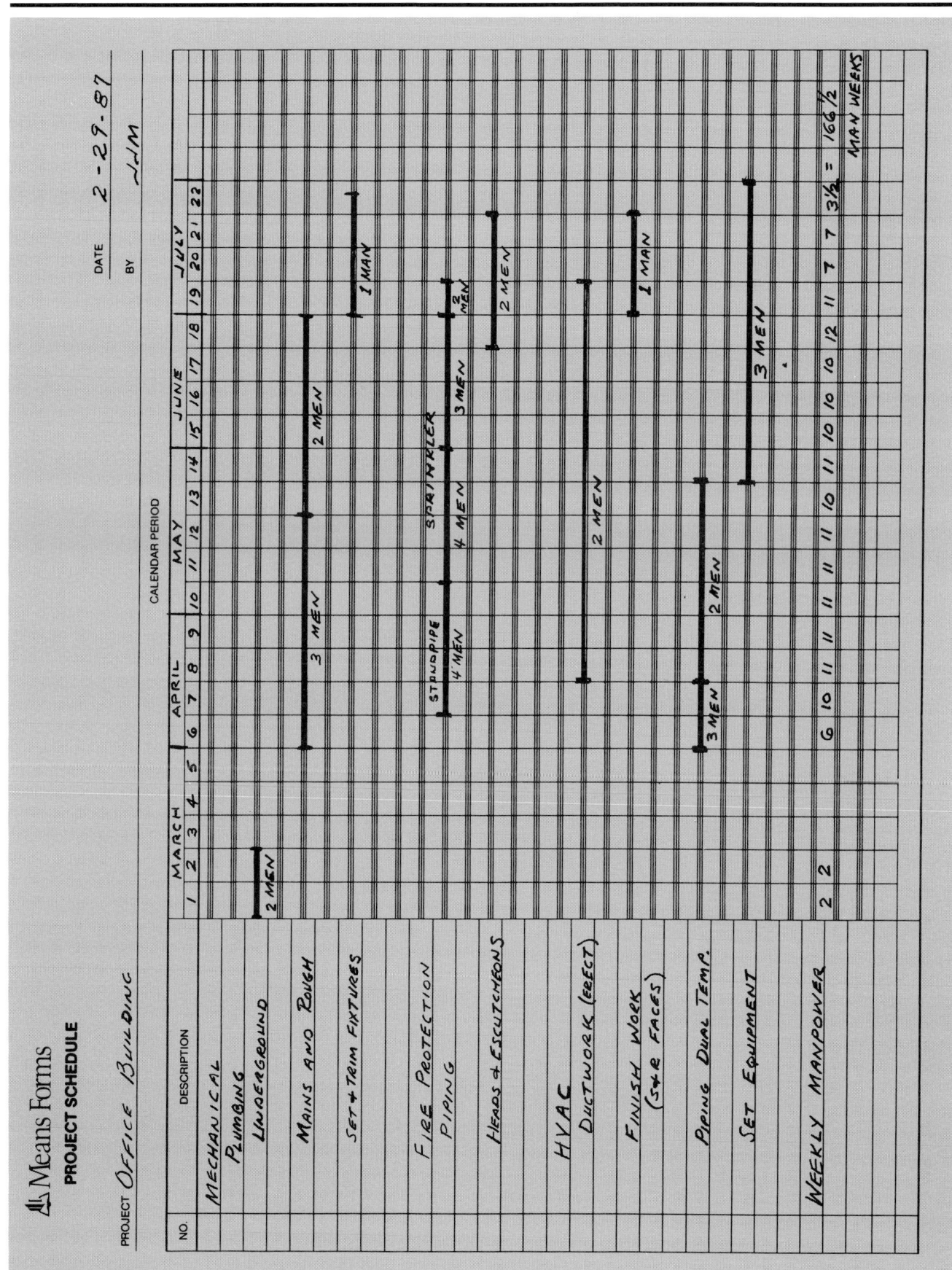

Figure 9.3

Chapter 10
BIDDING STRATEGIES

The goal of most contractors is to make as much money as possible on each job, but more importantly, to maximize return on investment on an annual basis. Often, this can be done by taking *fewer* jobs at a *higher* profit and by limiting bidding to the jobs which are most likely to be successful for the company.

Resource Analysis

Since most contractors cannot physically bid every job in a geographic area, a selection process must determine which projects to bid. This process should begin with an analysis of the strengths and weaknesses of the contractor. The following items must be considered as objectively as possible:
- Individual strengths of the company's top management
- Management experience with the type of construction involved, from top management to project superintendents
- Cost records adequate for the appropriate type of construction
- Bonding capability and capacity
- Size of projects with which the company is "comfortable"
- Geographic area that can be managed effectively
- Unusual corporate assets such as:
    - Specialized equipment availability
    - Reliable and timely cost control systems
    - Strong balance sheet
    - Familiarity with designer, owner, or general contractor

Market Analysis

Most contractors tend to concentrate on a few particular kinds of projects. From time to time, the company should step back and examine the portion of the industry they are serving. During this process, the following items should be carefully analyzed:
- Historical trend of the market segment
- Expected future trend of the market segment
- Geographic expectations of the market segment
- Historical and expected competition from other contractors
- Risk involved in the particular market segment
- Typical size of projects in this market
- Expected return on investment from the market segment

If several of these areas are experiencing a downturn, then it is definitely appropriate to examine an alternate market. On the other hand, many managers would feel that "bad times" are the times when they should consolidate and narrow their market. These managers are only likely to expand or broaden their market when current activities are especially strong.

Bidding Analysis

Certain steps should be taken to develop a bid strategy within a particular market. The first is to obtain the bid results of jobs in the prospective geographic area. These results should be set up on a tabular basis. This is fairly easy to do in the case of public jobs since the bid results are normally published (or at least available) from the agency responsible for the project. In private work, this step is more difficult, since the bid results are not normally divulged by the owner.

Determining Risk in a New Market Area

One way to measure success in bidding is how much money is "left on the table", the difference between the low bid and next lowest bid. The contractor who consistently takes jobs by a wide margin below the next bidder is obviously not making as much money as possible. Information on competitive public bidding is used to determine the amount of money left on the table; this information serves as the basis for fine-tuning a future bidding strategy.

For example, assume a public market where all bid prices and the total number of bidders are known. For each "type" of market sector, create a chart showing the percentage left on the table versus the total number of bidders. When the median figure (percent left on the table) for each number of bidders is connected with a smooth curve, the usual shape of the curve is shown in Figure 10.1.

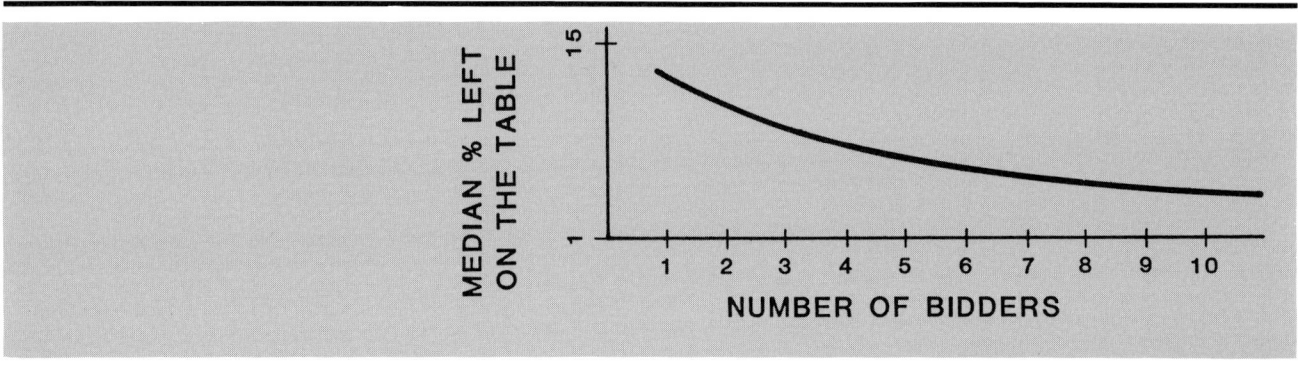

Figure 10.1

The exact shape and magnitude of the amounts left on the table will depend on how much risk is involved with that type of work. If the percentages left on the table are high, then the work can be assumed to be very risky — with a high profit or loss potential if the award is won. If the percentages are low, the work is probably neither as risky nor as potentially profitable.

Analyzing the Bid's Risk

If a company has been bidding in a particular market, certain information should be collected and recorded as a basis for a bidding analysis. First, the percentage left on the table should be tabulated (as shown in Figure 10.1), along with the number of bidders for the projects in that market on which the company was the low bidder. By probability, half the bids should be above the median line and half below. If more than half are below the line, the company is doing well; if more than half are above, the bidding strategy should be examined.

Maximizing the Profit-to-Volume Ratio

Once the bidding track record for the company has been established, the next step is to reduce the historical percentage left on the table. One method is to create a chart showing, for instance, the last ten jobs on which the company was low bidder, and the dollar spread between the low and second lowest bid. Next, rank the percentage differences from one to ten (one being the smallest and ten being the largest left on the table). An example is shown in Figure 10.2. This example is for a larger general contractor, but the principles apply to any size or type of contractor involved in bidding.

Job No.	"Cost"	Low Bid	Second Bid	Difference	% Diff.	% Rank	Profit (Assumed at 10%)
1	$ 918,000	$ 1,009,800	$1,095,000	$ 85,200	9.28	10	$ 91,800
2	1,955,000	2,150,500	2,238,000	87,500	4.48	3	195,500
3	2,141,000	2,355,100	2,493,000	137,900	6.44	6	214,100
4	1,005,000	1,105,500	1,118,000	12,500	1.24	1	100,500
5	2,391,000	2,630,100	2,805,000	174,900	7.31	8	239,100
6	2,782,000	3,060,200	3,188,000	127,800	4.59	4	278,200
7	1,093,000	1,202,300	1,282,000	79,700	7.29	7	109,300
8	832,000	915,200	926,000	10,800	1.30	2	83,200
9	2,372,000	2,609,200	2,745,000	135,800	5.73	5	237,200
10	1,681,000	1,849,100	2,005,000	155,900	9.27	9	168,100
	$17,170,000	$18,887,000		$1,008,000			$1,717,000 = 10% of Cost

Figure 10.2

The company's "costs" ($17,170,000) are derived from the company's low bids ($18,887,000) assuming a 10% profit ($1,717,000). The "second bid" is the next lowest bid. The "difference" is the dollar amount between the low bid and the second bid (money "left on the table"). The differences are then ranked based on the percentage of job "costs" left on the table for each. Figure the median difference by averaging the two middle percentages — that is, the fifth and sixth ranked numbers.

$$\text{Median \% Difference} = \frac{5.73 + 6.44}{2} = 6.09\%$$

From Figure 10.2, the median percentage left on the table is 6.09%. To maximize the potential returns on a series of competitive bids, a useful formula is needed for pricing profit. The following formula has proven effective.

$$\text{Normal Profit \%} + \frac{\text{Median \% Difference}}{2} = \text{Adjusted Profit \%}$$

$$10.00 + \frac{6.09}{2} = 13.05\%$$

Now apply this adjusted profit percentage to the same list of ten jobs as shown in Figure 10.3. Note that the job "costs" remain the same, but that the low bids have been revised. Compare the bottom line results of Figure 10.2 to those of Figure 10.3 based on the two profit margins, 10% and 13.05%, respectively.

Total volume *drops* from $18,887,000 to $17,333,900.

Net profits *rise* from $1,717,000 to $2,000,900.

Job No.	Company's "Cost"	Revised Low Bid	Second Bid	Adj. Diff.	Profit [10% + 3.05%]	Total
1	$ 918,000	$ 1,037,800	$1,095,000	$ 57,200	$ 91,800 + $28,000	$119,800
2	1,955,000	2,210,100	2,238,000	27,900	195,500 + 59,600	255,100
3	2,141,000	2,420,400	2,493,000	72,600	214,100 + 65,300	279,400
4	(1,005,000)	(1,136,100)	1,118,000(L)	—	100,500 + 30,600	0
5	2,391,000	2,703,000	2,805,000	102,000	239,100 + 72,900	312,000
6	2,782,000	3,145,100	3,188,000	42,900	278,200 + 84,900	363,100
7	1,093,000	1,235,600	1,282,000	46,400	109,300 + 33,300	142,600
8	(832,000)	(940,600)	926,000(L)	—	83,200 + 25,400	0
9	2,372,000	2,681,500	2,745,000	63,500	237,200 + 72,300	309,500
10	1,681,000	1,900,400	2,005,000	104,600	168,100 + 51,300	219,400
	$15,333,000	$17,333,900		$517,100		$2,000,900

Figure 10.3

Profits rise while volume drops! If the original volume is maintained or even increased, profits rise even further. Note how this occurs. By determining a reasonable increase in profit margin, the company has, in effect, raised all bids. By doing so, the company loses two jobs to the second bidder (jobs 4 and 8 in Figure 10.3).

A positive effect of this volume loss is reduced exposure to risk. Since the profit margin is higher, the remaining eight jobs collectively produce more profit than the ten jobs based on the original, lower profit margin. From where did this money come? The money "left on the table" has been reduced from $1,008,000 to $517,100. The whole purpose is to systematically lessen the dollar amount difference between the low bids and the second low bids. Caution: This is a hypothetical approach based upon the following assumptions.

- Profit must be assumed to be the bid or estimated profit, not the actual profit when the job is over.
- Bidding must be done within the same market in which data for the analysis was gathered.
- Economic conditions should be stable from the time the data is gathered until the analysis is used in bidding. If conditions change, use of such an analysis should be reviewed.
- Each contractor must make roughly the same number of bidding mistakes. For higher numbers of jobs in the sample, this requirement becomes more probable.
- The company must bid additional jobs if total annual volume is to be maintained or increased. Likewise, if net total profit margin is to remain constant, even fewer jobs need be won.
- Finally, the basic cost numbers and sources must remain constant. If a new estimator is hired or a better cost source is found, this technique cannot work effectively until a new track record has been established.

The accuracy of this strategy depends upon the criteria listed above. Nevertheless, it is a valid concept that can be applied, with appropriate and reasonable judgement, to many bidding situations.

Chapter 11
COST CONTROL AND ANALYSIS

An internal accounting system should be used by contractors and construction managers to logically gather and track the costs of a construction project. With this information, a cost analysis can be made about each activity — both during the installation process and at its conclusion. This information or "feedback" becomes the basis for management decisions through the duration of the project. This cost data is also helpful for future bid proposals.

The categories for a mechanical project are major items of construction (e.g., HVAC) which can be subdivided into component activities (e.g., ductwork, piping, controls, etc.). These activities should coincide with the system and methods of the quantity takeoff. Uniformity is important in terms of the units of measure and in the grouping of components into cost centers. For instance, cost centers for a mechanical project might include categories such as HVAC, insulation, and plumbing. Activities within one of these categories — plumbing, for example, might include such items as fixtures, water supply, waste, and vent piping.

The major purposes of cost control and analysis are as follows:
- To provide management with a system to monitor costs and progress
- To provide cost feedback to the estimator(s)
- To determine the costs of change orders
- To be used as a basis for progress payment requisitions to the owner, the general contractor, or his representative
- To manage cash flow
- To identify areas of potential cost overruns to management for corrective action

It is important to establish a cost control system that is uniform — both throughout the company and from job to job. Such a system might begin with a uniform Chart of Accounts, a listing of code numbers for work activities. The Chart of Accounts is used to assign time and cost against work activities for the purpose of creating cost reports. A Chart of Accounts should have enough scope and detail so that it can be used for any of the projects that the company may win. Naturally, an effective Chart of Accounts will also be flexible enough to incorporate new activities as the company takes on new or different kinds of projects. Using a cost control system, the various costs can be consistently allocated. The following information should be recorded for each cost component:

- Labor charges in dollars and man-days are summarized from weekly time cards and distributed by code.
- Quantities completed to date must also be recorded in order to determine unit costs.
- Equipment rental costs are derived from purchase orders or from weekly charges issued by an equipment company.
- Material charges are determined from purchase orders.
- Appropriate subcontractor charges are allocated.
- Job overhead items may be listed separately or by component.

Each component of costs — labor, materials, and equipment — is now calculated on a unit basis by dividing the quantity installed to date into the cost to date. This procedure establishes the actual installed unit cost to date.

At this point, it is also useful to calculate the percent complete to date. This is done in two steps, or levels. First, the percent complete for each activity is calculated based on the actual quantity installed to date divided by the total quantity estimated for the activity. Second, the project total percent complete is calculated. To arrive at this number, multiply the percent complete of each activity times the total estimated cost for that activity. Then sum these results to a total (sometimes called "earned" dollars) and divide this sum by the estimated total. The result is the project's overall percent complete.

The quantities that remain to be installed for each activity should be estimated based on the actual unit cost to date. This is the projected cost to complete each activity. Due to inefficiencies and mobilization costs as the work begins, it is not practical to use the actual units for projecting costs until an activity is 20% complete. Below 20%, the costs as originally estimated should be used. The actual costs to date are added to the projected costs to obtain the anticipated costs at the end of the project.

Typical forms that may be used to develop a cost control system are shown in Figures 11.1 to 11.6.

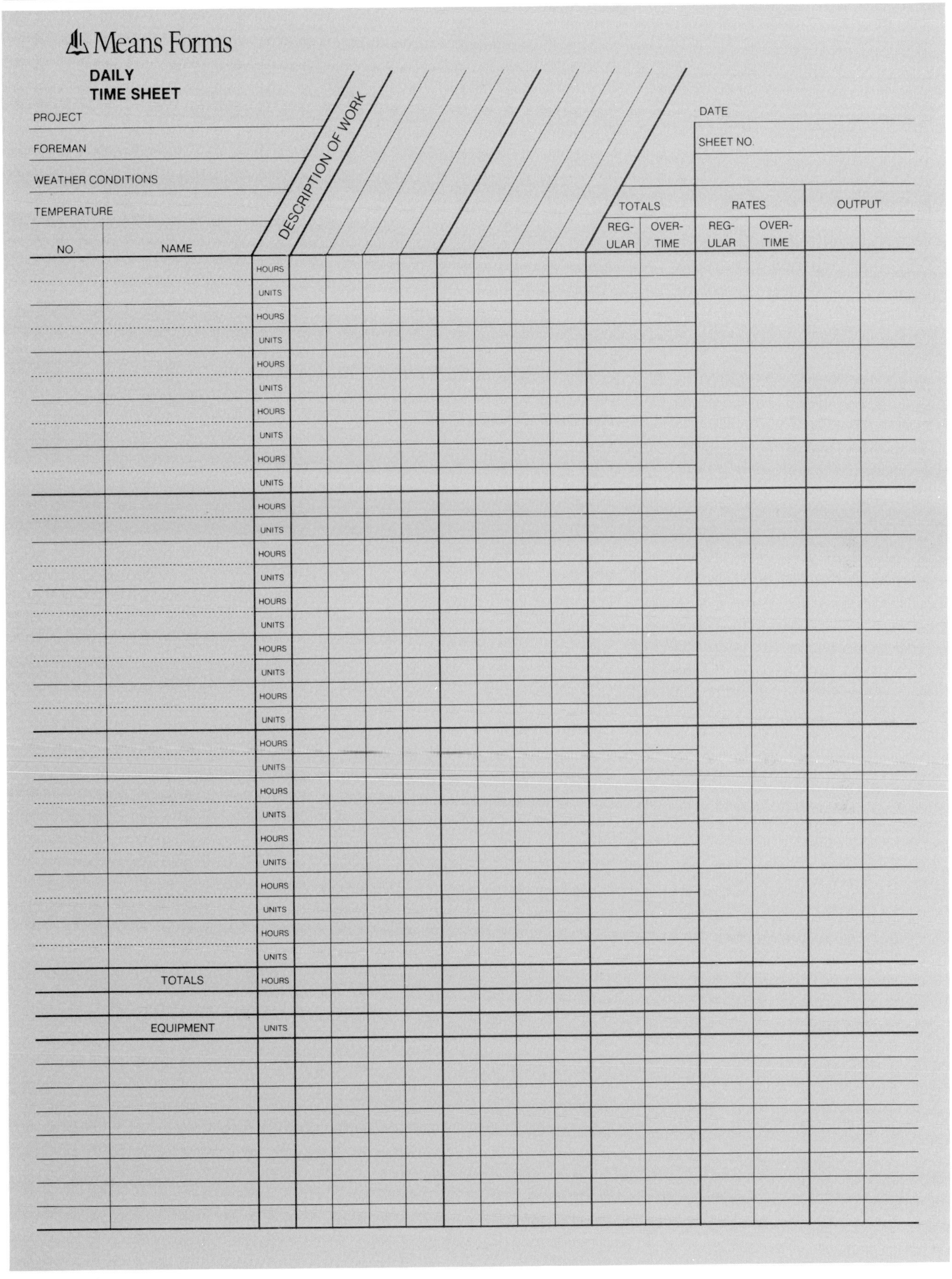

Means Forms

DAILY TIME SHEET

PROJECT

FOREMAN

WEATHER CONDITIONS

TEMPERATURE

DATE

SHEET NO.

NO.	NAME	DESCRIPTION OF WORK									TOTALS		RATES		OUTPUT	
											REG-ULAR	OVER-TIME	REG-ULAR	OVER-TIME		
		HOURS														
		UNITS														
		HOURS														
		UNITS														
		HOURS														
		UNITS														
		HOURS														
		UNITS														
		HOURS														
		UNITS														
		HOURS														
		UNITS														
		HOURS														
		UNITS														
		HOURS														
		UNITS														
		HOURS														
		UNITS														
		HOURS														
		UNITS														
		HOURS														
		UNITS														
		HOURS														
		UNITS														
		HOURS														
		UNITS														
		HOURS														
		UNITS														
		HOURS														
		UNITS														
	TOTALS	HOURS														
	EQUIPMENT	UNITS														

Figure 11.1

The analysis of categories serves as a useful management tool, providing information on a constant, up-to-date basis. Immediate attention is attracted to any center that is projecting a loss. Management can concentrate on this item in an attempt to make it profitable or to minimize the expected loss.

Means Forms

WEEKLY TIME SHEET

EMPLOYEE	EMPLOYEE NUMBER	WEEK ENDING
FOREMAN	HOURLY RATE	PIECE WORK RATE

PROJECT	NO.	DESCRIPTION OF WORK	DAY								TOTAL	HOURS		RATE	TOTAL COST	UNIT COST
			HOURS									REG.				
			UNITS									O.T.				
			HOURS									REG.				
			UNITS									O.T.				
			HOURS									REG.				
			UNITS									O.T.				
			HOURS									REG.				
			UNITS									O.T.				
			HOURS									REG.				
			UNITS									O.T.				
			HOURS									REG.				
			UNITS									O.T.				
			HOURS									REG.				
			UNITS									O.T.				
		TOTAL	HOURS									REG.				
			UNITS									O.T.				

PAYROLL DEDUCTIONS	F.I.C.A.	INCOME TAXES FEDERAL	INCOME TAXES STATE	INCOME TAXES CITY/COUNTY	HOSPITAL-IZATION			TOTAL DEDUCTIONS		LESS DEDUCTIONS
	$	$	$	$	$	$	$	$		NET PAY

Figure 11.2

Means Forms

LABOR
COST RECORD

SHEET NO.

DATE FROM:

PROJECT

DATE TO:

LOCATION

BY:

DATE	CHARGE NO.	DESCRIPTION	HOURS	RATE	AMOUNT	HOURS	RATE	AMOUNT	HOURS	RATE	AMOUNT

Figure 11.3

Means Forms

MATERIAL
COST RECORD

SHEET NO.

DATE FROM

PROJECT

DATE TO

LOCATION

BY

DATE	NUMBER	VENDOR/DESCRIPTION	QTY.	UNIT PRICE		QTY.	UNIT PRICE		QTY.	UNIT PRICE	

Figure 11.4

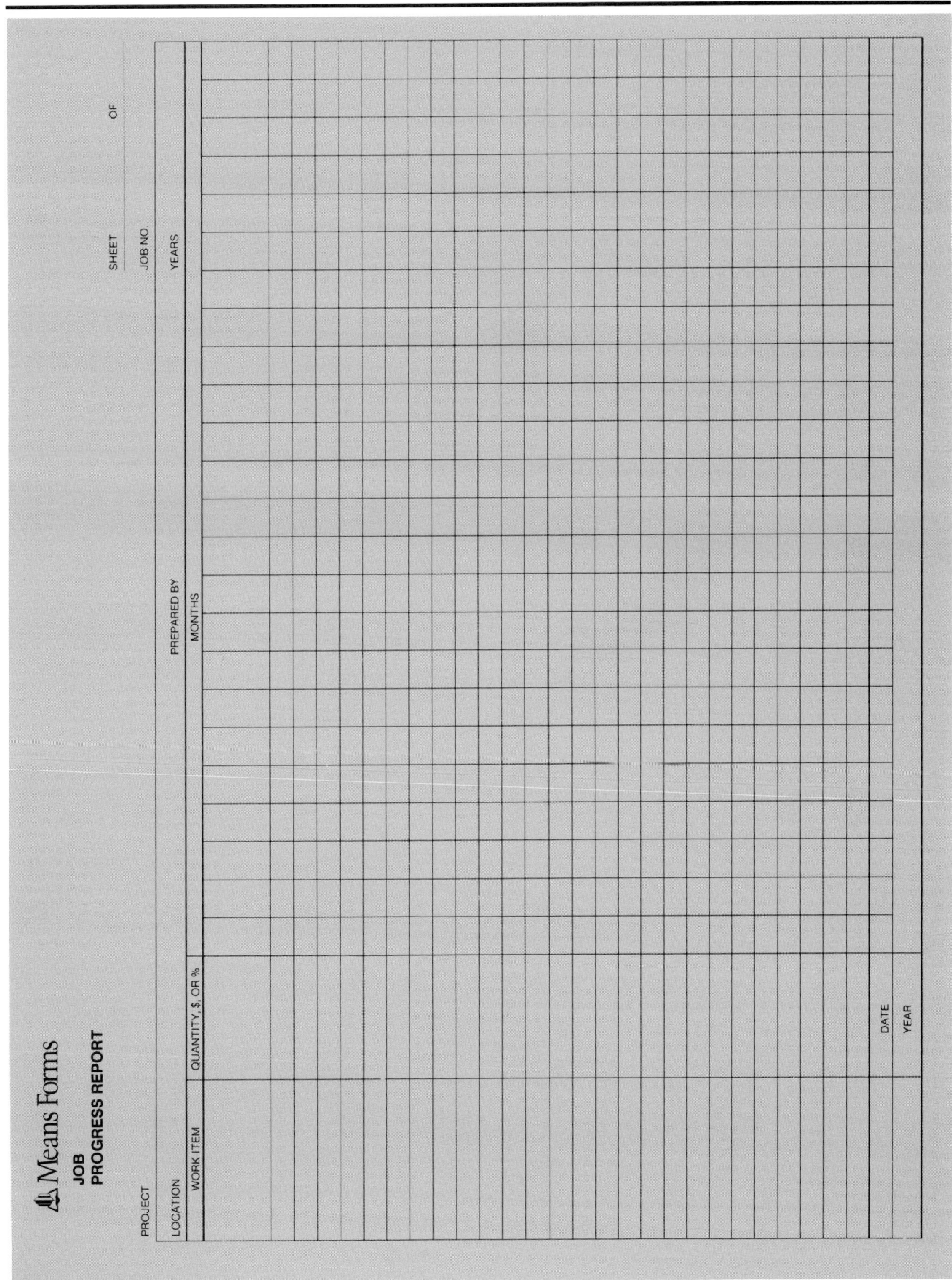

Figure 11.5

Means Forms
PERCENTAGE COMPLETE ANALYSIS

PAGE
PROJECT DATE
ARCHITECT BY FROM TO

NO	DESCRIPTION	ACTUAL OR ESTIMATED	TOTAL PROJECT	THIS PERIOD QUANTITY	%	QUANTITY	PERCENT TOTAL TO DATE 10	20	30	40	50	60	70	80	90	100
		ACTUAL														
		ESTIMATED														
		ACTUAL														
		ESTIMATED														
		ACTUAL														
		ESTIMATED														
		ACTUAL														
		ESTIMATED														
		ACTUAL														
		ESTIMATED														
		ACTUAL														
		ESTIMATED														
		ACTUAL														
		ESTIMATED														
		ACTUAL														
		ESTIMATED														
		ACTUAL														
		ESTIMATED														
		ACTUAL														
		ESTIMATED														
		ACTUAL														
		ESTIMATED														
		ACTUAL														
		ESTIMATED														
		ACTUAL														
		ESTIMATED														
		ACTUAL														
		ESTIMATED														
		ACTUAL														
		ESTIMATED														
		ACTUAL														
		ESTIMATED														
		ACTUAL														
		ESTIMATED														
		ACTUAL														
		ESTIMATED														

Figure 11.6

The estimating department can use the unit costs developed in the field as background information for future bidding purposes. Particularly useful are unit labor costs and unit man-hours (productivity) for the separate activities. This information should be integrated into the accumulated historical data. A particular advantage of unit man-hour records is that they tend to be constant over time. Current unit labor costs can be figured simply by multiplying these man-hour standards times the current labor rate per hour.

Frequently, items are added to or deleted from the contract either via field change orders or through contract amendments. Accurate cost records are an excellent basis for determining the cost changes that will result.

As discussed above, the determination of completed quantities is necessary in order to calculate unit costs. These quantities are used to figure the percent complete in each category. These percentages can, in turn, be used to calculate the billing for progress payment requisitions (invoices).

A cost system is only as good as the people responsible for coding and recording the required information. Simplicity is the key word. Do not try to break down the code into very small items unless there is a specific need. Continuous updating of reports is important so that operations which are not in control can be immediately brought to the attention of management.

Also, be sure to draft clear directions and instructions for each phase of the process. Adequate time must be spent to ensure that all who are involved (especially the foremen and supervisors) clearly understand the program.

Productivity and Efficiency

When using a cost control system such as the one described above, the unit costs should reflect standard practices. Productivity should be based on a five-day, eight-hour-per-day (during daylight hours) workweek. Exceptions can be made if a company's requirements are particularly and most often unique. Installation costs should be derived using normal minimum crew sizes, under normal weather conditions, during the normal construction season.

All unusual costs incurred or expected should be recorded separately for each category of work. For example, an overtime situation might occur on every job and in the same proportion. In this case, it would make sense to carry the unit price adjusted for the added cost of premium time. Likewise, unusual weather delays, strike activity, owner/architect delays, or contractor interference should have separate, identifiable cost contributions; these are applied as isolated costs to the activities affected by the delays. This procedure serves two purposes:

- To identify and separate the cost contribution of the delay so that future job estimates will not automatically include an allowance for these "non-typical" delays, and
- To serve as a basis for an extra compensation claim and/or as justification for reasonable extension of the job.

Overtime Impact

The use of long-term overtime is counter-productive on almost any construction job; that is, the longer the period of overtime, the lower the actual production rate. There have been numerous studies conducted which come up with slightly different numbers, but all reach the same conclusion. Figure 11.7 tabulates the effects of overtime work on efficiency.

Days per Week	Hours per Day	Production Efficiency					Payroll Cost Factors	
		1 Week	2 Weeks	3 Weeks	4 Weeks	Average 4 Weeks	@ 1½ Times	@ 2 Times
5	8	100%	100%	100%	100%	100%	100%	100%
	9	100	100	95	90	96.25	105.6	111.1
	10	100	95	90	85	91.25	110.0	120.0
	11	95	90	75	65	81.25	113.6	127.3
	12	90	85	70	60	76.25	116.7	133.3
6	8	100	100	95	90	96.25	108.3	116.7
	9	100	95	90	85	92.50	113.0	125.9
	10	95	90	85	80	87.50	116.7	133.3
	11	95	85	70	65	78.75	119.7	139.4
	12	90	80	65	60	73.75	122.2	144.4
7	8	100	95	85	75	88.75	114.3	128.6
	9	95	90	80	70	83.75	118.3	136.5
	10	90	85	75	65	78.75	121.4	142.9
	11	85	80	65	60	72.50	124.0	148.1
	12	85	75	60	55	68.75	126.2	152.4

Figure 11.7

As illustrated in Figure 11.8, there can be a difference between the *actual* payroll cost per hour and the *effective* cost per hour for overtime work. This is due to the reduced production efficiency with the increase in weekly hours beyond 40. This difference between actual and effective cost results from overtime work over a prolonged period. Short-term overtime work does not result in as great a reduction in efficiency, and in such cases, effective cost may not vary significantly from the actual payroll cost. As the total hours per week are increased on a regular basis, more time is lost because of fatigue, lowered morale, and an increased accident rate.

					Payroll Cost per Hour		Effective Cost per Hour	
Days per Week	Hours per Day	Total Hours Worked	Actual Productive Hours	Production Efficiency	Overtime after 40 hrs.		Overtime after 40 hrs.	
					@ 1-1/2 times	@ 2 times	@ 1-1/2 times	@ 2 times
5	8	40	40.0	100.0%	100.0%	100.0%	100.0%	100.0%
	9	45	43.4	96.5	105.6	111.1	109.4	115.2
	10	50	46.5	93.0	110.0	120.0	118.3	129.0
	11	55	49.2	89.5	113.6	127.3	127.0	142.3
	12	60	51.6	86.0	116.7	133.3	135.7	155.0
6	8	48	46.1	96.0	108.3	116.7	112.8	121.5
	9	54	48.9	90.6	113.0	125.9	124.7	139.1
	10	60	51.1	85.2	116.7	133.3	137.0	156.6
	11	66	52.7	79.8	119.7	139.4	149.9	174.6
	12	72	53.6	74.4	122.2	144.4	164.2	194.0
7	8	56	48.8	87.1	114.3	128.6	131.1	147.5
	9	63	52.2	82.8	118.3	136.5	142.7	164.8
	10	70	55.0	78.5	121.4	142.9	154.5	181.8
	11	77	57.1	74.2	124.0	148.1	167.3	199.6
	12	84	58.7	69.9	126.2	152.4	180.6	218.1

Figure 11.8

As an example, assume a project where workers are working 6 days a week, 10 hours per day. From Figure 11.8 (based on productivity studies), the actual productive hours are 51.1 hours. This represents a theoretical production efficiency of 51.1/60 or 85.2%.

Depending upon the locale and day of week, overtime hours may be paid at time and a half or double time. For time and a half, the overall (average) *actual* payroll cost (including regular and overtime hours) is determined as follows:

For time and a half:

$$\frac{40 \text{ reg. hrs.} + (20 \text{ overtime hrs.} \times 1.5)}{60 \text{ hrs.}} = 1.167$$

Based on 60 hours, the payroll cost per hour will be (on average) 116.7% of the normal rate at 40 hours per week. However, because the actual production (efficiency) for 60 hours is reduced to the equivalent of 51.1 hours, the *effective* cost of overtime is calculated as shown below:

For time and a half:

$$\frac{40 \text{ reg. hrs.} + (20 \text{ overtime hrs.} \times 1.5)}{51.1 \text{ hrs.}} = 1.37$$

Installed cost will be 137% of the normal rate (for labor).

Thus, when figuring overtime, the actual cost per unit of work will be higher than the apparent overtime payroll dollar increase, due to the reduced productivity of the longer work week. These calculations are true only for those cost factors determined by hours worked. Costs that are applied weekly or monthly, such as equipment rentals, will not be similarly affected.

Retainage and Cash Flow

The majority of construction projects have some percentage of retainage held back by the owner until the job is complete and accepted. This retainage can range from 5% to as high as 15% or 20% in unusual cases. The most typical retainage is 10%. Since the profit on a given job may be less than the amount of withheld retainage, the contractor must wait longer before a positive cash flow is achieved than if there were no retainage.

Figures 11.9 and 11.10 are graphic and tabular representations of the projected cash flow for a small project. With this kind of projection, the contractor is able to anticipate cash needs throughout the course of the job. Note that at the eleventh of May, before the second payment is received, the contractor has paid out about $25,000 more than has been received. This is the maximum amount of cash (on hand or financed) that is required for the whole project. At an early stage of planning, the contractor can determine if there will be adequate cash available or if a loan is needed. In the latter case, the expense of interest could be anticipated and included in the estimate. On larger projects, the projection of cash flow becomes crucial, because unexpected interest expense can quickly erode profits.

A note on the subject of financing may be helpful here. In the above example, assume that the contractor has adequate cash resources to finance the project. Does this mean that the project need not bear any interest charges? No. At the very least, money withdrawn from savings or investments will result in unrealized interest (effective losses). These "losses" should be included in the project records for the purpose of determining actual job profit. For this reason, many companies assign interest charges to their jobs for negative cash flow. Likewise, the project is credited if a positive cash flow is realized. In this way, an owner or manager can readily assess the contribution (or liability) of each job to the company's "financial health". For example, if a project has a 5% profit at completion but has incurred 10% in "finance charges", it certainly has not helped the company to stay in business.

The General Conditions section of the specifications usually explains the responsibilities of both the owner and the contractor with regard to billing and payments. Even the best planning and projections are contingent upon the general contractor paying requisitions as anticipated. There is an almost unavoidable adversary relationship between the subcontractor and general contractor regarding payment during the construction process. However, it is in the best interest of the general contractor that the subcontractor be solvent so that delays, complications, and financial difficulties can be avoided prior to final completion of the project. The interest of both parties is best served if information is shared and communication is open. Both are working toward the same goal: the timely and successful completion of the project.

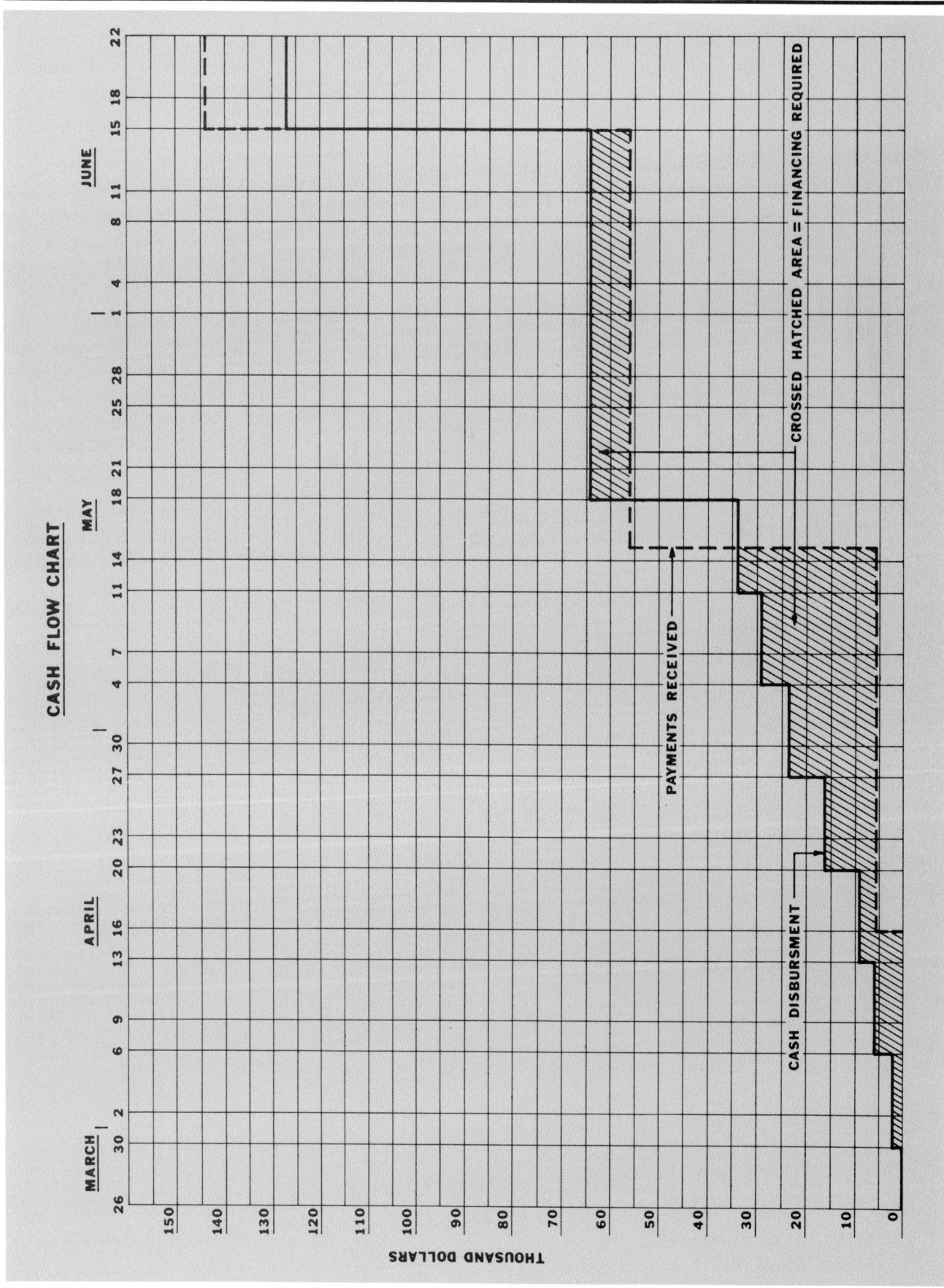

Figure 11.9

Date	Payroll Incl. Taxes	Workers Comp.	Monthly Billing & Payment	Retainage	Subs Billing & Payment	Retainage	Material Incl. Taxes	Equip. Incl. Taxes	Accumulated Costs
3-30	Payroll	11926	1148						— 11926
3-30	Monthly Billing			4558	506	708			
4-6	Payroll	3246	253						— 5172
4-13	Payroll	4197	311						— 9369
4-16	Payment			4558					— 4811
4-20	Pay Sub					708	77		— 5519
4-20	Pay Material							515	— 6304
4-20	Pay Equipment							965	— 6999
4-20	Payroll	3821	275						— 10826
4-27	Payroll	4525	333						— 15351
4-30	Monthly Billing			51330	5704	5245	582		
5-4	Payroll	5525	508						— 20876
5-11	Payroll	3631	370						— 24507
5-15	Payment			51330					— 24923
5-18	Payroll	2613	243						— 24310
5-18	Pay Sub					5245			— 19065
5-18	Pay Material							8112	779	— 3053
5-18	Pay Equipment									— 3832
6-1	Final Billing			88883	9876	48915	5435		
6-15	Payment			88883					+ 85051
	Pay Subs					48916			+ 36136
	Pay Material						12685		+ 23451
	Pay Equipment							738	+ 22713
		$ 29320	$ 2441	$ 144771	$ 16080	$ 54868	$ 6094	$ 35818	$ 2489

Cash 22713
Retainage 16080
Pay Workers Comp 2441 — 38719
Pay Subs Retainage 36358 — 6094
General Conditions 30264 — 7034 — 23240

General Conditions
Permit 150 300
Supervision Carpenter Foreman 35 days @ 170 5950
Temp. Power & Water 6 wks @ $30 450 200
Temp. Office & Storage 6 wks @ $30 180
Clean Up Laborer 2 days @ $122 244 — 7034

Figure 11.10

Life Cycle Costs

Life cycle costing is a valuable method of evaluating the total costs of an economic unit during its entire life. Regardless of whether the unit is a piece of excavating machinery or a manufacturing building, life cycle costing gives the owner an opportunity to look at the economic consequences of a decision. Today, the initial cost of a unit is often not the most important cost factor; the operation and maintenance costs of some building types far exceed the initial outlay. Hospitals, for example, may have operating costs within a three year period that exceed the original construction costs.

Estimators are in the business of initial costs. But what about the other costs that affect any owner: taxes, fuel, inflation, borrowing, estimated salvage at the end of the facility's lifespan, and expected repair or remodeling costs? These costs that may occur at a later date need to be evaluated in advance. The thread that ties all of these costs together is the time-value of money. The value of money today is quite different from what it will be tomorrow. One thousand dollars placed in a savings bank at 5% interest will, in four years, have increased to $1,215.50. Conversely, if $1,000 is needed four years from now, then $822.70 should be placed in an account today. Another way of saying the same thing is that at 5% interest, the value of $1,000 four years from now is only worth $822.70 today.

Using interest and time, future costs are equated to the present by means of a present worth formula. Standard texts in engineering economics have outlined different methods for handling interest and time. A present worth evaluation could be used, or all costs might be converted into an equivalent, uniform annual cost method.

The present worth of a piece of equipment (such as a trenching machine) can be figured, based on a given lifespan, anticipated maintenance cost, operating costs, and money borrowed at the current rate. Having these figures can help in determining whether to rent or purchase. The costs between two different machines can also be analyzed in this manner to determine which is a better investment.

Chapter 12
ESTIMATING WITH A COMPUTER

The estimating process discussed in the preceding chapters can be enhanced with the use of a computer and appropriate software. Computerization can add efficiency to virtually all phases of a contractor's management tasks — from accounting — to estimating — to project management. Since these functions are interrelated, any decision about computer application should be based on the whole operation.

The question of computer application should be considered only after a well-organized system of accounting, estimating, and management is in place. Too often, a computer and software are brought in to cure problems that demand a more basic solution. Be sure that your move to computerize is made to enhance a system with which you are already working comfortably. If the old process is familiar and running smoothly, the shock of learning a new procedure will be lessened. Remember that output from a computer is no better than the input which generates it.

With these cautions in mind, let us take a look at what the computer can do for your estimating process. As a first step, define how your estimates fit into your project management and business administration systems. Will you use the estimate to control the project? Will your job accounting system be based on your estimates as well? If so, look at ways to interrelate these functions. There are integrated software packages for either general or specific applications. These packages provide the framework to incorporate these varied functions under a common operating system. Integrated software packages may be desirable, but they tend to be quite structured and, as a result, may not meet all your individual needs.

Another option is a stand-alone specific applications program for estimating, scheduling, or other functions. These packages are capable of performing specified tasks. These programs may be capable of interchanging information, or they may support more than one function; for example an estimating program may also support scheduling.

A third option is the use of general application programs: spreadsheets, word processors, or data base managers. Because these programs are designed for general use, they are quite flexible. As a result, they require the user to create an outline of the system to be followed. For example, an estimating system can be set up on a spreadsheet, or an historical cost system can be set up on a data base manager. The advantage of increased flexibility is offset by the need to develop your own specific application. Creating one's own custom software — either internally or through an outside consultation — is also an option. With the ever-widening selection of software products, creating one's own custom software is not a cost-effective option for most companies. While learning to use many off-the-shelf programs requires a considerable time investment, it is insignificant compared to the worker-years required to fully develop a top notch package.

Deciding which of these options to pursue will depend largely upon your specific needs and the goals of your company. The choice will most likely involve a good deal of research. The rest of this chapter presents and examines guidelines and how specific application and general application can be applied to the estimating function in your company.

Specific Application Programs

Specific application programs are available to perform a variety of functions for the construction estimator. Programs vary from general estimating packages to integrated accounting, estimating, and scheduling packages. Estimating packages incorporating automatic quantity takeoff are also available. These programs use a digitizer board in conjunction with the computer to electronically read drawings and record quantities. Some of these programs have been developed specifically for the mechanical contractor and include routines to facilitate pipe and ductwork takeoff and pricing. In addition, some of these specialized programs can help to lay out ductwork for fabrication by graphically depicting pieces and their dimensions and efficiently laying them out for production. Computer controlled production equipment can be teamed with several of these packages to reduce fabrication labor and material waste. Specialty contractors should carefully investigate their options, as these type of programs are rapidly becoming more varied and sophisticated.

For the contractor looking for a partially integrated estimating program with automated takeoff capabilities, there are several options. An example of this program and its capabilities is the Means *Galaxy System*.

R.S. Means *Galaxy* is an estimating program for the micro-computer. The program allows the user to read drawings electronically, record the quantities in an estimate, and price the estimate based on cost records (Means annual construction cost data in the form of computer files). An available module also allows the user to develop a schedule as the project progresses.

The *Galaxy* quantity takeoff system (see Figure 12.1) uses a digitizer table (plan table with an electronic stylus, which can pinpoint locations on the grid). By inputting lineal dimensions from project plans into the program using the electronic pen, accurate readings are made based on lengths, perimeters, areas, volumes, and numeric count. The measured values are then recorded as estimate quantities. Using the paging feature of the software, similar, frequently used items can be grouped together to streamline the process. For the mechanical contractor, the quantity takeoff system could readily calculate lengths of pipe, area of duct, or fitting counts, for example.

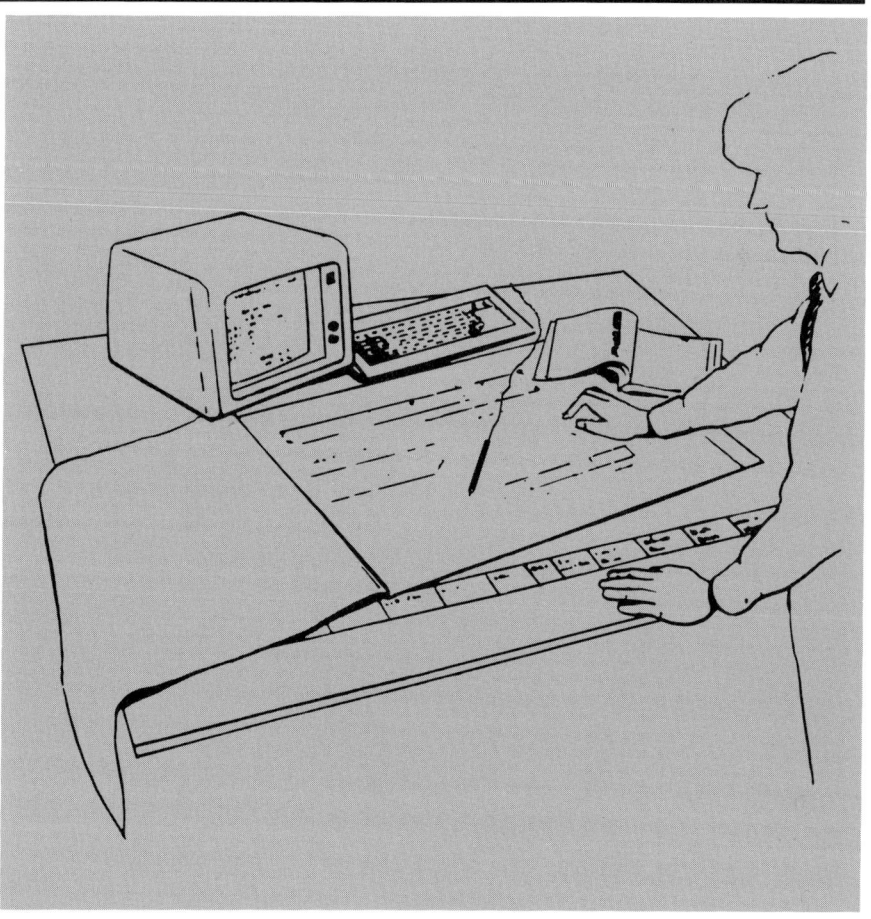

Figure 12.1

As the takeoff is made, data from the historical cost file is applied to the quantities in order to provide an item description and costs. Estimate items can then be listed separately and/or summarized by division to produce the estimate. As items are input, or upon later review, costs can be changed from those recorded in the historical data base. This feature allows the user to create an estimate that is responsive to the individual project.

A typical estimate also includes subcontracted items. These items can be listed in the takeoff and priced as a lump sum quotation in the estimate.

Standard Reports

When the takeoff and pricing are completed, estimate reports can be generated. There are 26 standard reports in the Galaxy program. The following discussion includes a brief description of the available reports and several examples from an estimate prepared on Galaxy.

Estimate Checklist: This report can be printed out as you are saving cost data information to your estimate or after your estimate is complete. It provides a hard copy of all cost data information stored on your diskette.

Burdened and Unburdened Reports: A Burdened Report uses cost figures that include national average overhead and profit costs of the installing contractor. An Unburdened Report uses only bare costs. There are three formats for each report, described below.

- A **Job Report Summary** (division totals only) provides a one-line summary by Construction Specifications (CSI) division. After printing each division, the system provides a menu for the user to enter mark-up factors for sales tax, material, labor, equipment, subcontractor, profit, and bond. These factors are reflected in the final portion of the report. An example of an Unburdened Job Report Summary is shown in Figure 12.2.
- A **Division Summary Report** (divisions and subdivisions) provide a one-line summary for each CSI subdivision and division contained in an estimate. An illustration of an Unburdened Division Summary is shown in Figure 12.3.
- An **Itemized Job Report** prints out each cost line contained in an estimate. All elements in the estimate are sorted and totalled by CSI subdivision and division. The user can select a report that presents only one specific division or all divisions covered in an estimate. A sample Unburdened Itemized Job Report is shown in Figure 12.4.

Percent Factor Report: Galaxy contains two fields for factoring labor, material, and quantity price by percentage. A separate report is generated to show the effect that these percentages have on the estimate. Items are listed by CSI subdivision with cost totals. The Percent 1 column is indicated by one asterisk (*); the Percent 2 column is indicated by two asterisks (**). A Percent 1 and 2 Report appears in Figure 12.5.

```
===============================================================================
NO. A-102                    UNBURDENED JOB REPORT SUMMARY           03-05-1985  01:11:19  PAGE  1
-------------------------------------------------------------------------------

PROJECT   : Means R & D Building         LOCATION : Kingston, MA
ARCHITECT : John Roth                    OWNER    : R.S. Means Co., Inc.
QUANTITIES: gfk                          ENTERED BY: gfk

===============================================================================
NO. DIVISION                             MANHOURS  MATERIAL   LABOR    EQUIP      SUB
                                                                               CONTRACT
-------------------------------------------------------------------------------
 1 GENERAL REQUIREMENTS                       4       135       90       0      4723
-------------------------------------------------------------------------------
 2 SITEWORK                                 593       829     9676    8547       163
-------------------------------------------------------------------------------
 3 CONCRETE                                  33     14878      665      41         0
-------------------------------------------------------------------------------
 4 MASONRY                                 1838    577651    34170     193         0
-------------------------------------------------------------------------------
 5 METALS                                    87      2180     1735       0         0
-------------------------------------------------------------------------------
 6 WOOD & PLASTICS                           45       725     1120      39         0
-------------------------------------------------------------------------------
 7 MOISTURE PROTECTION                      187      5376     3713       0         0
-------------------------------------------------------------------------------
 9 FINISHES                                  77       809     1878       0         0
-------------------------------------------------------------------------------
15 MECHANICAL                                41     29913      845       0         0
-------------------------------------------------------------------------------
16 ELECTRICAL                                26       200      580       0         0
-------------------------------------------------------------------------------
                        JOB TOTAL         2931    632696    54472    8820      4886
-------------------------------------------------------------------------------
                        SALES TAX           5.50%   34798
                        MATERIAL MARK-UP    7.00%   44289
                        LABOR MARK-UP      35.00%            19065
                        EQUIPMENT MARK-UP  10.00%                      882
                        SUBCONTRACTOR MARK-UP 9.00%                              440

  USER-DEFINED MARKUPS ──────────▶       711783    73537    9702      5326

                        TOTAL BEFORE PROFIT                                   800348
                        PROFIT             15.00%                             120052
                        BOND                0.02%                                160
                                                                          ==========
                        JOB TOTAL PROFIT & BOND INCLUDED              $       920560
-------------------------------------------------------------------------------
```

Figure 12.2

```
========================================================================================================
NO. A-102              UNBURDENED DIVISION SUMMARY                    03-05-1985  01:14:56  PAGE   1
--------------------------------------------------------------------------------------------------------

PROJECT   : Means R & D Building              LOCATION : Kingston, MA
ARCHITECT : John Roth                         OWNER    : R.S. Means Co., Inc.
QUANTITIES: gfk                               ENTERED BY: gfk

========================================================================================================

                                    1 GENERAL REQUIREMENTS

--------------------------------------------------------------------------------------------------------
SECTION NO.    DESCRIPTION                        MANHOURS  MATERIAL   LABOR   EQUIP   TOTAL    SUB
                                                                                        COST   CONTRACT
========================================================================================================
1.1            OVERHEAD                                  4       135      90       0     225     4723

========================================================================================================
                                 DIVISION TOTAL :        4       135      90       0     225     4723
========================================================================================================

                                         2 SITEWORK

--------------------------------------------------------------------------------------------------------
SECTION NO.    DESCRIPTION                        MANHOURS  MATERIAL   LABOR   EQUIP   TOTAL    SUB
                                                                                        COST   CONTRACT
========================================================================================================
2.1            EXPLORATION & CLEARING                  574         0    9377    8417   17794        0

2.3            EARTHWORK                                 7       135     121      65     321        0

2.6            ROADS & WALKS                            11       694     178      65     937      163

========================================================================================================
                                 DIVISION TOTAL :      593       829    9676    8547   19052      163
========================================================================================================

                                         3 CONCRETE

--------------------------------------------------------------------------------------------------------
SECTION NO.    DESCRIPTION                        MANHOURS  MATERIAL   LABOR   EQUIP   TOTAL    SUB
                                                                                        COST   CONTRACT
========================================================================================================
3.1            FORMWORK                                  7       104     173       1     278        0

3.2            REINFORCING STEEL                        11       225     229       0     454        0

3.3            CAST IN PLACE CONCRETE                   15     14549     263      40   14852        0

========================================================================================================
                                 DIVISION TOTAL :       33     14878     665      41   15584        0
========================================================================================================
```

Figure 12.3

```
===============================================================================================
NO. A-102                    UNBURDENED ITEMIZED JOB REPORT           03-05-1985  01:20:12  PAGE  1
-----------------------------------------------------------------------------------------------

PROJECT   : Means R & D Building              LOCATION : Kingston, MA
ARCHITECT : John Roth                         OWNER    : R.S. Means Co., Inc.
QUANTITIES: gfk                               ENTERED BY: gfk

===============================================================================================
                                        4 MASONRY

-----------------------------------------------------------------------------------------------
4.1 MORTAR & MASONRY ACCESSORIES
-----------------------------------------------------------------------------------------------
DESCRIPTION                                 CREW  QUANTITY  UNIT    M/H    MATERIAL   LABOR    EQUIP    TOTAL    SUB
LINE NO.              TAG
ASMBLY# RENUMBER1     RENUMBER2   PER1 PER2
-----------------------------------------------------------------------------------------------
CONTROL JOINT, PVC 4" AND WIDER WALL
  041 200 0050 00  M                        BRIC1  272.50  L.F.   0.013      1.20     0.27     0.00     1.47
  04-1200  BLD W       PLAN 12    98%  112%                       3.63     327.00    72.30     0.00   399.30

GROUTING CAVITY WALL 2" SPACE, .167@C@F/@S@F, PUMPED
  041 300 0500 00  M                        D4    2725.00  S.F.   0.016      0.36     0.28     0.07     0.71
  04-1200  BLD W       PLAN 12    98%  112%                      43.60     981.00   756.46   192.79  1930.25

WALL TIES FOR BRICK VENEER GALV. CORRG 7/8" X 7", 24 GAGE
  041 700 0010 00  M                        BRIC1   54.50  C      0.762      3.10    15.16     0.00    18.26
  04-1200  BLD W       PLAN 12    98%  112%                      41.52     168.95   826.32     0.00   995.27

===============================================================================================
                                        SUB TOTAL :          89      1477     1655      193     3325        0
-----------------------------------------------------------------------------------------------
4.2 BRICK MASONRY
-----------------------------------------------------------------------------------------------
DESCRIPTION                                 CREW  QUANTITY  UNIT    M/H    MATERIAL   LABOR    EQUIP    TOTAL    SUB
LINE NO.              TAG
ASMBLY# RENUMBER1     RENUMBER2   PER1 PER2
-----------------------------------------------------------------------------------------------
BRICK VENEER 4" THICK RUNNING BOND
  042 560 0010 00  U                        D2    5450.00  S.F.   0.200      1.68     3.75     0.00     5.43
  04-1200  BLD W       PLAN 12    98%  112%                    1090.00    9156.00 20432.54     0.00 29588.54

WASHING BRICK, SMOOTH BRICK
  042 680 0010 00  M                        BRIC1 5450.00  S.F.   0.014      0.04     0.28     0.00     0.32
  04-1200  BLD W       PLAN 12    98%  112%                      77.86     218.00  1549.36     0.00  1767.36

===============================================================================================
                                        SUB TOTAL :        1168      9374    21982        0    31356        0
```

Figure 12.4

```
===============================================================================================
NO. A-132                    Cost of Project Due to % Markups          05-20-1986 10:40:39 PAGE 6
-----------------------------------------------------------------------------------------------

PROJECT     : Means R & D Building              LOCATION : Kingston, MA
ARCH/ENGR   : John Roth                         OWNER    : R.S. Means Co., Inc.
QUANTITIES BY: gfk                              ENTERED BY: gfk

===============================================================================================

                                    4 MASONRY

-----------------------------------------------------------------------------------------------
 4.1 MORTAR & ACCESSORIES
-----------------------------------------------------------------------------------------------
DESCRIPTION                                 CREW  QUANTITY UNIT    M/H   MATERIAL    LABOR    EQUIP     TOTAL     SUB
 LINE NO.            TAG        *     **                                    *                  **
ASMBLY# RENUMBER1   RENUMBER2  PER1  PER2
-----------------------------------------------------------------------------------------------
CONTROL JOINT, PVC 4" AND WIDER WALL
   041 200 0050 00  M                       BRIC1  272.50 L.F.    0.013      1.18     0.30     0.00      1.47
 04-1200  BLD W      PLAN 12    98%  112%                          3.63    320.46    80.98     0.00    401.44

GROUTING CAVITY WALL 2" SPACE, .167@C@F/@S@F, PUMPED
   041 300 0500 00  M                       D4    2725.00 S.F.    0.016      0.35     0.31     0.08      0.74
 04-1200  BLD W      PLAN 12    98%  112%                         43.60    961.38   847.24   215.93   2024.54

WALL TIES FOR BRICK VENEER GALV. CORRG 7/8" X 7", 24 GAGE
   041 700 0010 00  M                       BRIC1   54.50 C       0.762      3.04    16.98     0.00     20.02
 04-1200  BLD W      PLAN 12    98%  112%                         41.52    165.57   925.48     0.00   1091.05

===============================================================================================
                                         SUB TOTAL :     89      1447      1854      216      3517        0
-----------------------------------------------------------------------------------------------
 4.2 BRICK MASONRY
-----------------------------------------------------------------------------------------------
DESCRIPTION                                 CREW  QUANTITY UNIT    M/H   MATERIAL    LABOR    EQUIP     TOTAL     SUB
 LINE NO.            TAG        *     **                                    *                  **
ASMBLY# RENUMBER1   RENUMBER2  PER1  PER2
-----------------------------------------------------------------------------------------------
BRICK VENEER 4" THICK RUNNING BOND
   042 560 0010 00  U                       D2    5450.00 S.F.    0.200      1.65     4.20     0.00      5.85
 04-1200  BLD W      PLAN 12    98%  112%                       1090.00   8972.88 22884.45     0.00  31857.33

WASHING BRICK, SMOOTH BRICK
   042 680 0010 00  M                       BRIC1 5450.00 S.F.    0.014      0.04     0.32     0.00      0.36
 04-1200  BLD W      PLAN 12    98%  112%                         77.86    213.64  1735.28     0.00   1948.92

===============================================================================================
                                         SUB TOTAL :   1168      9187     24620        0     33806        0
-----------------------------------------------------------------------------------------------
 4.3 BLOCK & TILE MASONRY
-----------------------------------------------------------------------------------------------
DESCRIPTION                                 CREW  QUANTITY UNIT    M/H   MATERIAL    LABOR    EQUIP     TOTAL     SUB
 LINE NO.            TAG        *     **                                    *                  **
ASMBLY# RENUMBER1   RENUMBER2  PER1  PER2
-----------------------------------------------------------------------------------------------
8" X 16" X 8" CONCRETE BLOCK PARTITIONS (INC SCAFFOLDING)
   043 520 0010 00  U                       D8    5450.00 S.F.    0.107    101.92     2.16     0.00    104.08
 04-1200  BLD W      PLAN 12    98%  112%                        581.33 555464.00 11797.81     0.00 567261.81
```

Figure 12.5

Renumber Report: Renumber reports summarize the estimate by using the code field Renumber #1 or Renumber #2. Items are listed by CSI division with cost totals. These reports are available with Galaxy. An example of a Renumber 1 Report is shown in Figure 12.6.

Job Crews Report: This report prints a listing of each crew being used in an estimate.

User Crews Report: This report is a listing of all user-built crews. It is similar to the Job Crews Report.

Trade Report: This report prints a listing of all trades required for the job, the man-hours involved for each trade, the hourly base rate, the total base pay, the hourly rate with overhead and profit, and total pay with overhead and profit. The man-hours, total base pay, and total pay with overhead and profit are totalled for all the trades. An illustration of a Trade Report is shown in Figure 12.7.

Assembly File Report: The Assembly Report puts together a collection of up to 59 related tasks per assembly under one number. The report includes the CSI line number, quantity, description, man-hours, material, installation, and total cost of each component of the assembly, as well as for the total assembly. A sample Assembly Report is shown in Figure 12.8.

User File Report: This report prints all the line items that have been created and stored in the User File, sorted by CSI subdivision. The line items in this file are modified Means lines or unique user lines. An example of a User File Report is shown in Figure 12.9.

Labor Cost Report: This report prints all the labor trades. The listing indicates the hourly base rate and hourly rate including the subcontractor's overhead and profit. It includes both the standard equipment codes and rates, plus any non-standard equipment codes and rates created by the user.

The preceding descriptions and example of available reports provides a good idea of how *Galaxy* can organize and expedite the estimating process. Once the estimate has been put together, the real power of the computer can be employed. The reporting features can be used to obtain the required information with the appropriate amount of detail to serve the needs of the end user.

Once created, the estimate can be modified to meet changing project requirements, and re-used as the basis for new projects. The ability to analyze different alternatives (What if I change this price, material, or method of construction?) is greatly improved when the estimate is prepared on the computer.

```
===============================================================================================================
NO. A-102                        All Items Located in Building W                    05-20-1986 11:50:42 PAGE  5
---------------------------------------------------------------------------------------------------------------

PROJECT     : Means R & D Building              LOCATION : Kingston, MA
ARCH/ENGR   : John Roth                         OWNER    : R.S. Means Co., Inc.
QUANTITIES BY: qfk                              ENTERED BY: qfk
RENUMBER1   : BLD W                             RENUMBER2 :

===============================================================================================================

                                  7 MOISTURE PROTECTION

---------------------------------------------------------------------------------------------------------------
  7.1 WATERPROOFING
---------------------------------------------------------------------------------------------------------------
DESCRIPTION                                  CREW  QUANTITY UNIT   M/H   MATERIAL   LABOR   EQUIP   TOTAL   SUB
LINE NO.            TAG
ASMBLY# RENUMBER1   RENUMBER2   PER1 PER2
---------------------------------------------------------------------------------------------------------------
CAULKING & SEALANTS, POLYETHYLENE BACKER ROD, 1/4" DIAMETER
   071 200 0030 00  M                        BRIC1  54.50 C.L.F.  1.739    3.50    34.61    0.00    38.11
 04-1200 BLD W      PLAN 12      98% 112%                        94.78  190.75  1886.17    0.00  2076.92

CAULKING COMPOUND BUTYL BASE, BULK IN PLACE, 1/4"X1/2" BEAD 54LF/GAL
   071 200 1700 00  M                        BRIC1 681.25 L.F.   0.035    0.22     0.69    0.00     0.91
 04-1200 BLD W      PLAN 12      98% 112%                        23.70  149.88   471.54    0.00   621.42

===============================================================================================================
                                             SUB TOTAL :          118     341     2358       0     2698      0
---------------------------------------------------------------------------------------------------------------
  7.2 INSULATION
---------------------------------------------------------------------------------------------------------------
DESCRIPTION                                  CREW  QUANTITY UNIT   M/H   MATERIAL   LABOR   EQUIP   TOTAL   SUB
LINE NO.            TAG
ASMBLY# RENUMBER1   RENUMBER2   PER1 PER2
---------------------------------------------------------------------------------------------------------------
CMU INSUL POURED PERLITE 8" WALL
   072 250 0300 00  M                        D1    5450.00 S.F.   0.007    0.85     0.12    0.00     0.97
 04-1200 BLD W      PLAN 12      98% 112%                         36.33 4632.50   642.19    0.00  5274.69

===============================================================================================================
                                             SUB TOTAL :           36    4633      642       0     5275      0
---------------------------------------------------------------------------------------------------------------
  7.6 SHEET METAL WORK
---------------------------------------------------------------------------------------------------------------
DESCRIPTION                                  CREW  QUANTITY UNIT   M/H   MATERIAL   LABOR   EQUIP   TOTAL   SUB
LINE NO.            TAG
ASMBLY# RENUMBER1   RENUMBER2   PER1 PER2
---------------------------------------------------------------------------------------------------------------
FLASHING ALUMINUM MILL FINISH .019" THICK
   076 250 0060 00  M                        SHEE1  545.00 S.F.   0.055    0.63     1.20    0.00     1.83
 04-1200 BLD W      PLAN 12      98% 112%                         30.07  343.35   652.50    0.00   995.85

===============================================================================================================
                                             SUB TOTAL :           30     343      652       0      996      0
===============================================================================================================
                                             DIVISION TOTAL :     185    5316     3652       0     8969      0
===============================================================================================================
```

Figure 12.6

```
===============================================================================
NO. A-102                      TRADE REPORT                    05-20-1986  16:44:44  PAGE  1
-------------------------------------------------------------------------------
PROJECT        : Means R & D Building        LOCATION : Kingston, MA
ARCH/ENGR      : John Roth                   OWNER    : R.S. Means Co., Inc.
QUANTITIES BY  : gfk                         ENTERED BY: gfk
===============================================================================
```

TRADE	DESCRIPTION	MANHOURS	HOURLY BASE	TOTAL BASE	HOURLY O & P	TOTAL O & P
BRHE	BRICKLAYER HELPERS	668.864	15.45	10333.94	21.90	14648.11
BRIC	BRICKLAYERS	1301.105	19.90	25891.98	28.20	36691.15
CAIR	LABORER AIR TOOL	145.241	15.70	2280.28	22.67	3292.61
CARP	CARPENTERS	88.295	25.00	2207.38	30.00	2648.86
CEFI	CEMENT FINISHERS	396.937	18.60	7383.04	26.00	10320.37
ELEC	ELECTRICIANS	27.266	22.10	602.57	31.65	862.96
ELEV	ELEVATOR CONSTRUCTORS	11.900	21.10	251.09	30.50	362.95
EQHV	EQUIPMENT OPERATORS CRANE OR S	96.971	20.15	1953.96	29.10	2821.85
EQMD	MEDIUM EQUIPMENT OPERATORS	1.043	19.70	20.55	28.45	29.68
EQMM	EQUIPMENT OPERATORS MASTER MEC	6.250	20.85	130.31	30.10	188.13
EQOL	EQUIPMENT OPERATORS OILERS	99.352	16.80	1669.11	24.25	2409.29
RODM	RODMEN (REINFORCING)	11.032	20.80	229.47	31.85	351.38
ROFP	ROOFER , PRECAST	8.000	18.65	149.20	28.32	226.56
ROFP (a)	ROOFER , PRECAST	8.000	14.92	119.36	22.66	181.25
SHEE	SHEET METAL WORKERS	30.069	21.70	652.50	31.55	948.68
SSWK	STRUCTURAL STEEL WORKERS	16.806	20.90	351.25	32.85	552.08
SSWK (a)	STRUCTURAL STEEL WORKERS	6.250	16.72	104.50	26.28	164.25
TILF	TILE LAYERS (FLOOR)	8.000	18.55	148.40	25.95	207.60
TOTALS FOR 18 TRADE(S):		2,931.381 hrs.		$54,478.90		$76,907.77

Figure 12.7

```
================================================================================================================
                                           ASSEMBLY REPORT              RUN: 07-10-1985    11:28:58    PAGE 1
    12" DIA. ROUND TIED COLUMN, 4KSI,100K MAX, 10' STORY                 CREATED: 07-10-1985  11:21:00
----------------------------------------------------------------------------------------------------------------
    NUMBER    UNIT         HOURS      MATERIAL      LABOR      EQUIPMENT     TOTAL
    03-0000   V.L.F.        0.38        5.96         7.20        0.50        13.66
----------------------------------------------------------------------------------------------------------------
    COMPONENT    QUANTITY UNIT   DESCRIPTION                                              HOURS  MATERIAL  INSTAL  TOTAL
    0312501600    1.00 L.F.      FORMS IN PLACE, COLUMNS, ROUND FIBER TUBE, 12" DIAMETER   0.21    2.85    4.10    6.95
    0320400200U   5.22 LB.       REINFORCING IN PLACE, COLUMNS, #3 TO #7, @A615 GRADE 60   0.06    1.30    1.19    2.49
    0320400200U   1.39 LB.       REINFORCING IN PLACE, COLUMNS, #3 TO #7, @A615 GRADE 60   0.01    0.35    0.32    0.66
    0331200010U   0.03 C.Y.      CONCRETE, READY MIX REGULAR WEIGHT 3000 PSI               0.00    1.30    0.00    1.30
    0333800400    0.03 C.Y.      PLACING CONCRETE, INCL. VIBRATING, 12" SQ./ROUND COLUMNS, PUMPED  0.04  0.00  1.05  1.05
    0332800050    3.14 S.F.      FINISHING WALL, BREAK TIES, PATCH VOIDS,BURLAP RUB W/GROUT 0.06    0.16    1.05    1.21
```

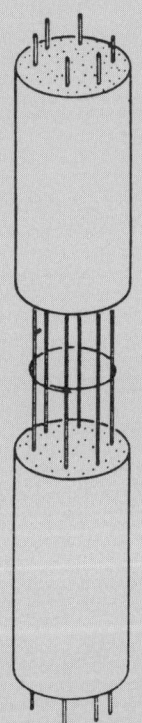

```
                        ASSEMBLY COMPONENTS

            ASSEMBLY 03-0000
            ROUND TIED COLUMNS, 4 KSI CONCRETE, 100K MAX. LOAD, 10' STORY, 12" SIZE
                Forms in place, columns, round fiber tube, 12" diameter
                Reinforcing in place, columns, #3 to #7
                Reinforcing in place, column ties
                Concrete ready mix, regular weight, 3000 PSI
                Place and vibrate concrete, 12" sq./round columns, pumped
                Finish, burlap rub w/grout

                                                         TOTAL
```

This ROUND TIED COLUMN represents an ASSEMBLY in the ASTRO program. By definition an assembly is a collection of unit prices originating from the R. S. Means cost file, the User's own cost file, or any combination. The assembly may contain up to 50 components. To call the above assembly in an estimate the estimator would key in the code 03-0000. Astro then retrieves the unit price components of the assembly. Once the estimator has created a library of assemblies they are permanently stored on the computer's hard disk.

Figure 12.8

```
================================================================================
                          USER FILE REPORT            03-06-1985  01:30:19  PAGE   1
================================================================================

                              2 SITEWORK
--------------------------------------------------------------------------------
2.1 EXPLORATION & CLEARING
--------------------------------------------------------------------------------
DESCRIPTION
LINE NO.                      CREW    M/H   UNIT   MATERIAL   LABOR   EQUIP   INSTALL   TOTAL   OANDP
--------------------------------------------------------------------------------
CLEAR MEDIUM TREES TO 10"DIA CUT & CHIP
021 100 0200                  B7    53.333  ACRE     0.00    870.67  781.56  1652.22  1652.22  2112.83

================================================================================
                              3 CONCRETE
--------------------------------------------------------------------------------
3.3 CAST IN PLACE CONCRETE
--------------------------------------------------------------------------------
DESCRIPTION
LINE NO.                      CREW    M/H   UNIT   MATERIAL   LABOR   EQUIP   INSTALL   TOTAL   OANDP
--------------------------------------------------------------------------------
4000 PSI CONCRETE READY MIX, REGULAR WEIGHT
033 120 0010                   1     0.000  C.Y.    49.25     0.00    0.00     0.00    49.25    0.00

CONCRETE IN PLACE INCL.FORMS AND REINFORCING, SLAB ON GRADE, 6" THICK
033 140 4700                  C17C   0.800  C.Y.    55.00    16.00    3.28    19.28    74.28   87.34

================================================================================
                              4 MASONRY
--------------------------------------------------------------------------------
4.1 MORTAR & MASONRY ACCESSORIES
--------------------------------------------------------------------------------
DESCRIPTION
LINE NO.                      CREW    M/H   UNIT   MATERIAL   LABOR   EQUIP   INSTALL   TOTAL   OANDP
--------------------------------------------------------------------------------
WALL TIES TO BRICK VENNEER CORRUGATED GALVANIZED 16 GAUGE
041 700 0010                  BRIC1  0.762   C      9.00    15.16    0.00    15.16    24.16   31.39

--------------------------------------------------------------------------------
4.2 BRICK MASONRY
--------------------------------------------------------------------------------
DESCRIPTION
LINE NO.                      CREW    M/H   UNIT   MATERIAL   LABOR   EQUIP   INSTALL   TOTAL   OANDP
--------------------------------------------------------------------------------
BRICK VENEER 4" THICK RUNNING BOND
042 560 0010                  D2     0.200  S.F.    1.68     3.75    0.00     3.75    5.43    7.06
```

Figure 12.9

Galaxy is an example of a special application program for estimating. It has the capability to perform quantity takeoffs, price estimate items, summarize the estimate, and print a variety of reports. As an added feature, it can also support a scheduling module to generate proper schedules from the estimate. Like any computer base system, it becomes more powerful the more it can be re-used for similar tasks. The program is representative of special application software currently available in the marketplace.

Acquiring and learning to use *Galaxy* involves a significant investment of resources. It is, therefore, a good idea to investigate your options carefully before committing to any system. You may have to adapt your present estimating system for this type of program. However, the many built-in features can organize the estimating process, as well as saving a great deal of time.

General Application Programs

General application programs can be adapted to meet many of the construction estimator's needs. As noted in preceding chapters, it is very important for the estimator to set up a system of forms to record information and organize the estimating process. General application programs are well suited to this purpose. Once created, computer-generated forms can be readily updated, modified, and re-used to facilitate the recording and analysis of construction project information. Understanding this application of general business software provides an idea of its overall capabilities.

Electronic Forms

Forms created on the computer offer the user a high degree of flexibility. The electronic form can be easily modified to suit specific requirements of individual preferences. Once an original copy of the form has been input, multiple variations can be made using the editing capabilities of the selected program. The original form can be "saved" and duplicated; thus the duplicate can be modified to create the desired variations, while the original remains intact. The ability to save and duplicate is a powerful feature of the computer.

The computer's ability to make changes rapidly adds to the flexibility of the electronic form. Once the form has been set up, data values can be input. Calculations can then be made automatically on the input data. If changes in data values are made, recalculations are automatically performed to determine new results. This feature allows the user to readily update information about ongoing events and makes it easier to perform a "what if" analysis.

Spreadsheet Programs: Spreadsheet programs are ideal for electronic forms. These programs were designed to function much like a pencil, a sheet of paper, and an electronic calculator. With any of the available spreadsheet programs, the user starts with a blank electronic sheet. The sheet is organized as a grid of columns and rows (a matrix). The intersecting lines of the columns and rows define thousands of entry positions. At each position, you can enter a Label (alphabetic title, description, etc.) and a Value, or a Formula to be calculated. Formatting commands allow you to individualize the appearance of your entries.

The power of the spreadsheet is that values can be changed and recalculations performed easily and rapidly. Once the sheet has been created, it can be saved for future use or for further expansion and modification.

Quantity takeoff, estimating, project tracking, and scheduling forms could be adapted to the spreadsheet. A detailed example of such an adaptation will be presented later in this chapter.

Word Processing Programs: Word processing programs can also be used to create electronic forms. These programs allow the user to create an electronic copy of a document that can be corrected or modified before it is committed to paper. By "saving" the document, it can be re-used in total or in part. Forms can be set up with a predetermined layout to be filled in as required. Forms for correspondence could be set up in this manner.

Data-base Management (DBM) Programs: These programs can also be used to create electronic forms and to perform operations on the information entered in them. With DBM programs, information can be recorded about specific items in defined categories. The information can then be listed, sorted, searched, indexed, edited, formatted, printed, added to, deleted from, and selected from based on specific criteria.

Using the editing and print formatting capabilities of the DBM, the stored information can be organized and printed in the same manner as it appears on a preprinted paper form. Consequently, the DBM can be set up to perform storage and analysis functions on data gathered with paper forms. Historical record keeping systems to track job costs and labor production can be developed with a data base manager.

Graphics Programs: Graphics programs can create forms as well. These programs can be used to draw lines and to make text entries. Using a graphics program, preprinted forms can be duplicated, or new forms created. The graphics form is best used when it is printed on a printer or plotter or duplicated on a copy machine for manual data entry. Unlike other types of general application programs, graphics programs will not allow the user to efficiently use the form on the computer.

Integrated General Application Software Packages: This type of program can be used to create electronic forms. These software packages combine the functions of the four types of individual programs discussed above. For example, a spreadsheet can be created using information contained in a data base. This spreadsheet can then be inserted into a written report. Because the different basic programs are combined into a single package, a simple combination of commands will invoke various functions. With a fully integrated software package, an estimating and project administration system can be developed sharing common data. As the programs become more sophisticated, they can be more challenging. However, once the forms are created, they can be re-used easily. Time is also saved by sharing information without the need to re-input to a different program.

Each of the general application programs discussed above has its particular strengths in pursuing the goal of information management. As a result, some programs are more appropriate for developing certain types of forms. Integrated programs with spreadsheet capabilities are the most generally adaptable because they combine the functions of several program types. Of the individual function programs, the spreadsheet with its calculating and editing capabilities is easily the most versatile.

Applications

The process of reproducing forms on a microcomputer is facilitated by the proper matching of the software to the form. This is particularly important if you are using a single function software package. Since integrated software packages incorporate a variety of functions and can be used in virtually all situations, the following discussion is limited to the adaptation of various form types to single function packages.

Estimating Forms

Estimating and bidding forms can be readily adapted to the computer using a spreadsheet program. Quantity takeoff and pricing forms used to record data and perform calculations are especially well suited to the spreadsheet format. Electronic versions of these forms can take advantage of the computer's recalculation and storage capabilities, thereby reducing time and margin for error in measuring and pricing work quantities. Performing a "what if" analysis also becomes an easier task. All types of estimating forms from the preliminary estimate to the final bid could be set up in a similar manner. An example of a bid summary sheet (set up with a computer spreadsheet) will be presented later in this section.

The task of storing subcontractor lists and historical bidder records is simplified with a data base management (DBM) program. The DBM can be set up to enter information and store it for later use. When a bid is upcoming, the appropriate subcontractors and competitors can be selected from the data base by specifying selection criteria. The resulting information is then printed in an arrangement of rows and columns similar to the preprinted forms from the book. This method can save time otherwise spent in researching and writing lists. The risk of overlooking a key subcontractor is also reduced as long as the appropriate selection criteria are specified.

Administration Forms

Administration forms used to document expenses over time can be handled with a spreadsheet program. Row and column layouts can be set up in a format similar to the preprinted form. With the appropriate formulas entered for the spreadsheet, the necessary calculations can be made on the information. Once the form is created, data can be input periodically and recalculations performed quickly to keep track of the project. Time sheets, job progress reports, percentage complete analyses, and labor and material cost records could all be set up in this manner. manner.

Administration forms that document communications are natural applications of word processing forms. A word processor can be used to replicate the preprinted forms. If a blank copy is saved, it can be filled in — when needed — with the specific information. Editing and correcting are expedited by the functions of the word processor. Any of the communications or reporting forms presented in this book could be duplicated using this type of software package. As the forms are recreated on the computer, they can be customized to the user's specifications. As requirements change, the ease of changing the electronic copy ensures that the right form or document is always available.

Schedule

Bar chart scheduling forms could be easily adapted to a spreadsheet. Project schedule forms, for example, consist of rows and columns for plotting activities over time. Such a form could easily be duplicated by formatting a spreadsheet with the appropriate column widths and headings. Activity descriptions could be input in the proper sequence. Activity durations could then be resequenced, descriptions rewritten, or durations changed easily. This flexibility expedites the task of performing a "what if" analysis when developing project schedules.

If estimating and/or project administration forms have also been set up on a spreadsheet, this information could be transferred to the project schedule. Activity descriptions, costs, and durations could be input from an existing estimate in order to set up the schedule. Then, as the project progresses, records of costs incurred could be used to update the schedule and to plot expenditures over time. Setting up a schedule template capable of handling these operations involves a sophisticated use of the program's capabilities. However, the result is a powerful project management system.

Estimate Summary Example

Perhaps the most critical phase of the estimating process is the final summary of costs that takes place immediately before the contractor's bid is submitted. During these last hours, the subcontractors and material suppliers phone in their prices, scopes of work, and any revisions to the same. The estimator must compare and analyze this information quickly and efficiently. At some point, prices must be plugged into the estimate summary to cover all phases of the work. With the prices incorporated, the estimate overhead can be adjusted and the summations checked. From this point on, all changes are handled as external adjustments. This process, while essential and unavoidable, leaves a margin for error. In fact, the sometimes frantic nature of the final minutes before the bid deadlines can even invite mistakes. Mathematical or judgmental errors can be very dangerous.

The microcomputer and an electronic spreadsheet program can be used to alleviate some of the problems encountered on bid day. By creating an electronic estimate summary sheet to replace the typical paper form, the problems associated with closing the estimate and having it checked can be virtually eliminated. Figure 12.10 displays an estimate summary sheet developed for a general contractor as it appears before the cost information has been input.

In addition to totalling costs, the spreadsheet program can also be used to calculate overhead items that are typically figured on early cost projections. As figures change, the overhead is automatically updated by formulas programmed into the Overhead Adjustments section of the spreadsheet.

The spreadsheet is created by first setting the widths of the columns into which information will be input. The column headings and other permanent text entries are then input. Formatting commands can be used to establish the appearance and location of these entries. When the sheet is arranged in the desired format, the formulas to perform extensions and summations can be input into the appropriate locations. For example, to total the labor dollars, a formula to add up all column entries is input at the bottom of that column. Other formulas for extensions and summations are entered in a like manner. The formulas are written so that they will work with any values input to the sheet. This advantage allows the spreadsheet (with formulas) to be "saved" before numbers are input and then re-used for any project.

Analysis of subcontractor/material supplier bids can be handled with the Subcontractor Bid Summary section of the spreadsheet (shown in Figure 12.11). Located adjacent to the Estimate Spreadsheet are a series of smaller spreadsheets set up for the entry of subcontractor/material supplier bid information. If an item is to be subcontracted, it can be indicated with a "Y" or "N" in the "Sub" column of the Estimate Summary sheet. A program command can be used to move the cursor to the corresponding subcontractor summary. Prices and scopes of work can then be entered. Upon choosing the appropriate subcontract/material supply price, the amount can be transferred to the estimate summary, which is then appropriately updated to reflect the new price. A sufficient number of summary locations can be created to handle even the largest projects.

After the appropriate profit margin has been input, the spreadsheet program can calculate the bid — based on this information and current adjusted costs. The flexibility of the electronic spreadsheet allows for continual change in the prices and markup rates until virtually seconds before the bid must be submitted, with the assurance that calculations on inputted numbers are accurate. Figure 12.12 shows the Estimate Summary with values input for an example bid.

```
========================
     ESTIMATE SPREADSHEET
========================

PROJECT:
ARCHITECT:
LOCATION:
BID DATE:
PROJECT DURATION (MOS.):     1.0
```

ESTIMATE SECTIONS	SUB.	MATERIAL	LABOR
1. GENERAL REQUIREMENTS			
2A. SITE WORK-EXCAV., ETC.			
2B. BIT. PAVING, LANDSCAPING			
2C. DEMOLITION			
3A. CONCRETE FOUNDATIONS			
3B. CONCRETE FLOORS, ETC.			
3C. STRUCT. CONCRETE, PRECAST			
3D. RE-STEEL, MISC. ADJ.			
4. MASONRY			
5A. STRUCT. & MISC. STEEL			
5B. METAL DECKING			
6A. ROUGH CARPENTRY, HDW.			
6B. FINISH CARPENTRY, ETC.			
7A. WATERPROOFING, INSUL.			
7B. PREFORMED METAL PANELS			
7C. ROOFING			
7D. SEALANTS			
8A. HOLLOW METAL, FIN. HDW.			
8B. SPECIALTY DOORS			
8C. GLASS & GLAZING, WINDOWS			
9A. DRYWALL, PLASTER, ACOUST.			
9C. PAINTING, WALLCOVERING			
10. MISC. SPECIALTIES			
11. EQUIPMENT			
12. FURNISHINGS			
13. SPECIAL CONSTRUCTION			
14. CONVEYING SYSTEMS			
15. MECHANICAL SYSTEMS			
16. ELECTRICAL SYSTEMS			
TOTALS		$0.00	$0.00

```
OVERHEAD ADJUSTMENTS
  BASED ON:
          $0.00
------------------------------------------
  BUILDER'S RISK INS.              $0.00
  OWNER'S PROT. INS.               $0.00
  UMBRELLA INS.                    $0.00
  INS. & SOC. SECURITY             $0.00
  PERFORMANCE BOND                 $0.00
  STATE SALES TAX                  $0.00
  CONTINGENCY
------------------------------------------
  ADJUSTMENTS TOTAL                $0.00

  ADJUSTED COST                    $0.00

  MARGIN (ENTER BELOW)
       0%                          $0.00
                           ==================
  *BASE BID*                       $0.00
                           ==================
```

Figure 12.10

```
                    SUBCONTRACTOR
                         BID
                      SUMMARYS
--------------------------------------------------------------------

SECT.:
         SUBCONTRACTOR         BID AMOUNT              SCOPE
      ------------------     ------------------    ------------------




                        ------------------------------------------

SECT.:
         SUBCONTRACTOR         BID AMOUNT              SCOPE
      ------------------     ------------------    ------------------




                        ------------------------------------------

SECT.:
         SUBCONTRACTOR         BID AMOUNT              SCOPE
      ------------------     ------------------    ------------------




                        ------------------------------------------

SECT.:
         SUBCONTRACTOR         BID AMOUNT              SCOPE
      ------------------     ------------------    ------------------




                        ------------------------------------------
```

Figure 12.11

```
=======================
    ESTIMATE SPREADSHEET
=======================

PROJECT: RSM Warehouse Addition
ARCHITECT: C. Linde Assoc.
LOCATION: Kingston, MA
BID DATE: 4 FEB. 1986
PROJECT DURATION (MOS.):      7.0
```

ESTIMATE SECTIONS	SUB.	MATERIAL	LABOR
1. GENERAL REQUIREMENTS		$17406.00	$36055.00
2A. SITE WORK-EXCAV., ETC.	Y	$28386.00	$135.00
2B. BIT. PAVING, LANDSCAPING		$0.00	$0.00
2C. DEMOLITION		$0.00	$0.00
3A. CONCRETE FOUNDATIONS	N	$18000.00	$32429.00
3B. CONCRETE FLOORS, ETC.	N	$15378.00	$14681.00
3C. STRUCT. CONCRETE, PRECAST	N	$1016.00	$2096.00
3D. RE-STEEL, MISC. ADJ.	Y	$9398.00	$5355.00
4. MASONRY	Y	$7000.00	$0.00
5A. STRUCT. & MISC. STEEL	Y	$6985.00	$1866.00
5B. METAL DECKING		$0.00	$0.00
6A. ROUGHH CARPENTRY, HDW.	N	$1577.00	$2571.00
6B. FINISH CARPENTRY, ETC.		$0.00	$0.00
7A. WATERPROOFING, INSUL.		$0.00	$0.00
7B. PREFORMED METAL PANELS		$0.00	$0.00
7C. ROOFING	Y	$7480.00	$0.00
7D. SEALANTS	N	$60.00	$160.00
8A. HOLLOW METAL, FIN. HDW.	Y	$1348.00	$160.00
8B. SPECIALTY DOORS	Y	$625.00	$0.00
8C. GLASS & GLAZING, WINDOWS		$0.00	$0.00
9A. DRYWALL, PLASTER, ACOUST.		$0.00	$0.00
9C. PAINTING, WALLCOVERING		$0.00	$0.00
10. MISC. SPECIALTIES		$0.00	
11. EQUIPMENT	Y	$1250.00	$1120.00
12. FURNISHINGS		$0.00	$0.00
13. SPECIAL CONSTRUCTION		$0.00	$0.00
14. CONVEYING SYSTEMS		$0.00	$0.00
15. MECHANICAL SYSTEMS	Y	$18209.00	$0.00
16. ELECTRICAL SYSTEMS		$0.00	$0.00
TOTALS		$134118.00	$96628.00

```
OVERHEAD ADJUSTMENTS
  BASED ON:
        $298876.34
```

BUILDER'S RISK INS.	$4.70
OWNER'S PROT. INS.	$343.71
UMBRELLA INS.	$388.54
INS. & SOC. SECURITY	$16049.17
PERFORMANCE BOND	$3586.52
STATE SALES TAX	$2268.45
CONTINGENCY	
ADJUSTMENTS TOTAL	$22641.08
ADJUSTED COST	$253387.08
MARGIN (ENTER BELOW) 7%	$17737.10
BASE BID	$271124.18

Figure 12.12

This example of an estimate summary is based on an Estimate Summary form for a general contractor. The mechanical subcontractor could create a similar form listing only the relevant items of work for the trade. The flexibility of the spreadsheet would enable this "template" to be easily adapted to the needs of a specialty contractor. The method used to develop this form applies to other types as well. Any of the forms recommended for use with a spreadsheet program are set up similarly. Forms used with other software programs also follow the same general pattern.

Many of the forms needed by the subcontractor can be set up on the computer. Used with the appropriate software, the computer can add efficiency to the information management functions for which the forms were created. Many forms are still needed in their paper and pencil versions for a variety of functions. Using both tools may be the ideal solution to the contractor's information management needs.

Summary and Conclusions

The computer can serve a wide variety of functions from general business management to specific project management. The wide variety of applications, many of which have been touched on in this chapter, are supported by an equally wide selection of programs. Of special interest to the estimator are specific application programs for estimating. These can be further specialized for use by the mechanical subcontractor to aid in layout and fabrication of materials. Estimating programs can be part of a complete "integrated" software package for the contractor, or they may stand alone, performing only the estimating function.

Also available are general business programs that can be used to complement specific application programs. These general application programs can also be used on their own to facilitate information management.

The viability of computers for construction estimating and management is evident. The proper system of hardware and software should be flexible, and can only be selected after a thorough analysis of the company's needs and goals.

Part II
COMPONENTS OF MECHANICAL SYSTEMS

Chapter 13
PIPING

Piping is the common thread that ties plumbing, fire protection, heating, and cooling together into the category known as "the mechanical trades". This chapter contains descriptions of different types of pipe, valves, fittings, and other appurtenances which make up the piping systems assembled by the mechanical trades. The appropriate units of measure, step-by-step takeoff procedures for recording materials, appropriate labor units, and any pertinent cost modifications are indicated for each component of the piping system.

The fluid conveyed, location used, and the pressure contained usually determine what piping materials and joining methods are required. Steel pipe, for example, is often used for handling steam, chilled water, hot water, compressed air, gas, fuel oil, storm drains, and sprinkler systems. This chapter describes the cost estimating procedure for each type of pipe, (e.g., steel pipe) rather than repeat the pricing procedure for every function of that particular pipe.

Fittings are used in piping systems to change direction (elbows, bends, tees, and crosses), to divert or merge (wyes, tees, and laterals), to connect (couplings, unions, flanges, hubs, and sleeves), to terminate (plugs, caps, and blind flanges), to change from one joining method or material to another (adapters and transitions) or to change size (reducers, increasers, and bushings).

Nipples are short pieces of threaded pipe with a maximum length of 12 inches, usually included in the overall pipe totals as increased footage, rather than being taken off and priced separately. An estimator may prefer to take them off individually, however this can be a time consuming procedure.

Valves in piping systems control the flow of the conveyed fluid or regulate its pressure. Valves are produced in a wide variety of materials, end connections, and configurations, to be compatible with the rest of the piping system. Valves are manufactured for both manual and automatic (motorized) operation.

Check valves prevent reversal of flow. *Gate, globe, ball,* and *butterfly valves* and *plug cocks* stop or start the flow. A gate valve should be either fully open or fully closed. Globe, ball, and butterfly valves and plug cocks are also designed to throttle the flow. Specially constructed *reducing* and *relief valves* are available to regulate the pressure within a piping system.

Supports, anchors, and guides support and direct the expansion and contraction of pipe lines subject to the temperature changes of the conveyed fluid. The type of material and hardware used to support a piping system can significantly affect the installed cost. As a general rule, supports (or hangers) are made of the same material as the piping system.

Steel supports are used with ferrous piping; copper or copper-clad supports are used with non-ferrous piping. Plastics are an exception, because of the many forms of plastic available and the lack of a complete range of plastic hangers. Another exception is glass pipe for acid waste where metal or plastic supports are used.

Anchors and guides are usually indicated on the plans. Pipe hangers are only described in the specifications. Spacing distance is determined by the piping material and size as well as the fluid conveyed, and in the case of plastic pipe, the ambient temperature.

Steel Pipe

Steel pipe is often referred to as "black steel", "black iron", "carbon steel", "full weight", "merchant pipe", or "standard pipe" in the construction industry. Although such terms are regional in origin and usage, they refer to the same product.

Steel pipe usually comes from the mill with a black external coating for protection during shipment and storage. Galvanized steel pipe is shipped with a zinc coating on both the interior and exterior surfaces for protection against corrosion. Galvanized pipe is used for critical duties, such as drainage, waste, vent, water service, and cooling towers (condenser water), where the pipe will be exposed to the atmosphere.

There are two forms of steel pipe, welded and seamless, according to American Society of Testing and Materials (ASTM) specifications. Three ASTM designations will be discussed here: A106, A53, and A120. Seamless pipe is primarily made to A106 and A53 specifications; welded pipe is made to both the A53 and A120 specifications. The ASTM designation and manufacturer's data will be stencilled along the exterior wall of the pipe, or in smaller diameters, on the bundle tag.

A120 pipe is traditionally for non-critical* use, when the pipe is not intended to be bent or coiled. Pipe made to the A53 specification is also used in the above mentioned situations, but this type of pipe can be bent or coiled. In recent years, dual stenciled or dual certified pipe has been made available at the same low cost of the A120, but with all the requisites for A53 up through 4" diameter. Seamless A106 pipe is used for higher pressures and temperatures, with the advantage of being flexible. A106 pipe is available in plain end only (must be field threaded).

*Note: Non-critical use of pipe referred to in this book means low pressure gas, steam, or water systems normally found in plumbing, heating, cooling, and fire protection use.

American Standards Association (ASA) schedule numbers classify wall thicknesses of steel pipe up through 24 inches O.D. (outside diameter). In diameters 1/8" through 10", schedule 40 is synonymous with standard weight, and schedule 80 with extra heavy (XH), or extra strong (XS) through 8". For larger sizes, the wall thickness varies depending on the anticipated pressures, temperatures, or stress conditions.

Lighter wall thickness pipe is used in non-critical plumbing, heating, cooling, or fire protection applications. With the advent of roll grooving, and to compete with plastic and thin wall copper piping, light wall steel pipe (e.g., schedule 10) has come into accepted use, particularly in fire protection systems.

Weights and Dimensions of Carbon Steel Pipe—Seamless and Welded
A.S.A. Pipe Schedule

Pipe Size	O.D. in Inches	5	10	20	30	40	Std.	60	80	XS	100	120	140	160	X.X.S.
1/8	.405	.035 / .1383	.049 / .1863			.068 / .2447	.068 / .2447		.095 / .3145	.095 / .3145					
1/4	.540	.049 / .2570	.065 / .3297			.088 / .4248	.088 / .4248		.119 / .5351	.119 / .5351					
3/8	.675	.049 / .3276	.065 / .4235			.091 / .5676	.091 / .5676		.126 / .7388	.126 / .7388					
1/2	.840	.065 / .5383	.083 / .6710			.109 / .8510	.109 / .8510		.147 / 1.088	.147 / 1.088				.187 / 1.304	.294 / 1.714
3/4	1.050	.065 / .6838	.083 / .8572			.113 / 1.131	.113 / 1.131		.154 / 1.474	.154 / 1.474				.218 / 1.937	.308 / 2.441
1	1.315	.065 / .8678	.109 / 1.404			.133 / 1.679	.133 / 1.679		.179 / 2.172	.179 / 2.172				.250 / 2.844	.358 / 3.659
1¼	1.660	.065 / 1.107	.109 / 1.806			.140 / 2.273	.140 / 2.273		.191 / 2.997	.191 / 2.997				.250 / 3.765	.382 / 5.214
1½	1.900	.065 / 1.274	.109 / 2.085			.145 / 2.718	.145 / 2.718		.200 / 3.631	.200 / 3.631				.281 / 4.859	.400 / 6.408
2	2.375	.065 / 1.604	.109 / 2.638			.154 / 3.653	.154 / 3.653		.218 / 5.022	.218 / 5.022				.343 / 7.444	.436 / 9.029
2½	2.875	.083 / 2.475	.120 / 3.531			.203 / 5.793	.203 / 5.793		.276 / 7.661	.276 / 7.661				.375 / 10.01	.552 / 13.70
3	3.5	.083 / 3.029	.120 / 4.332			.216 / 7.576	.216 / 7.576		.300 / 10.25	.300 / 10.25				.437 / 14.32	.600 / 18.58
3½	4.0	.083 / 3.472	.120 / 4.973			.226 / 9.109	.226 / 9.109		.318 / 12.51	.318 / 12.51					.636 / 22.85
4	4.5	.083 / 3.915	.120 / 5.613			.237 / 10.79	.237 / 10.79	.281 / 12.66	.337 / 14.98	.337 / 14.98		.437 / 19.01		.531 / 22.51	.674 / 27.54
5	5.563	.109 / 6.349	.134 / 7.770			.258 / 14.62	.258 / 14.62		.375 / 20.78	.375 / 20.78		.500 / 27.04		.625 / 32.96	.750 / 38.55
6	6.625	.109 / 7.585	.134 / 9.289			.280 / 18.97	.280 / 18.97		.432 / 28.57	.432 / 28.57		.562 / 36.39		.718 / 45.30	.864 / 53.16
8	8.625	.109 / 9.914	.148 / 13.40	.250 / 22.36	.277 / 24.70	.322 / 28.55	.322 / 28.55	.406 / 35.64	.500 / 43.39	.500 / 43.39	.593 / 50.87	.718 / 60.93	.812 / 67.76	.906 / 74.69	.875 / 72.42
10	10.75	.134 / 15.19	.165 / 18.70	.250 / 28.04	.307 / 34.24	.365 / 40.48	.365 / 40.48	.500 / 54.74	.593 / 64.33	.500 / 54.74	.718 / 76.93	.843 / 89.20	1.000 / 104.1	1.125 / 115.7	
12	12.75	.165 / 22.18	.180 / 24.20	.250 / 33.38	.330 / 43.77	.406 / 53.53	.375 / 49.56	.562 / 73.16	.688 / 88.51	.500 / 65.42	.844 / 107.2	1.000 / 125.5	1.125 / 139.7	1.312 / 160.3	
14	14.0		.250 / 36.71	.312 / 45.68	.375 / 54.57	.437 / 63.37	.375 / 54.57	.593 / 84.91	.750 / 106.1	.500 / 72.09	.937 / 130.7	1.093 / 150.7	1.250 / 170.2	1.406 / 189.1	
16	16.0		.250 / 42.05	.312 / 52.36	.375 / 62.58	.500 / 82.77	.375 / 62.58	.656 / 107.5	.843 / 136.5	.500 / 82.77	1.031 / 164.8	1.218 / 192.3	1.437 / 223.5	1.593 / 245.1	
18	18.0		.250 / 47.39	.312 / 59.03	.437 / 82.06	.562 / 104.8	.375 / 70.59	.750 / 138.2	.937 / 170.8	.500 / 93.45	1.156 / 208.0	1.375 / 244.1	1.562 / 274.2	1.781 / 308.5	
20	20.0		.250 / 52.73	.375 / 78.60	.500 / 104.1	.593 / 122.9	.375 / 78.60	.812 / 166.4	1.031 / 208.9	.500 / 104.1	1.280 / 256.1	1.500 / 296.4	1.750 / 341.1	1.968 / 379.0	
24	24.0		.250 / 63.41	.375 / 94.62	.562 / 140.8	.687 / 171.2	.375 / 94.62	.968 / 238.1	1.218 / 296.4	.500 / 125.5	1.531 / 367.4	1.812 / 429.4	2.062 / 483.1	2.343 / 541.9	

Figure 13.1

Steel pipe, regardless of wall thickness, retains its designated outside diameter. For example, 3" pipe, regardless of wall thickness or schedule, is always 3-1/2" in diameter, as shown in Figure 13.1. This is necessary for threading, socket welding, and other fitting and mating purposes. The inside diameter decreases or increases depending on the wall thickness specified.

Steel pipe is produced in three lengths: a "uniform" length of 21 feet; "random" length, which varies from 16 to 22 feet; or cut specifically to the customer's required length. Double lengths are available by special order only.

Steel piping is shipped in bundles for sizes 1-1/2 inches and smaller. Figure 13.2 is a bundling schedule for 21 foot uniform lengths. Piping is also available from local suppliers in individual lengths.

Bundling Schedule			
Number of Pieces — Average Feet and Weight of Steel Pipe Per Bundle			
Size (Inches)	Pieces per Bundle	Average per Bundle	
		Feet	Weight (Lbs.)
Standard Weight Pipe			
1/8	30	630	151
1/4	24	504	212
3/8	18	378	215
1/2	12	252	214
3/4	7	147	166
1	5	105	176
1¼	3	63	144
1½	3	63	172
Extra Strong Pipe			
1/8	30	630	195
1/4	24	504	272
3/8	18	378	280
1/2	12	252	275
3/4	7	147	216
1	5	105	228
1¼	3	63	189
1½	3	63	229

Figure 13.2

The pipe end finishes are determined by the proposed joining or coupling method. End finishes can be plain (or bald) end, bevelled end for welding, threaded and coupled (T & C), cut grooved, or roll grooved. Cut or roll grooved ends are used with grip type bolted or hinged fittings. Piping larger than 12 inches is threaded on special order only.

Units of Measure: Steel piping is measured in linear feet (L.F.). Associated fittings, supports, and joints, are recorded as each.

Labor Units: The following procedures are generally included in the per foot labor cost of piping.

- Unloading and relocating from receiving or storage area to the point of installation.
- Set up of scaffolding, staging, ladders, or other work platforms.
- Lay out of piping route.

These additional procedures are also recorded and included in the piping estimate.

- Joining of straight lengths every 20 feet.
- Installation of fittings, valves, and appurtenances.
- Installation of supports, anchors, and guides.
- Testing and cleaning individual systems.

Takeoff Procedure: A takeoff sheet should be set up for each piping system to record the various footages required for each diameter of pipe (shown in Figure 13.9). To figure the footages, the piping runs should be scaled or otherwise measured, beginning at the source of each system. By starting at the source of each system, the larger, more costly pipe will be taken off first. This does not hold true, however, when taking off condensate return, drain, or waste piping, where the larger pipe is installed at the termination point.

Each system should be marked with a colored pencil as it is taken off. A color code should be developed, to distinguish between piping materials or piping systems, which can be standardized for use throughout the company or throughout the estimating department. Horizontal piping

Height Modifications	
10-15 feet	add 10 percent
15-20 feet	add 20 percent
20-25 feet	add 25 percent
25-30 feet	add 35 percent
30-35 feet	add 40 percent
35-40 feet	add 50 percent
over 40	add 55 percent

Figure 13.3

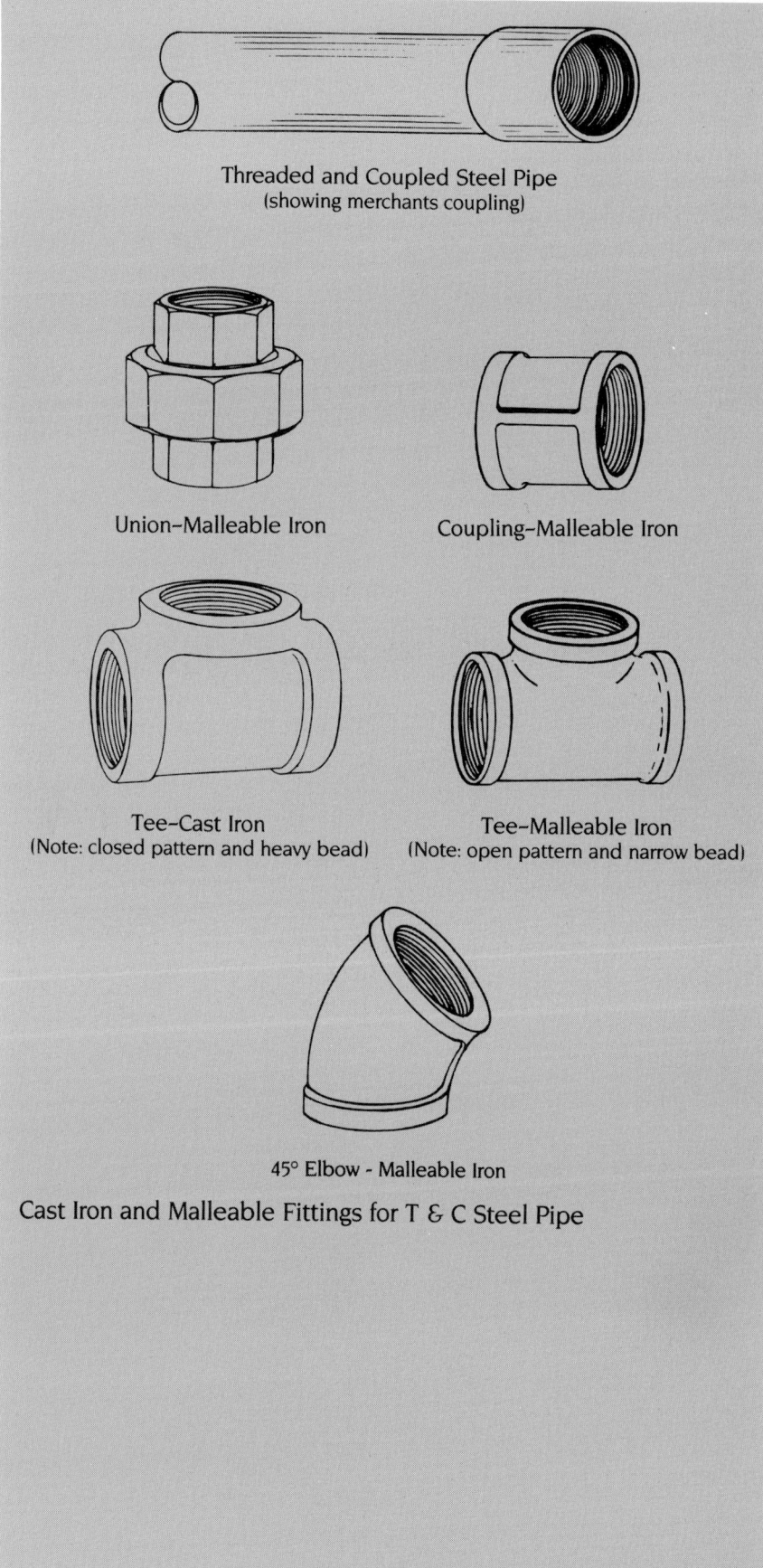

Figure 13.4

runs should be recorded and measured first, then vertical risers or drops. The total should be rounded off to coincide with the average lengths available from local suppliers (e.g., 20 feet) for steel pipe.

Height Modifications: Additional labor costs are incurred when piping is installed over 10 feet above the work area. For anticipated work over 10 feet, add the percentages shown in Figure 13.3 to compensate for the difference between actual cost and usual labor cost figures.

Screwed Fittings and Flanges for Steel Pipe

Fittings for use with steel pipe come in a variety of shapes and materials to accommodate the end finish of the piping and the fluid conveyed. Fittings are available for any end finish found on steel pipe, including threaded, welded, flanged, grooved, or plain end. Figures 13.4 and 13.5 show different fittings used with steel pipe.

Cast Iron Screwed Fittings and Flanges are produced in standard black (125 lb.) or extra heavy (250 lb.) pressure classes and in galvanized finish for corrosive atmospheres. Cast iron (also known as "gray iron") is non-ductile, therefore it is not used in systems subject to excessive shock or hammering. (Malleable fittings are used for these more critical conditions.) The non-ductability of cast iron, however, becomes an asset when dismantling or replacing a fitting that is "frozen" or "cooked" in place by prolonged heat. In this case, the fitting will shatter after several hammer blows if the joint does not submit to a pipe wrench.

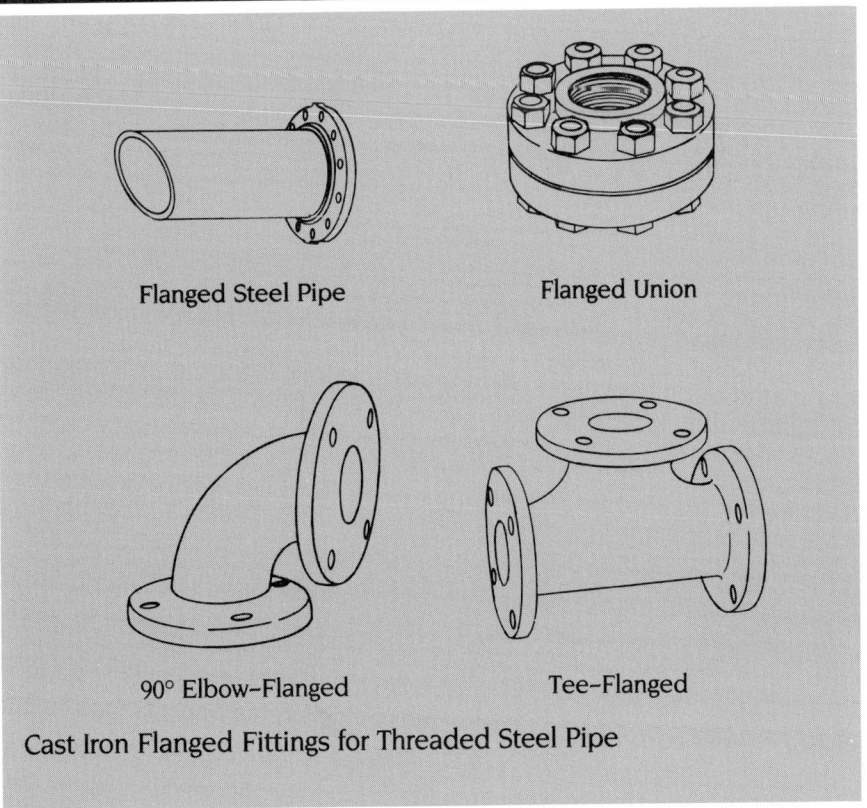

Figure 13.5

Special drainage or pitched fittings are designed to give an unobstructed flow by having an inside diameter (I.D.) the same as the I.D. of the pipe. These fittings are used for plumbing drains, vents, and vacuum cleaning systems. Tees and elbows have pitched threads to assure a 1/4" per foot pitch of the drain line. These fittings do not have a pressure rating and are normally shipped with a coating of black enamel. Plain or galvanized coating is also available by special order.

Some local plumbing code requirements call for drainage fittings with combinations of bell and spigot (soil pipe fittings), and threaded ports. Combination fittings are not stock items, and are available by special order only.

Malleable Iron Screwed Fittings are used in gas piping, smaller sizes of sprinkler piping, fuel oil piping, gasoline, and any system subject to shock or physical abuse. Malleable iron fittings are produced in standard (150 lb.) and extra heavy (300 lb.) pressure classes, except for unions, which come in classes of 150, 200, 250, and 300 lb. These fittings are available in most of the same sizes and configurations as cast iron fittings, except the unions, which are exclusively malleable iron.

Malleable iron screwed fittings are similar in appearance to cast iron fittings, except for the bead on the fitting ends. Cast iron fittings have a heavy bead and are of a closed pattern. Malleable fittings have a narrow bead and an open pattern. The 300 lb. malleable iron fitting is the exception, bearing the pattern usually found on cast iron fittings (a heavy bead and a closed pattern). Malleable iron fittings are available with a black or galvanized finish.

Flanged Cast Iron Fittings, Threaded Cast Iron Flanges, and *Flanged Unions* are used with threaded steel pipe. Flanged fittings are most often used in large diameter piping systems, although they are made in smaller sizes, from 1-1/2" diameter and larger. These fittings are used to mate with flanged valves, pumps, and other mechanical equipment provided with flanged ports. The flanges and fittings have 125 or 250 lb. ratings. A 250 lb. flange has a larger O.D. than a 125 lb. flange, and also has a larger bolt circle, requiring more bolts to make up the flanged joint.

Flanging is frequently used for offsite prefabricated piping (e.g., fire protection sprinkler systems). Flanged elbows are available in standard or long radius configurations. The advantage of flanged fittings is the comparative ease with which a flanged joint may be assembled or disassembled.

Ductile Iron Fittings are not usually encountered in building construction. Both 300 lb. screwed and 150 lb. flanged are available in limited sizes for use in more critical systems.

Units of Measure: All fittings are taken off and recorded as each. The number of joints will be determined by this fitting count.

Material Units: In addition to the cost of the fitting, material is needed to prepare and make up each joint. Threaded joints require cutting oil, cotton waste or wipes, pipe dope (pipe joint compound), or a suitable tape or paste lubricant or sealant. Joints in screwed piping usually require a certain quantity of pipe nipples. These are not shown on the drawings, but must be included in the material and labor costs.

Flanged joints require bolts, nuts, and gaskets. Threaded steel flanges can be substituted for cast iron flanges, but the cost of a steel flange is considerably higher than cast iron. In some flanged installations it is possible to estimate the number of companion flanges or sets of bolts, nuts, and gaskets (B.N.&G.) by doubling the count of flanged valves or strainers.

Each assembly should be checked carefully to determine the exact number of flanges or joints required. An example is shown in Figure 13.6. The sketch indicates two butterfly valves, one check valve, two flexible connections, and two flanged reducers. If the flanged items are doubled, it would seem that fourteen flanges and fourteen sets of B.N.&G. are required. A closer look, however, reveals that eight companion flanges and thirteen sets of B.N.&G. would be sufficient to button up this assembly. If the discharge side butterfly valve and the check valve were buttoned together, two more flanges, two joints, and one set of B.N.&G. could be eliminated.

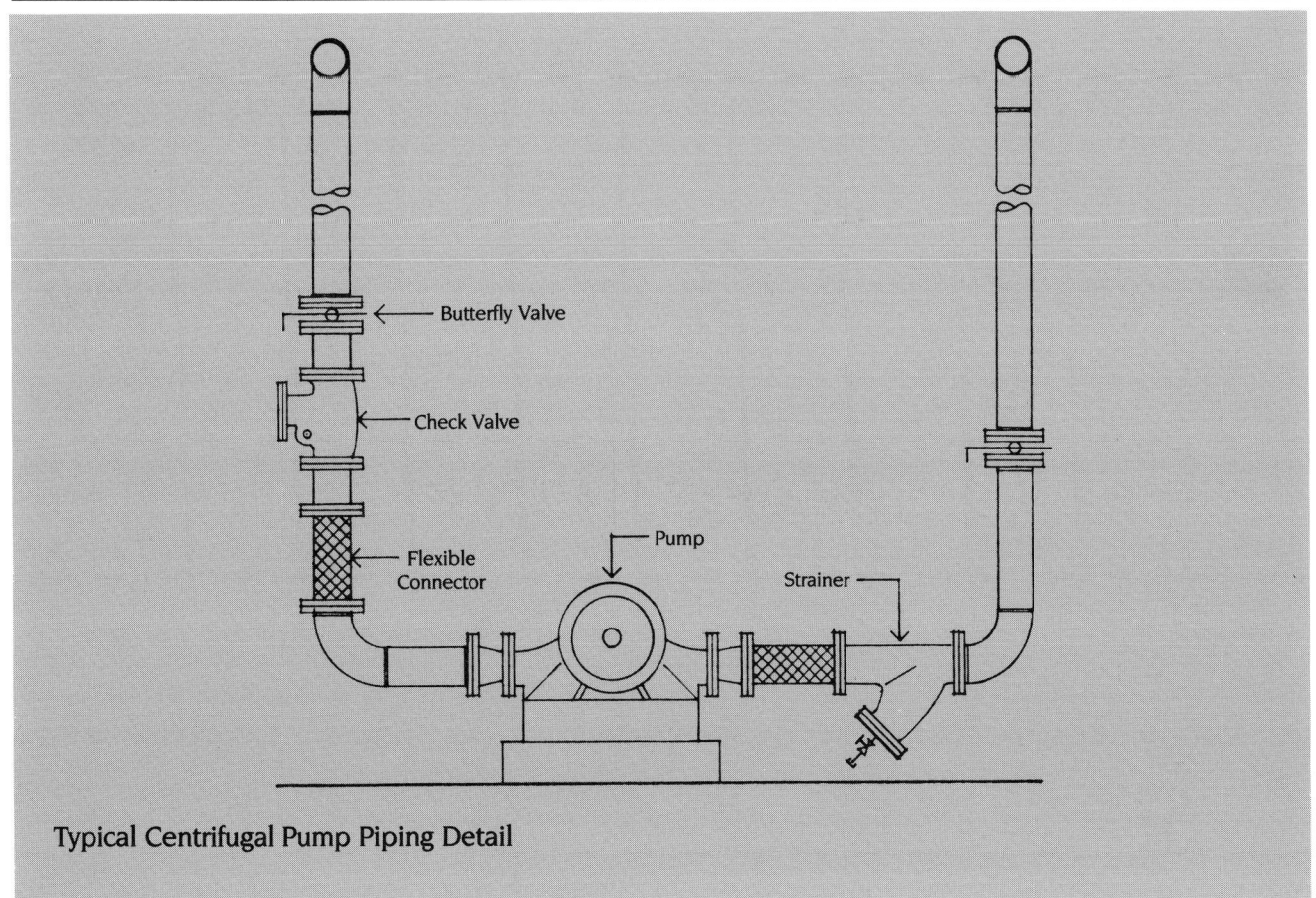

Typical Centrifugal Pump Piping Detail

Figure 13.6

Labor Units: The labor units for fittings usually involve making up a tight joint. For larger fittings, weight becomes an added factor in handling, installing, and supporting. The field or shop cutting and threading must also be recorded.

When estimating by the joint method, each fitting, valve, or other piping accessory is counted and recorded by pipe size, material, and the total number of joints per fitting; each elbow having two joints, tees having three, crosses having four, etc.

Rather than take off all of the screwed fittings and nipples, some estimators have an allowance built into the piping labor factor based on historical labor data for smaller sizes of T & C piping (two inches and less). In this case, an allowance based on a certain percentage of the T & C pipe material cost is added for the screwed fittings and nipple material costs. Until an estimator's data base includes information on past labor and fitting costs, the time consuming individual takeoff and pricing procedure must be followed.

Takeoff Procedure: The fittings may be recorded below the accumulated footage on the takeoff sheet for steel piping systems. If the job is of significant size, separate takeoff sheets for fittings are necessary, and should be attached to the pipe takeoff sheets for each system. The same procedure as that used for pipe should be followed, treating larger sizes first, using the color code to indicate that each fitting and intermediate joint has been recorded. Intermediate joints are couplings connecting straight sections of pipe. In this case, threaded and coupled (T & C) pipe is supplied with a steel coupling attached to each length. Therefore, only the labor to make up both joints of the coupling need be recorded. For gas piping, the wrought steel or merchants coupling is removed and replaced with a malleable iron coupling. In a flanged piping system, a flanged union or a pair of companion flanges couple the straight lengths of pipe and still allow the capability for future dismantling.

Steel Weld Fittings and Flanges

Steel welding fittings (shown in Figure 13.7) are produced in standard weight and extra heavy weight for normal piping systems encountered in the building construction industry. Welding fittings are manufactured in higher ratings for industrial or process use. A light weight fitting is also available by special order that is compatible with schedule 10 pipe wall thickness. This type of fitting is most often used for gas distribution piping.

The standard weight line is used for general purposes. Most elbows are of the long radius pattern, where the center-to-face dimension is one and one half times the nominal pipe diameter.

Steel welding flanges (shown in Figure 13.7) are produced in standard weight (150 lb.) and extra heavy (300 lb.), as well as the higher ratings for power and process service. The 150 lb. weld neck flanges are most commonly used. Despite the differences in pressure ratings, the 150 lb. series weld flanges mate with 125 lb. cast iron or steel threaded flanges and the 300 lb. series are compatible with the 250 lb. flanges, because they have the same bore, O.D., and bolt circle.

Units of Measure: All fittings are taken off and recorded as each. The number of joints and the inches of welding will be determined by this fitting count plus the number of intermediate welds required.

Material Units: Additional material is necessary to prepare and make up each joint. This cost must be estimated in addition to the actual purchase cost of a fitting or flange. Welded joints require welding rod, oxygen, and acetylene for gas welding. For arc welding, electrodes (coated rod) and a power source are needed. These are the two most commonly used welding procedures for mild carbon steel pipe and fittings in the field. Other materials to include in the welding estimate are backing rings (chill rings) if specified. These are counted per joint. Chill rings allow for uniform spacing of the two pipe ends to be welded, resulting in a smooth interior surface and eliminating the necessity of tack welding. Backing rings are not normally specified for non-critical piping used in mechanical systems.

An experienced estimator has learned that there are some welding fittings never used unless space conditions mandate: reducing weld elbows and welding tees which reduce on the run.

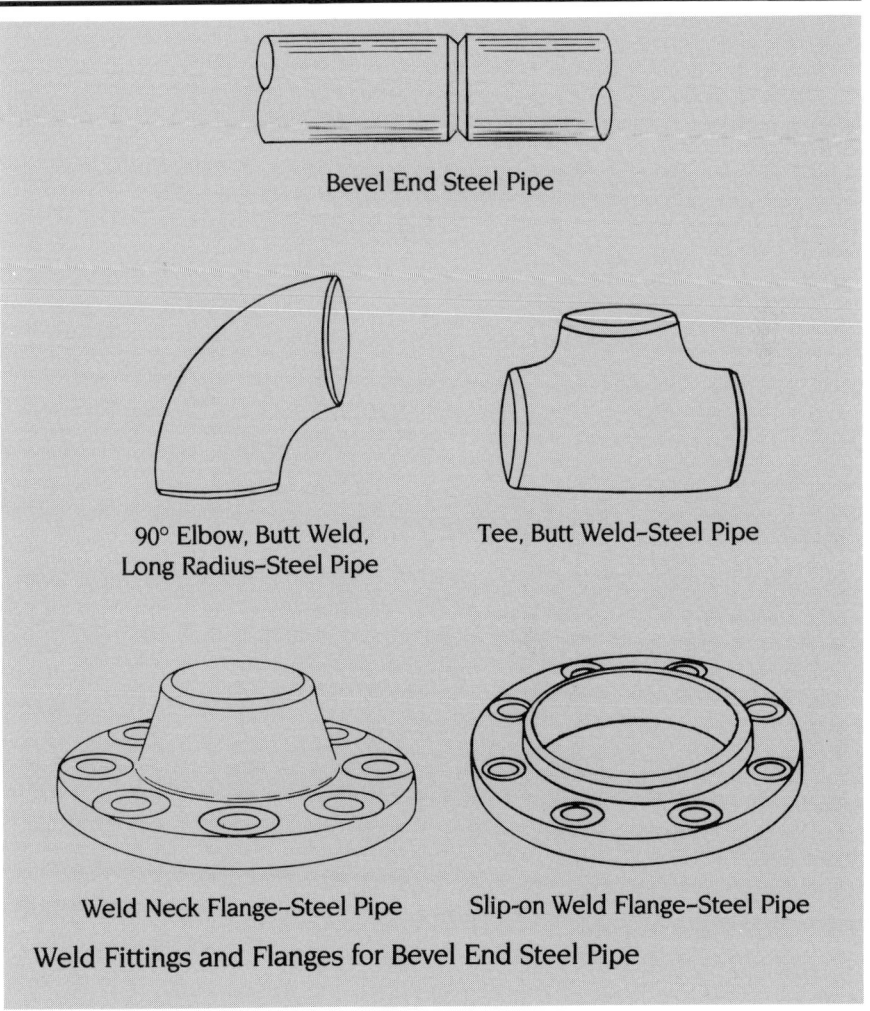

Figure 13.7

Reducing Weld Elbows are not usually stock items because they are prohibitively priced. It is cheaper, both material-wise and labor-wise, to use a straight size elbow and a concentric weld reducer.

Welding Tees reducing on the run are also prohibitively priced and should be used with straight size on the run plus a reducer, either concentric or eccentric, depending on the fluid being piped and the position of the fitting. When a reducer is used in vertical piping, it should be concentric, which is less expensive than eccentric. For horizontal piping, an eccentric reducer is necessary. For steam piping, the eccentric reducer is installed with the flat on the bottom to maintain a flat surface for the condensate flow. For forced hot and cold water and condensate return, the flat is installed at the top to prevent air pockets from forming.

Welding tees reducing more than two sizes on the outlet are extremely expensive and sometimes not available at all. For example, a 10" main requiring a 3/4" outlet would require a 10" by 4" tee plus a 4" by 1-1/2" reducer plus a 1-1/2" by 3/4" reducer. This would result in an outlet 6" long and 26-1/4 diameter inches of welding. A preferable alternative would be to use outlet fittings which are made to be welded directly into an opening cut in the pipe itself. Some trade names for those fittings are weldolets, threadolets, and sockolets, depending on the type of pipe connection required at the outlet. A weldolet for the above example would require two diameter inches of welding. Field expedients for these outlets are often nipples cut in half for a male outlet, or any of the surplus wrought steel merchant pipe couplings that accumulate on a job or in a shop for a female outlet. These field expedients should be used only for non-critical piping systems. The contract specifications should be checked, as they may prohibit the use of these field expedients.

Welding material costs are estimated by daily consumption; the pounds of wire or rod per day and the number of cubic feet of oxygen and acetylene, or gallons of fuel for engine driven generators. The material costs for gas or arc welding should usually be the same or close enough for estimating purposes.

Labor Units: The labor for weld fittings entails more than just the actual time required to gas or arc weld the two pieces together. Joint preparation, cutting, clamping, bevelling, chipping, cleaning, and measuring are all entailed. Daily output of a welder is generally accepted as the best way to estimate the welding labor for a piping system. The welder's output is determined by the diameter inches of welding accomplished per day. The term "diameter inches" refers to the nominal diameter of the pipe and considers the number of passes per joint as well as the circumference. These two factors determine the labor per joint. For example, a four inch diameter pipe weld is four "diameter inches". For branch welds, the diameter inches are recorded and doubled to allow for the cutting or burning of the hole. For example, a 1-1/2" diameter branch would be recorded as three diameter inches of weld.

The labor for weld flanges would include aligning, welding, and bolting up each joint. Bolting requirements are shown in Figure 13.8. Slip on flanges require two welds, an internal and an external. In low pressure heating and cooling applications it is not uncommon to tack weld the hub and complete the normal interior weld. This practice, although not recommended, may save a considerable amount of time.

The daily output of a welder varies from mechanic to mechanic. Thirty diameter inches per day for a welder and a layout man working together is considered a good overall average for an eight hour day (sixteen man-hours). Do not anticipate that a welder will spend eight hours under the hood. Preparation and positioning of the assembly to be welded is time consuming. A piping fit, prefabricated on the bench or floor and hoisted into place, increases daily output.

150 Lb. Steel Flanges				
Pipe Size	Diam. of Bolt Circle	Diam. of of Bolts	No. of Bolts	Bolt Length
1/2	2⅜	1/2	4	1¾
3/4	2¾	1/2	4	2
1	3⅛	1/2	4	2
1¼	3½	1/2	4	2¼
1½	3⅞	1/2	4	2¼
2	4¾	5/8	4	2¾
2½	5½	5/8	4	3
3	6	5/8	4	3
3½	7	5/8	8	3
4	7½	5/8	8	3
5	8½	3/4	8	3¼
6	9½	3/4	8	3¼
8	11¾	3/4	8	3½
10	14¼	7/8	12	3¾
12	17	7/8	12	4
14	18¾	1	12	4¼
16	21¼	1	16	4½
18	22¾	1⅛	16	4¾
20	25	1⅛	20	5¼
22	27¼	1¼	20	5½
24	29½	1¼	20	5¾
26	31¾	1¼	24	6
30	36	1¼	28	6¼
34	40½	1½	32	7
36	42¾	1½	32	7
42	49½	1½	36	7½

Steel, cast iron, bronze, stainless, etc. Flanges have identical bolting requirements per pipe size and pressure rating.

Figure 13.8

Means Forms
PIPING SCHEDULE

JOB: Office Building
SYSTEM: Dual Temp. – Hot/Chill Water – 2 1/2" & up welded 2" & down T&C
HVAC
PAGE: 1 of 1
DATE: 3-17-87
BY: JJM

PIPE DIAMETER IN INCHES

	12	10	8	6	5	4	3 1/2	3	2 1/2	2	1 1/2	1 1/4	1	3/4	1/2	
Black Steel						26 50 23 74 33 29 ⟨235⟩		14 14 ⟨28⟩	14 14 ⟨28⟩	28 180 ⟨208⟩			20 ⟨20⟩			**Air Vents** ⁂ |||
Ells						⁂ ⁂ ⁂ ⁂ ⁂	⁂	"	"	"						
Tees						⁂ ⁂ ⁂ ⟨17⟩	⁂	"	"	"						**Thermometers** ⁂ ⁂ |||
Reducers						⁂ ⟨4⟩		⟨2⟩	⟨2⟩							
Gate Valves						⁂ ⟨8⟩		⟨2⟩	⟨2⟩							**Gauges** ⁂ |||
Check Valves						⁂ ⟨4⟩		"								
Balancing Cocks						⁂ ⟨4⟩		"								**Drain Valves** ⁂ ⁂ ⁂	
Strainers						⁂ ⟨4⟩		"								
Flexible Connectors						⁂ ⟨8⟩		"		⁂ ⟨12⟩						
Butt Welds						107 18	⁂	⁂								C.I. Flanged Elbows 4–4"
Weld Flanges						10										
Branch Welds														⁂		
Bolts Nuts Gaskets						40										

Figure 13.9

Takeoff Procedure: The fittings, flanges, nozzles, branch welds, and butt welds needed for intermediate joints should be recorded by size on the takeoff sheet for steel piping (shown in Figure 13.9). From this fitting list, the diameter inches are determined for the total welding labor and material cost.

Grip Type Mechanical Joint Fittings

The grip type of fitting is used in the fire protection field, building service systems (i.e., storm and sanitary drains), potable, heating, cooling, and condenser water systems. (Not available for steam or gas piping.)

Grip type fittings (shown in Figure 13.10) are designed for use with plain end, grooved, roll grooved, or bevel end steel pipe, and are available with

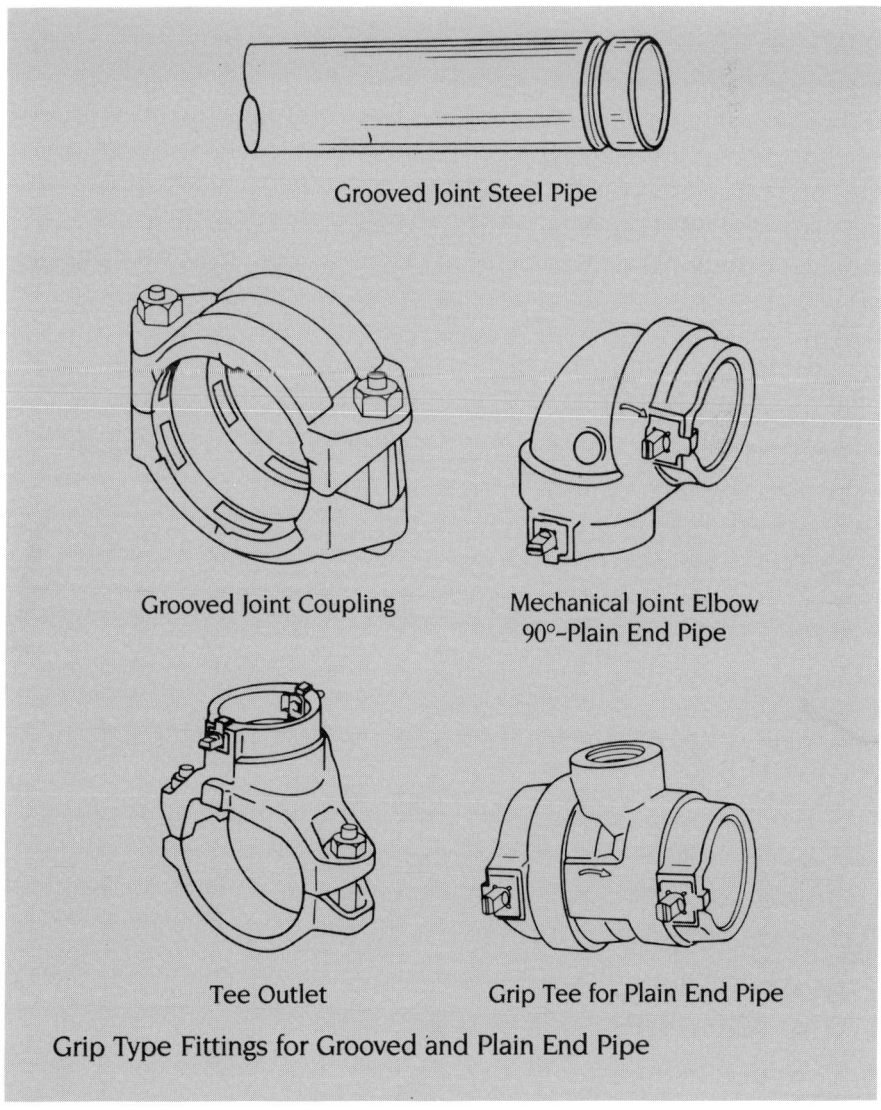

Figure 13.10

plain or galvanized finish. This kind of fitting is produced for use with lighter schedules of piping, as well as with standard weight (schedule 40). The wide variety of sizes and configurations of grip fittings allows a complete system to be assembled using only these fittings.

Special tools are required for field or shop cutting, rolling of grooves, and drilling outlets in the pipe wall where "hook type" or other mechanically joined outlets are substituted for tees or welded nozzles.

Units of Measure: All fittings are taken off and recorded as each. The number of joints, grooves, or holes to be cut is determined by this fitting count.

Material Units: Little else is required, other than the special tools previously mentioned, to prepare the grip type joint. The only exception is adapter flanges (available in standard or extra heavy); the installer should furnish the normal quantity of nuts and bolts, but the gasket is supplied with the adapter flange.

Labor Units: The labor units for grip type fittings include pipe end preparation. Even plain end pipe requires cleaning and deburring. The fittings are then mechanically joined to the pipe using factory supplied bolts, pins, or quick disconnect lever handles. A system based on diameter inches, similar to welding, is often used with a claimed productivity of 90 to 100 inches per day.

Takeoff Procedure: Grip joint systems are recorded by size for each fitting or outlet on the takeoff sheet for steel piping. Straight lengths of pipe are joined by one coupling, but each pipe end must be prepared per joint. Intermediate joints for long runs of straight pipe are recorded separately because the grooving labor is not required (assuming that the pipe is purchased already grooved). Therefore, only the labor to bolt on the coupling is included in the estimate.

Cast Iron Pipe

Cast iron pipe, also known as gray iron pipe, is used in soil pipe systems for sanitary waste, vent systems, and storm drain piping systems. Cast iron soil pipe is manufactured in both service weight and extra heavy, or full weight. Extra heavy soil pipe is generally used in underground installations. A corrosion-resistant high silicon content soil pipe is available from certain manufacturers for use in acid type waste systems and corrosive atmospheres.

Cast iron pipe is available with bell and spigot joints, which are made up or poured with hot lead, or by push on type gasketed compression joints. Plain end, or no hub cast iron soil pipe is joined by applying bolted clamp type couplings and neoprene gaskets. The clamp type couplings are produced in rigid cast iron or flexible stainless steel bands, depending on use and local acceptance.

Soil pipe is produced in both 5' and 10' lengths, in diameters of 2", 3", 4", 5", 6", 8", 10", 12" and 15". The 5' lengths are manufactured in double or single hub (hub being another term for the bell shaped end).

Units of Measure: Cast iron pipe is measured in linear feet (L.F.). Associated fittings, supports, joints, and other appurtenances are recorded as each.

Labor Units: The following procedures are generally included in the per foot labor cost:
- Unloading and relocating from receiving or storage area to point of installation.
- Set up of scaffolding, staging, ladders, or other work platform.
- Lay out of piping route.

These additional procedures should also be recorded and included in the piping estimate:
- Joining of straight lengths every 5 or 10 feet.
- Installation of fittings, valves, supports, and anchors.
- Leak and flow testing of the assembled system.

Takeoff Procedure: A takeoff sheet should be set up for each piping system and footages recorded for each diameter of pipe, beginning with the waste system at its termination point. This will be the largest diameter pipe found in the system. If a portion of the system is to be tarred or coated, it should be taken off separately. Underground pipe should also be recorded separately, as its joining, supporting, and laying will be estimated differently.

Each system should be marked on the plans in a different color pencil. The pipe totals should be rounded off to coincide with 5' or 10' lengths available for each category, type, and size of pipe.

Cast Iron Soil Pipe Fittings

Fittings for cast iron soil pipe, because of their designated use in drainage and venting, have many more configurations (shown in Figure 13.11) than do other types of fittings. A manufacturer's catalogue showing all the available fittings and patterns is a must for an estimator's reference library. Soil pipe fittings used for a change in direction are called bends rather than elbows. Other piping systems have 90 degree and 45 degree elbows whereas drain waste and vent systems have quarter, eighth, fifth, sixteenth, and, for special use, half and sixth bends. Quarter bends are also available in long turn and short turn pattern. Tees and wyes are used in these systems, plus variations peculiar to DWV use, such as sanitary tees for vertical use. The traditional pouring of lead and oakum joints is being replaced, where allowed by local code, with push-on gasketed joints or no hub coupled joints.

Combination fittings which conserve space and joint makeup labor can often be used in cast iron soil pipe systems. Examples are combination wye and eighth bends, single and double vent branches, offsets, running traps, and many more (available on special order). These fittings are all available with special cleanout or vent ports, and in straight or reducing sizes. Soil pipe fittings reduce on the spigot end only.

Cast iron soil pipe fittings are manufactured in bell and spigot type (in service or extra heavy weight) and no hub (in service weight).

Units of Measure: Soil pipe fittings and combination fittings are taken off and recorded as each.

Material Units: The number of joints or no hub couplings will be determined from the fitting count, as well as any special supports, cleanout access, or extension covers.

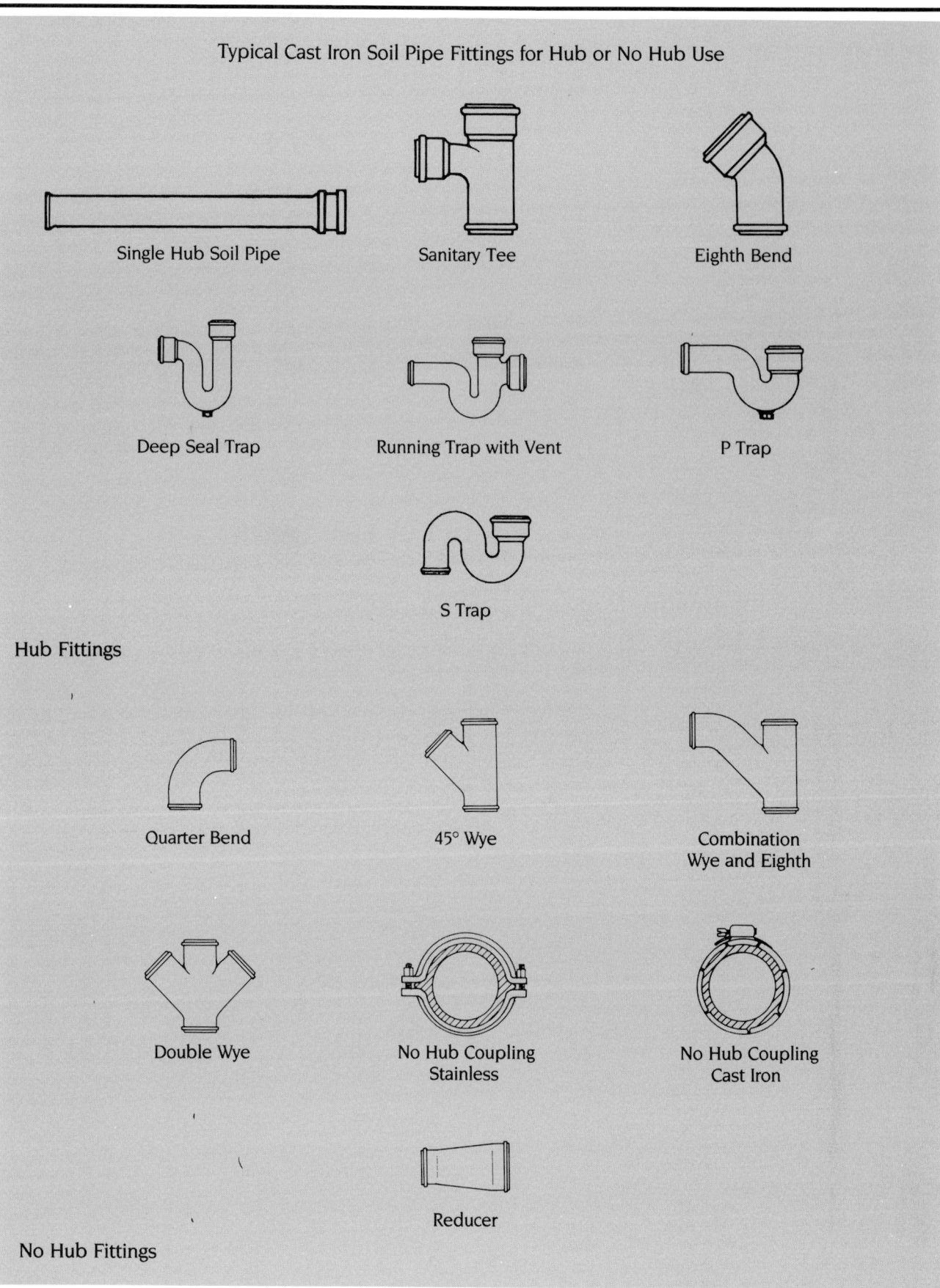

Figure 13.11

Labor Units: Unloading, sorting, and bringing the fittings to the point of installation should be included in the installation cost. Joint makeup, testing, and inspection by local municipal authorities, should also be considered.

Takeoff Procedure: Each fitting and combination should be recorded on the soil pipe estimate sheet (or one attached to it) by size and type. Line drawings usually do not show every fitting or possible combination of fittings. This is where knowledge of local codes and available fittings can be useful to the mechanical estimator. The plumbing fixture count will also aid in determining many fitting requirements.

Copper Pipe and Tubing

Copper pipe is not as commonly used in the construction industry today as it was in years past. Copper is now relegated to very special systems and trim around boiler controls and the like. Copper, yellow brass, and red brass (85% percent copper) pipe have, for the most part, been replaced by copper tubing. These materials are still available, should the need for them arise, with threaded brass fittings 125 lb. MIP (malleable iron pattern), and 250 lb. CIP (cast iron pattern).

Copper pipe, having the same O.D. as steel pipe, will accept threaded valves and fittings. Labor units for threaded copper pipe are similar to those for threaded steel.

Because copper tubing has a thinner wall and smaller O.D. than pipe, it cannot be threaded and must instead be assembled using soldered, flared or brazed fittings and valves. The tubing comes in several classes and types as well as tempers.

The first class is *water tubing*, which is produced in types K, L, M, and DWV. Type K heavy wall tubing is used in interior and underground systems, available in soft or hard temper. Type L medium wall tubing is used in interior systems, available in hard or soft temper. Type M light wall tubing is used in low pressure systems, and cannot be used in refrigeration systems. Type M is available in hard temper only. Type DWV light wall tubing is used in plumbing drainage, waste, and vent systems, and is available in hard temper only. The Copper Development Association and the BOCA National Plumbing Code allow types L and M in underground systems, types M and DWV in chilled water systems and forced hot water systems excluding high rise and other pressure restrictions.

The second class of copper tubing is type ACR, produced for use in *air conditioning and refrigeration systems*. Type ACR tubing has the same wall thickness as type L, and is shipped cleaned, capped, and charged with nitrogen, available in hard or soft drawn.

Water tubing is sized nominally. Types K and L are produced in 1/8" through 12" diameters, Type M in 1/4" through 12", DWV in 1-1/4" through 8" diameters. ACR tubing is sized by outside diameter (O.D.), produced in 1/8" through 4-1/8" O.D. Hard drawn tubing is manufactured in 20' lengths; soft, or annealed, is manufactured in straight lengths or coils. Figure 13.12 shows commercially available lengths of copper tubing.

Units of Measure: Copper pipe and tubing is measured in linear feet (L.F.). Associated fittings, supports, and joints are recorded as each.

Labor Units: The following procedures are generally included in the per foot labor cost:

- Unloading and relocating from receiving or storage area to point of installation.
- Set up of scaffolding, staging, ladders, or other work platforms.
- Lay out of piping route.

Tube	Commercially Available Lengths			
	Hard Drawn		Annealed	

Type K	Straight lengths up to:
	  8-inch diameter
	  10-inch diameter
	  12-inch diameter
		20 ft.
		18 ft.
		12 ft.
			Straight lengths up to:
			  8-inch diameter
			  10-inch diameter
			  12-inch diameter
			
			Coils up to:
			  1-inch diameter
			
			1¼ and 1½-inch
			  diameter
			2-inch diameter
				20 ft.
				18 ft.
				12 ft.
				
				60 ft.
				100 ft.
				
				60 ft.
				40 ft.
				45 ft.

Type L	Straight lengths up to:
	  10-inch diameter
	  12-inch diameter
		20 ft.
		18 ft.
			Straight lengths up to:
			  10-inch diameter
			  12-inch diameter
			
			Coils up to:
			  1-inch diameter
			
			1¼ and 1½-inch
			  diameter
			2-inch diameter
				20 ft.
				18 ft.
				
				60 ft.
				100 ft.
				
				60 ft.
				40 ft.
				45 ft.

Type M	Straight lengths:
	  All diameters
		20 ft.	Straight lengths up to:
			  12-inch diameter
			
			Coils up to:
			  1-inch diameter
			
			1¼ and 1½-inch
			  diameter
			2-inch diameter
				20 ft.
				
				60 ft.
				100 ft.
				
				60 ft.
				40 ft.
				45 ft.

DWV	Straight lengths: All diameters	20 ft.	Not available	—
ACR	Straight lengths:	20 ft.	Coils:	50 ft.

Figure 13.12

These additional procedures must be recorded and included in the piping estimate:
- Joining straight lengths every 20 feet (or 10 feet in larger sizes).
- Installation of fittings, valves, appurtenances, supports, anchors, and guides.
- Testing and cleaning of systems (special emphasis on cleaning and chlorinating potable systems.)

Takeoff Procedure: A takeoff sheet should be set up for each piping system and record the various footages for each diameter and type of pipe. Hot, cold, drain, and vent systems should be listed separately. Underground pipe should also be listed separately due to differences in support and installation procedures. Each system should be marked on the plans as it is taken off, beginning with the larger diameter pipe and proceeding in the direction of the flow (except for drain piping, where the larger pipe is at the termination point). Round off individual totals to coincide with 20', lengths, (60', or 100' coils) available from local suppliers. The height modifications given for steel pipe apply to copper pipe and tubing as well (refer back to Figure 13.3).

Copper Fittings and Flanges

Threaded copper and brass pipe both require the same fitting types; cast bronze threaded fittings, available in the 125 lb. rating in the open pattern (with a light bead). These fittings (shown in Figure 13.13) are often referred to as "malleable pattern bronze fittings". Extra heavy cast bronze fittings (250 lb.) are also available for use with copper pipe, in the cast iron pattern (closed pattern with a heavy bead). Drainage and flanged fittings are also made which are similar to their cast iron counterparts. For special requirements, threadless fittings are produced for brazing to pipe.

Fittings and flanges for copper tubing are soldered. Solder end (sweat) fittings are produced for use in both pressure and drainage systems. In soldered pressure systems only one standard, or line of fittings, is produced for types K, L, and M tubing. More critical pressures and temperatures are managed by using solder with higher melting points or by brazing. For most plumbing, heating, and cooling systems, soft solder is more than adequate for these non-critical applications.

At this writing, the use of lead in solder for potable water systems has been banned in most areas of the country. Instead, solder comprised of a mixture of tin and antimony or other substitutes should be used.

In general, solder type fittings are produced in wrought copper and cast bronze. Both are priced equally and have similar installation costs. In the larger sizes, some manufacturers do not make fittings in wrought copper, and cast bronze must be used. Cast bronze, however, requires more heat to make the joint but does not cool down quickly (which can be important when soldering large diameter pipe). Drainage fittings are also produced in wrought copper and cast bronze. Since drainage fittings do not have a pressure factor, soft solder can be used in making up the joints.

Flared fittings are used with annealed copper tubing (soft drawn) in water, refrigeration, and gas systems. These fittings are used when soldering or brazing is not practical, and ease of disassembly is a requirement. Special tools are needed, in this case, to flare the tubing end, which becomes an integral part of the joint. These fittings are limited in size from 3/8" to 2" nominal.

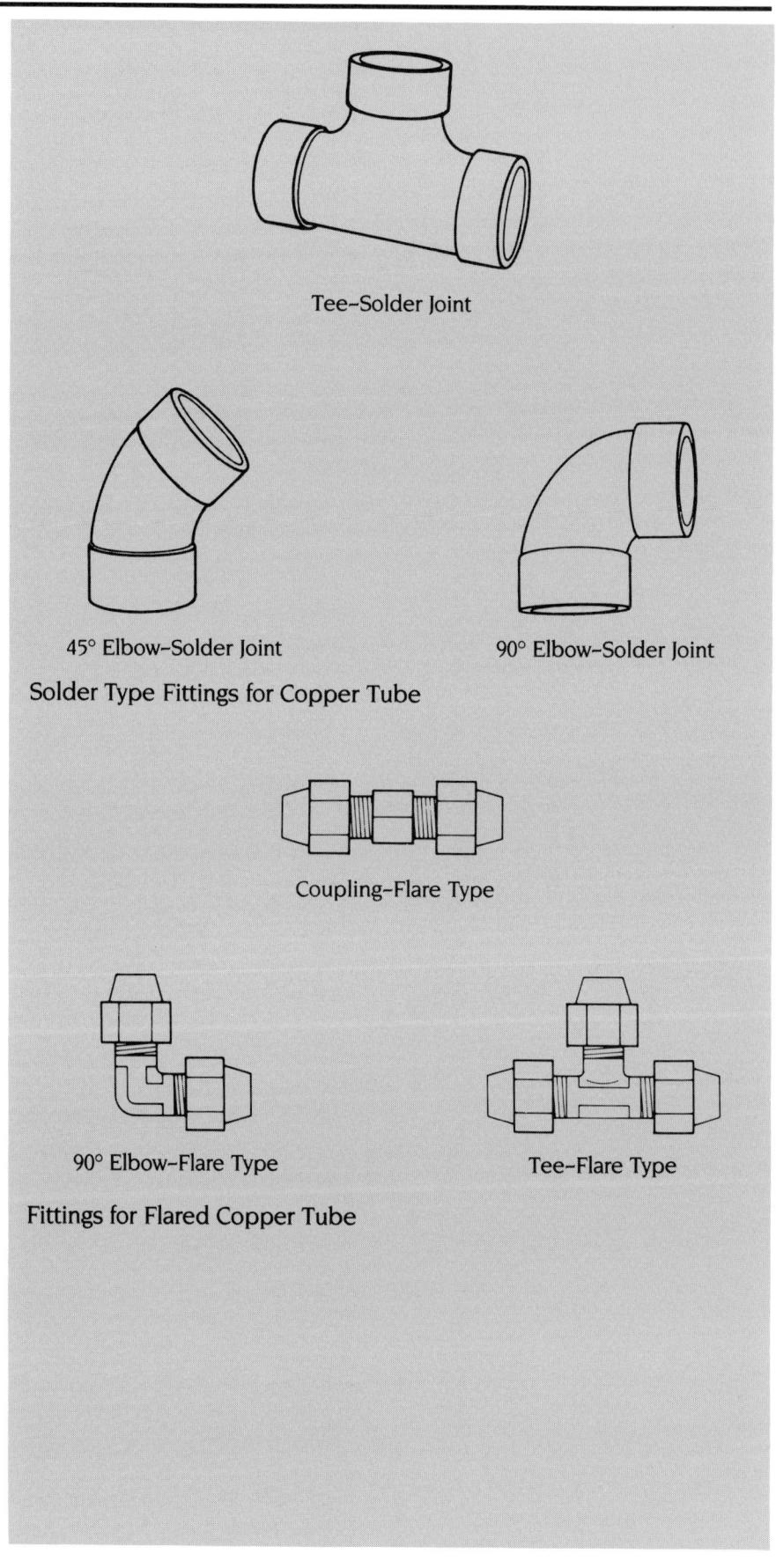

Figure 13.13

Compression type fittings are also used with soft drawn tubing for some oil burner work, domestic refrigeration, and instrumentation.

Units of Measure: All fittings are taken off and recorded as each. The number of joints is determined by this fitting count. The number of fittings and joints may be greatly reduced by the judicious use of bends, end swaging (expanding), and drilled and formed branch outlets. These items should be recorded in the same manner as fittings (as each) to anticipate the labor units necessary to perform the work.

Material Units: In addition to the actual cost of the fittings are the materials necessary to prepare the joints. For threaded bronze fittings, material unit costs include nipples, cutting oil, and joint compound, similar to threaded steel joints. For solder type fittings, sandcloth, flux, solder, and a heating source (gas) are needed. A chart of solder and flux consumption for these joints is shown in Figure 13.14

Estimated Quantity of Soft Solder Required to Make 100 Joints	
Size	Quantity in Pounds
3/8"	.5
1/2"	.75
3/4"	1.0
2"	1.5
1¼"	1.75
1½"	2.0
2"	2.5
2½"	3.4
3"	4.0
3½"	4.8
4"	6.8
5"	8.0
6"	15.0
8"	32.0
10"	42.0

1. The quantity of hard solder used is dependent on the skill of the operator, but for estimating purposes, 75% of the above figures may be used.
2. Two oz. of Solder Flux will be required for each pound of solder.
3. Includes an allowance for waste.
4. Drainage fittings consume 20% less.

Figure 13.14

Labor Units: The labor units for threaded bronze fittings include:
- Tightening and sealing each threaded joint.
- Cutting and threading, when required for intermediate joints in the shop or field.

These considerations are identical to cast iron and malleable iron fittings for steel pipe, both threaded and flanged. The total number of joints is determined from the fitting and valve counts, plus any intermediate joints in long straight runs of pipe.

Solder joint fittings require a unique method of joint preparation: a square cut tube end, deburred, cleaned with a sandcloth or commercial (acid) cleaner; application of flux; insertion into the similarly cleaned hub of the fitting or valve; application of heat; and filling the space between the hub and tube with solder. Allowances should also be considered for cooling or quenching a soldered assembly if it is to be handled immediately.

Flared fitting joint assembly consists of a square cut of the tube end, deburring, and slipping the flare nut onto the tubing *before* flaring the end of the tube. After the flare is completed, the nut is mechanically tightened to the male thread of the fitting. Flared joints require no other materials besides proper cutting and flaring tools and wrenches.

Takeoff Procedure: Fittings, valves, appurtenances, bends, swaged joints, field drawn tees, and test nipples should be recorded on the takeoff sheet for copper pipe or tubing, by size and system. From this count, the number of joints and the labor for each system can be determined. By recording the systems separately, the same takeoff sheets can be used for purchasing and scheduling during construction.

Plastic Pipe and Tubing

Plastic and thermoplastic (material which softens when subjected to heat, rehardening when cooled) pipe and tubing have revolutionized mechanical piping systems over the past few decades, and will continue to do so in the years to come as the plastics industry grows. Because of its relative "newness" to the market, a wide variety of materials, types, and joining procedures are available, all seeking their share of this expanding market. For example, the most familiar plastic piping material found in mechanical systems is polyvinyl chloride (PVC) pipe, used in drain, waste, and vent (DWV) and cold water systems. PVC competes with acrylonitrile butadiene styrene (ABS) piping, especially in DWV applications. Polypropylene is another competitor in DWV applications. Chlorinated PVC (CPVC) pipe has been developed for use in hot water systems; polyethylene is used in cold water systems; and polybutylene is used in hot and cold water systems. CPVC and polybutylene are recently approved for use in wet type fire protection systems, while polyethylene is commonly used in gas and water services.

As with other types of piping systems, there are both advantages and disadvantages to the use of plastic pipe. The light weight of plastic pipe is a major installation labor saving advantage over more traditional heavier materials. The flexibility of some types of plastic is another labor saving advantage. This flexibility eliminates some bends and elbows, and reduces the need for expansion control or water hammer eliminators. The disadvantages of some plastics are vulnerability to sunlight, ambient temperatures, abrasion, and piercing by sharp objects.

Plastic pipe comes in rigid 10' and 20' lengths, or in coils of 1,500', 1,000', 600', 500', 400', 300', 200', and 100' lengths, depending on diameter. Plastic pipe and tubing are also produced in piping schedules 40 and 80, plus in the O.D. sizes comparable to those of copper, and in a wall thickness rating method beyond scheduled pipe classifications, known as SDR (standard dimension ratio). The SDR method was developed to standardize the inside and outside dimensions of plastic piping which does not conform to the pipe schedule method. It is determined by dividing the average O.D. by the minimum wall thickness.

Pipe diameters are generally comparable to those found in steel, copper, and cast iron systems. Some manufacturers, however, do not produce a full line, skipping some intermediate sizes due to closeness in diameter or infrequency of use.

Units of Measure: Plastic pipe and tubing are taken off and recorded per linear foot. Associated fittings, joints, supports, valves, and appurtenances are counted as each. In some instances, constant or continuous support may be required, which should also be recorded by linear foot.

Labor Units: The following procedures are generally included in the per foot installation labor cost of plastic pipe and tubing:

- Unloading, protecting, relocating from receiving or storage area to point of installation.
- Setup of scaffolding, staging, ladders, or other work platforms.
- Lay out of piping route.

These additional procedures should also be recorded and included in the piping estimate:

- Joining or coupling intermediate lengths.
- Curing of solvent cement or otherwise preparing the joints for fusion welding, threading, etc.
- Installation of fittings, valves, appurtenances, supports, anchors, and guides.
- Testing and flushing of systems (especially portions concealed below grade in the building construction).

Takeoff Procedure: A takeoff sheet for each piping system should be set up. If more than one type of plastic pipe or joining method is to be used in the same system, each type should be listed separately. The various footages required should be recorded for each pipe diameter. Color coding the systems as they are taken off helps track the progress of the takeoff, as well as serving to define each system so it can be quickly identified on the plans later on. The largest size pipe should be taken off first, and the totals rounded off to coincide with pipe lengths available. Height modifications (shown in Figure 13.3) may need to be considered. Underground or structurally supported lengths should be listed separately because they have unique support considerations.

Fittings for Plastic Pipe and Tubing

There are at least three times as many joining methods for plastic pipe as there are different types of plastic. There are fittings available to accommodate any end finish or joining procedure required for the various types and sizes of plastic pipe or tubing. This may include threaded, flanged, compression, socket (solvent) weld, fusion weld, butt weld, brass insert, plastic insert, clamp type, grip type, flared, gasketed fittings, and many combinations thereof. Some of the common fitting and joining methods for plastic pipe are shown in Figure 13.15.

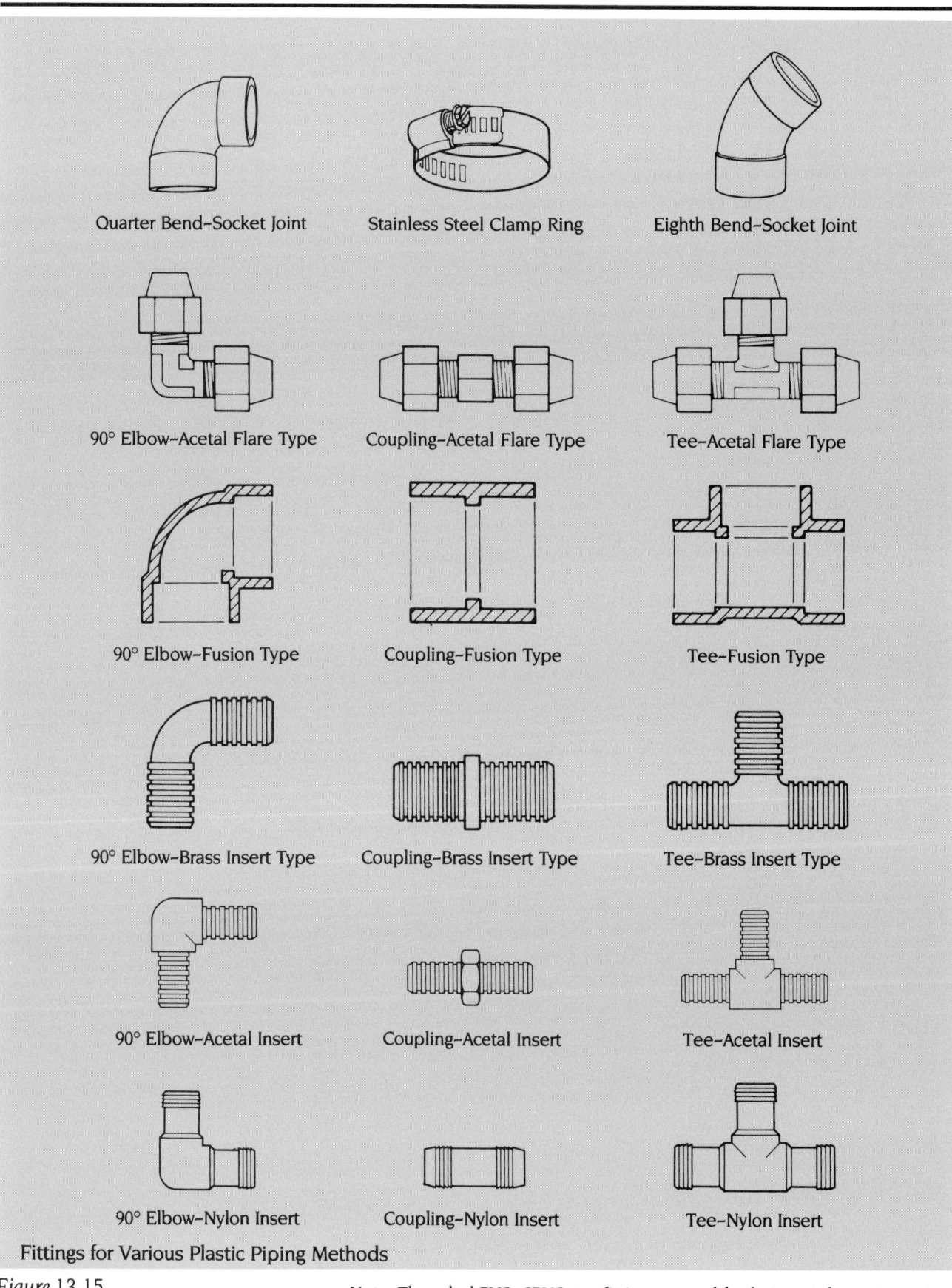

Fittings for Various Plastic Piping Methods

Figure 13.15   Note: Threaded PVC, CPVC etc. fittings resemble their metal counterparts.

Units of Measure: Fittings and flanges are taken off and recorded as each. The number of joints, cuts threads, and welds will be determined by this fitting count.

Material Units: Unique tools for cutting, deburring, and preparing are required for plastic fittings. Solvent welds also require primers, solvent cements, and applicators. Fusion welds require special torches and tips. Electrical resistance fusion kits consisting of power units and fusion insert coils are needed for some polypropylene drain waste and vent systems for use with acids and other corrosives. Threaded joints require pipe joint compound or tape; flanged joints require bolts, nuts, and gaskets; and some compression or clamp type joints require gaskets.

Labor Units: Labor units for plastic fittings are similar to those of metal pipe for flanged or threaded joints. For solvent weld fittings, the simple but essential procedure of priming and applying the solvent must be adhered to. Fusion welding procedures vary widely with each manufacturer. It is important to follow the manufacturer's instructions and to verify that the fittings are compatible with the pipe or tubing being installed. Local codes should be consulted before selecting the piping material and joining method. It is also critical in plastic pipe joint makeup that dirt or any foreign matter be removed. A bad plastic joint usually cannot be redone; it must be removed and replaced.

Takeoff Procedure: Plastic fittings, valves, appurtenances, bends, and intermediate joints should be recorded by size, system, and joining method on the takeoff sheet for plastic pipe or tubing. From this count, the number of joints or diameter inches can be determined, and the labor units can be estimated. It is advantageous to record the various systems on separate takeoff sheets for use during construction planning and purchasing.

Preinsulated Pipe or Conduit

Conduit, or preinsulated pipe, is available in various combinations of pipe and insulation materials. It consists of a prefabricated core of pipe (one or more) coated with insulation. The core is then protected with an outer jacket of heavy gauge sheet metal or a rigid plastic casing (shown in Figure 13.16). Conduits have been developed for underground piping systems, overhead lines between buildings, along waterfront docks and piers, spanning rivers, railroad tracks, etc., by using a bridge for support. Conduit systems may contain one or more pipe, (either nested or individually insulated) and prefabricated elbows, tees, expansion loops, anchors, and building entry leakplates.

Units of Measure: All conduit is taken off by individual sizes per linear foot. Prefabricated fittings, manholes, and the like should be recorded as each.

Material Units: The conduit supplier should furnish all material required to make a complete section (including slings for handling) and any unique tool for connecting. All the material for field joint assembly, insulation and outer closure should also be part of the supplier's job quotation.

Labor Units: The labor units required here are to physically place the sections of conduit in the trench or on the proper supports. Diameter and weight will determine whether the installation is to be by manual labor or the use of a crane or if powered lifting equipment is required. The pipe joints are prepared and made up in accordance with manufacturers' instructions and are pressure tested prior to insulation.

After testing and insulation are completed the outer closure or casing is installed. Whether steel or plastic, the linear inches of welding this closure must be taken off and recorded for material and labor costs. The exterior closure itself is then tested with a pressure test (using compressed air).

In addition to the manufacturer's drawings and instructions, a field service engineer is normally provided to direct the installer and certify that the installation conforms to the manufacturer's recommendations.

Takeoff Procedure: The conduit is normally taken off by the manufacturer's representative, but the estimator should also do a takeoff as both a double check and to determine labor units, incidental supports needed, etc. This takeoff should be listed on its own takeoff sheet and the labor should be listed separately for future observation.

Valves

A valve is a mechanical device which controls the flow, pressure, level, or temperature of the fluid being conveyed or contained in a piping system. Valves can be operated manually or automatically. An automatic valve may be operated by a motor, or by the temperature, pressure, liquid level, or direction of the liquid flow. A manual valve may have a wheel, lever, wrench, or gear operator.

Valves are used in virtually all piping systems with the exception of drain, waste, and vent; backwater valves which are sometimes used in drain or waste piping systems being the exception.

There are valves for each pipe and tubing size and for each method of connection, whether it be flanged, screwed, weld, solder, compression, grip type, or mechanical joint installation. In mechanical systems valves are usually iron body, bronze, copper, or plastic. Valve sizes are rated as standard (in the 100 to 175 lb. range), or extra heavy (in the 200 to 300 lb. range). Valves described in the following paragraphs are shown in Figure 13.17.

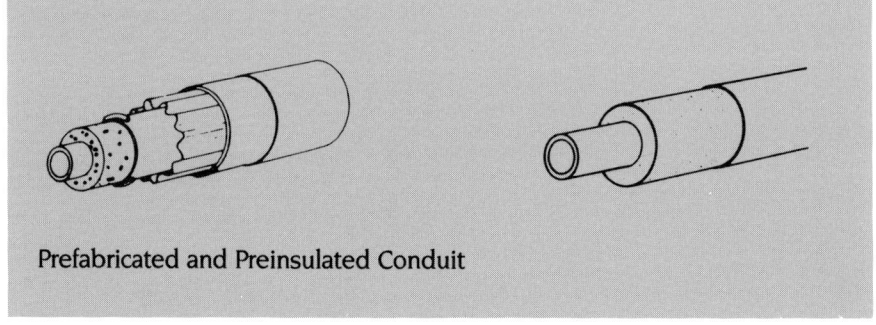

Prefabricated and Preinsulated Conduit

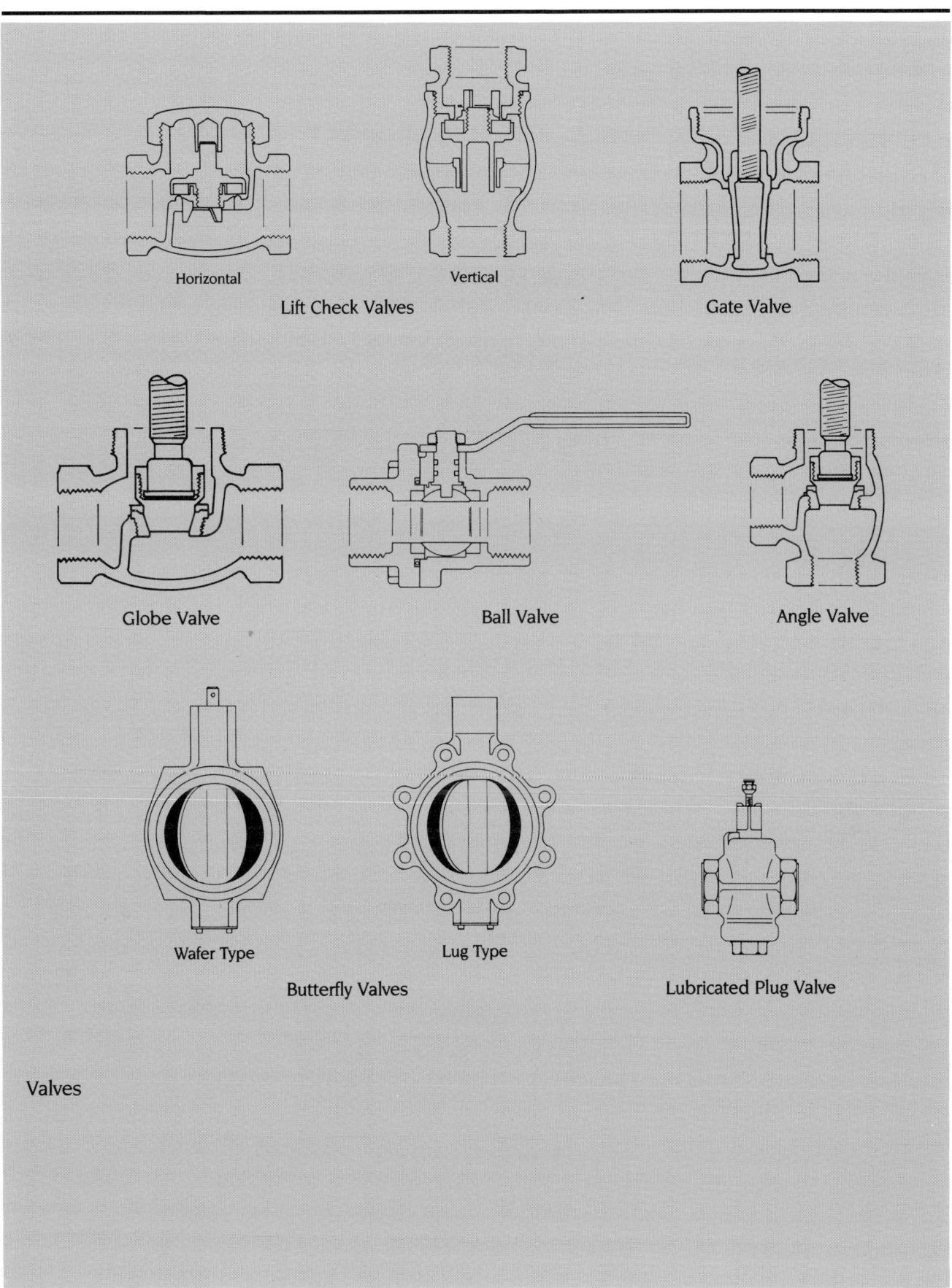

Valves

Figure 13.17

Gate Valves provide full flow, minute pressure drop, minimum turbulence and minimum fluid trapped in the line. The flow is controlled by raising or lowering a gate via a screw mechanism. Gate valves should always be fully opened or fully closed. They are normally used when operation is infrequent.

Globe Valves are designed primarily for throttling flow. A plug is raised or lowered via a screw mechanism into a seat. The configuration of the globe valve causes restricted flow which results in increased pressure drop. Besides throttling, globe valves can also shut off the flow.

The fundamental difference between the *Angle Valve* and the globe valve is the fluid flow through the angle valve. When a 90 degree turn is required, an angle valve offers less resistance to flow than a globe valve and elbow combination. Use of an angle valve also reduces the number of joints and installation time.

Swing Check Valves are designed to prevent backflow by automatically seating when the direction of the fluid is reversed. Swing check valves are usually installed with gate valves, as they provide comparable full flow. Usually recommended for lines where flow velocities are low, they should not be used on lines with pulsating flow. Swing checks valves are recommended for horizontal installation, or in vertical lines where flow is upward.

Lift Check Valves have diaphragm seating arrangements similar to globe valves and are recommended for preventing backflow of steam, air, gas, and water, and on vapor lines with high flow velocities. For horizontal lines, horizontal lift checks should be used. Vertical lift checks should be used for vertical lines.

Ball Valves are light and easily installed, and because of modern elastomeric seats, provide tight closure. Flow is controlled by rotating up to 90 degrees a drilled ball which fits tightly against resilient seals. This ball seats with the flow in either direction, and the valve handle indicates the degree of opening. Recommended for frequent operation, readily adaptable to automation, it is ideal for installation where space is limited.

Butterfly Valves provide bubble tight closure with excellent throttling characteristics. They can be used for full-open, closed, and throttling applications. The butterfly valve consists of a disc within the valve body which is controlled by a shaft. In its closed position, the valve disc seals against a resilient seat. The disc position throughout the 90 degree rotation is visually indicated by the position of the operator. A butterfly valve is only a fraction of the weight of a gate valve and, in most cases, requires no gaskets between flanges. Recommended for frequent operation, it is adaptable to automation. Wafer and lug type bodies, when installed between two pipe flanges, can be easily removed from the line. The pressure of the bolted flanges holds the valves in place. Locating lugs makes installation easier.

Lubricated *Plug Valves* are similar to ball and butterfly valves in operation and configuration. Because of the wide range of services to which they are adapted, they may be classified as all-purpose valves. Plug valves can be safely used at all pressure and vacuums, and at all temperatures up to the limits of available lubricants. They are the most satisfactory valves for the handling of gritty suspensions and many other destructive, erosive, corrosive, and chemical solutions.

Units of Measure: Valves are taken off and recorded as each.

Material Units: The material cost of valves usually involves only the cost of the valve itself, except in electric or pneumatic operated systems. Electric or pneumatic operation requires the addition of a motor or actuator, installed after the valve body has been piped. Remote chain wheel operators and gear operators are also purchased and installed separately.

Labor Units: The labor involved in installing a valve is usually the makeup and handling of one, two, or three piping connections. Valves which have only one piping connection include safety or relief valves, drain cocks, sill cocks, hose bibbs, vacuum breakers, air vents, and ball cocks. Valves having two piping connections are the type most often encountered. Two connection valves are gates, globes, checks, plug cocks, butterfly, ball, stop and waste, balancing, radiator control, flushometer, fire, fusible oil, backflow preventer, pressure regulating, angle stop, etc. Three way or mixing valves include shower valves, motor operated temperature control valves, and blending valves for fuel.

As with fittings and piping, weight is always a contributing factor to the labor cost of installing valves, as well as height above the floor or work platform (for height modifications, see Figure 13.3)

Takeoff Procedure: The valves and accessories for each system should be recorded by size and type on the takeoff sheet used for pipe and fittings (shown in Figure 13.9). An estimator may have to include valves not shown on the plans, but which may be necessary to isolate portions of a piping system for testing and draining.

Pipe Hangers and Supports

Piping for mechanical installations is supported by a wide variety of hangers and anchoring devices. The location of piping supports is an important consideration, and many different building components may be used for anchoring pipe hangers. Some of the commonly used locations for anchoring devices (shown in Figure 13.18) include the roof slab or floor slab, structural members, side walls, another pipe line, machinery, or building equipment.The pipe hanger material is usually black or galvanized steel. For appearance or in corrosive atmospheres, chrome or copper plated steel, cast iron, or a variety of plastics may also be used.

Another consideration is the method of anchoring the hanger assembly to the building structure. The method selected for anchoring the hanger depends on the type of material to which it is being secured, usually concrete, steel, or wood. If the roof or floor slab is constructed of concrete, formed, and placed at the site, then concrete inserts may be nailed in the forms at the required locations prior to the placing of concrete. These inserts (shown in Figure 13.18) may be manufactured from steel or malleable iron, and they are either tapped to receive the hanger rod machine thread or contain a slot to receive an insert nut. Because the slotted type of insert requires separate insertable nuts, for the various rod diameters, only one size of insert need be warehoused. For multiple side by side runs, long, slotted insert channels of up to 10' increments are available with several types of adjustable insert nuts.

When precast slabs are used, the inserts may be installed on-site in the joints between slabs, drilled, or shot into the slab itself. Electric or pneumatic drills and hammers are available to drill holes for anchors,

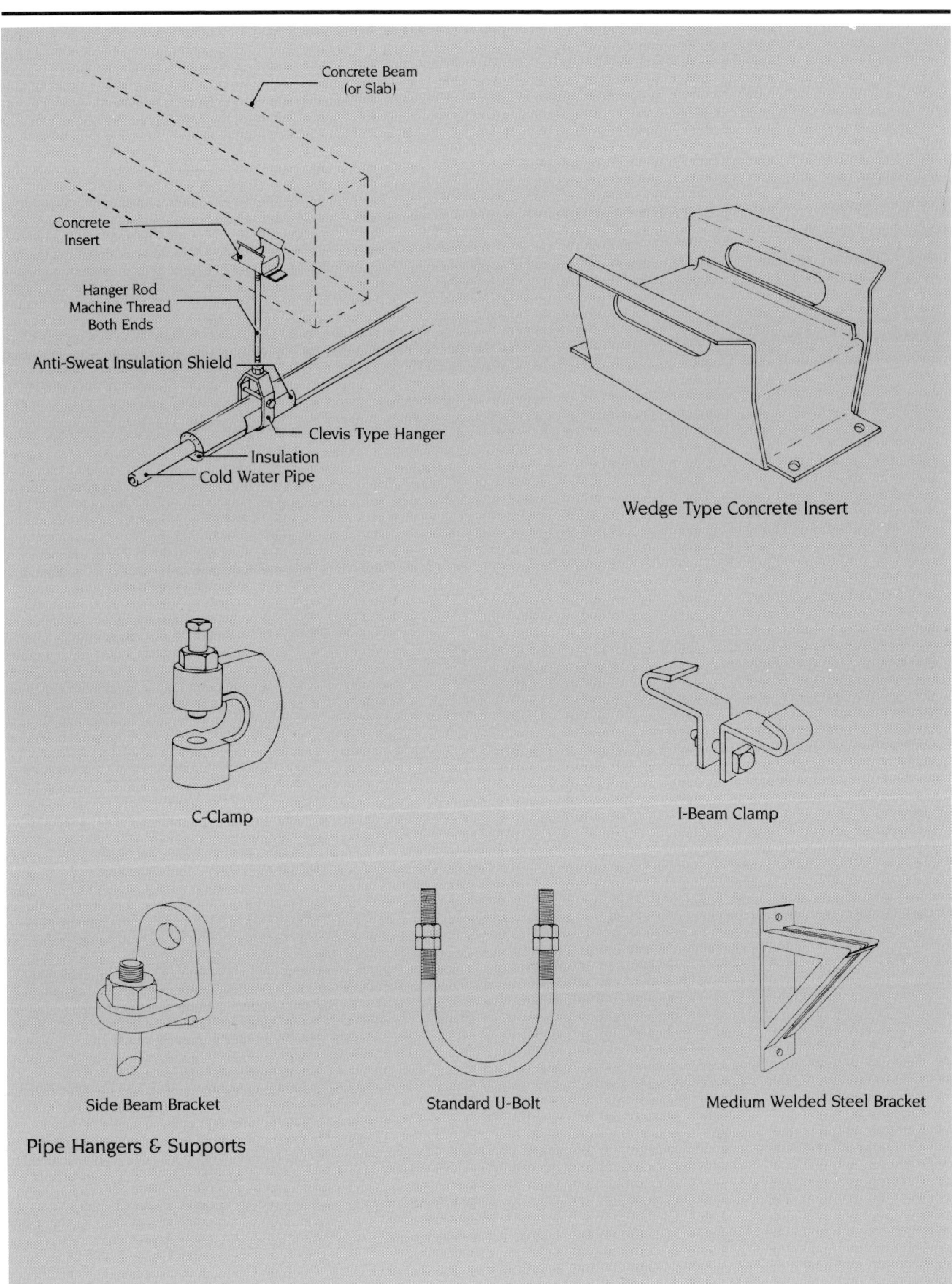

Pipe Hangers & Supports

Figure 13.18

shields, or expansion bolts. Another method of installing anchors on site utilizes a gunpowder actuated stud driver, which partially embeds a threaded stud into the concrete.

If the piping is to be supported from the sidewalls, similar methods of drilling, driving, or anchoring to those mentioned above are used for concrete walls. Where hollow-core masonry walls are to be fitted, holes may be drilled for toggle or expansion-type bolts or anchors.

Attaching the pipe supports to steel or wood requires different methods from those used for concrete. When the piping is to be supported from the building's structural steel members, a wide variety of beam clamps, fish plates (rectangular steel washers), and welded attachments can be employed. If the piping is being run in areas where the structural steel is not located directly overhead, then intermediate steel is used to bridge the gap. This intermediate steel is usually erected at the piping contractor's expense. If the building is constructed of wood, then lag screws, drive screws, or nails are used to secure the support assembly.

From the anchoring device, a steel hanger rod, threaded on both ends, extends to receive the pipe hanger. One end of this rod is threaded into the anchoring device, and the other is fastened to the hanger itself by a washer and a nut. For cost effectiveness and convenience, continuous thread rod may be used. The pipe hanger itself (shown in Figure 13.19) may be a ring, band, roll, or clamp, depending on the function and size of the piping being supported. Spring-type hangers are also used when it is necessary to cushion or isolate vibration.

Piping systems subject to thermal expansion must often absorb this expansion by use of piping loops, bends, or manufactured expansion joints. The piping must be anchored to force the expansion movement back to the joint. In order to prevent distortion of the piping joint itself, alignment guides are installed. Figure 13.20 shows alignment guides and roll hangers used in these types of installations.

Specifically designed pipe hangers are used for fire protection piping with underwriters and factory mutual approvals.

In wood frame residential construction, holes drilled through joists or studs have often been the sole support for certain pipes. These pipes may occasionally be reinforced with wedges cut from a 2" x 4". A piece of wood cut and nailed between two studs is often used to support the plumbing stubouts to a fixture. Prepunched metal brackets are available to span studs for both 16" and 24" centers. These brackets can be used with both plastic or copper pipe and maintain supply stub spacing 4", 6", or 8" on centers, giving perfect alignment and a secure time saving support. Plastic support and alignment systems are also available for any possible plumbing fixture rough-in. Some typical pipe supports for wood construction are shown in Figure 13.21.

Units of Measure: Pipe hangers and supports are taken off, recorded, and priced as each. As a general rule, the number of pipe hanger assemblies can be approximated from the piping totals and the required hanger spacing (e.g., one hanger every 10 feet). The fixture count will give an accurate measure for bracket supports or alignment systems.

Material Units: Hanger assemblies should be pre-priced for each different size and type of piping. The assembly price should include the insert or beam clamp, vertical rod, nuts, hanger, and, if necessary, an insulation

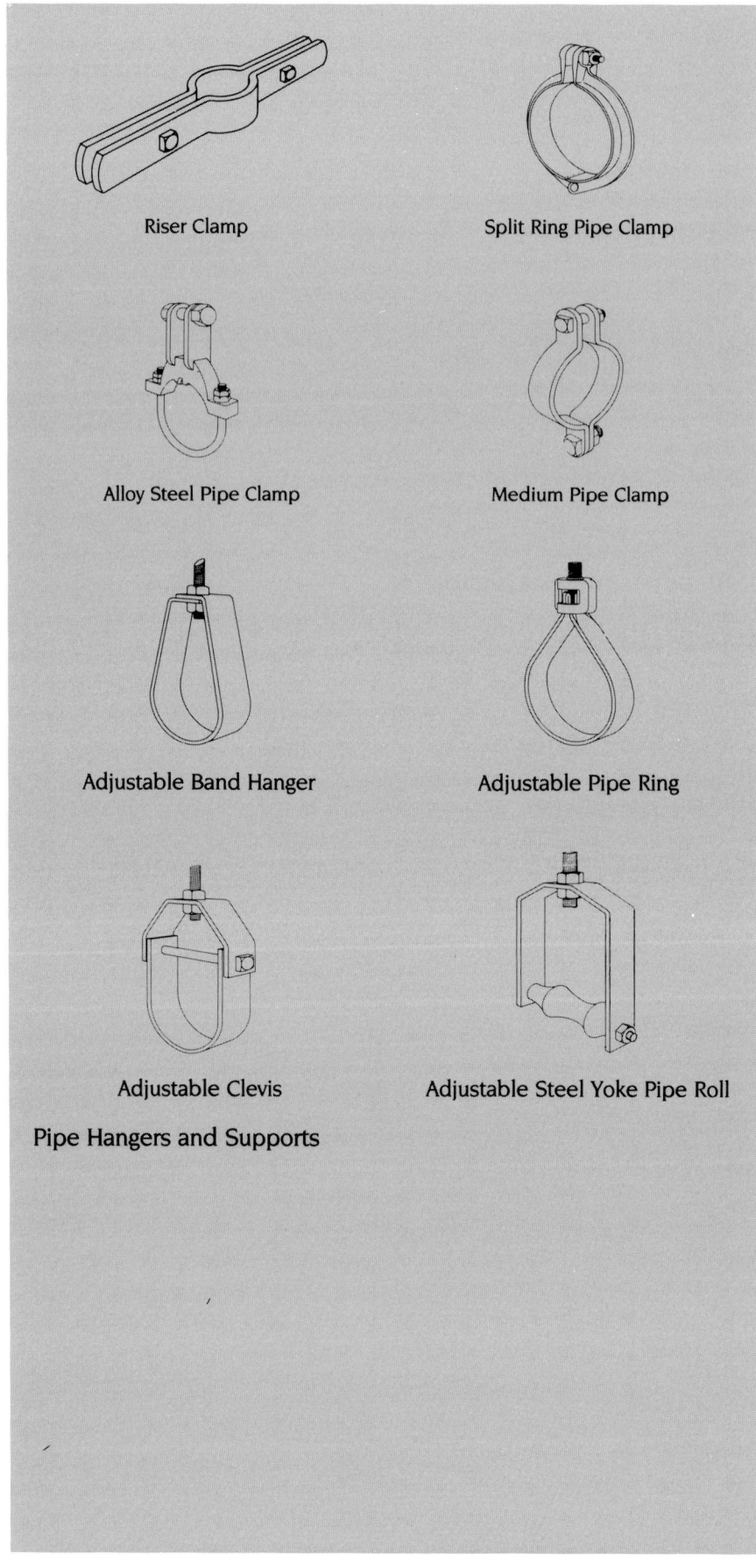

Figure 13.19

shield. The estimator will decide, from experience, which hanger assemblies should be pre-priced, depending on the type of work or system a firm might specialize in. This method can save an estimator many hours of repetitive pricing each nut, bolt, and washer for each and every size of pipe and for each estimate prepared.

Labor Units: The following procedures must be considered and included in the labor unit per hanger assembly, depending on the job conditions:

- Sleeving and inserting for each concrete slab when it is being formed, otherwise, drilling or driving expansion shields or studs.
- Cutting the hanger rods (cutting and threading if labor conditions do not permit the use of threaded rod).
- Installing and adjusting the height of each assembly to maintain the desired pitch.

The type of hanger selected can significantly influence the time and cost of piping installation.

Residential construction (wood frame) permits the installation of simple supports or hangers after the piping is in place. Underground piping may not require supports depending on the type of pipe or tubing used and the condition of the pipe bed. Masonry blocks or bricks are commonly used if the pipe bed is not satisfactory. Steel pipe (insulation protection) saddles will have the added labor of tack welding the saddles to the pipe.

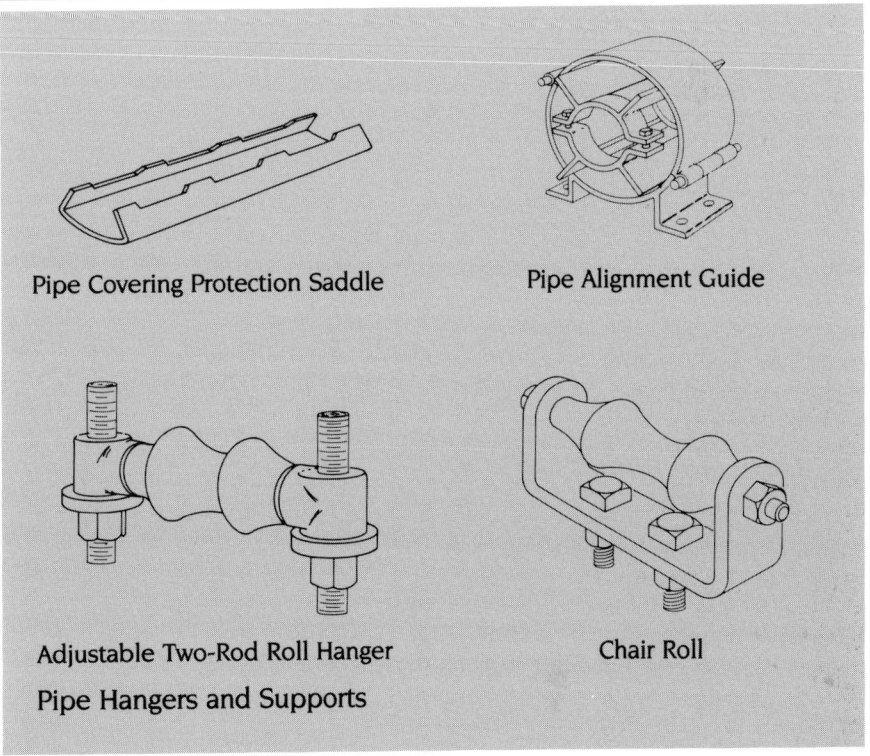

Figure 13.20

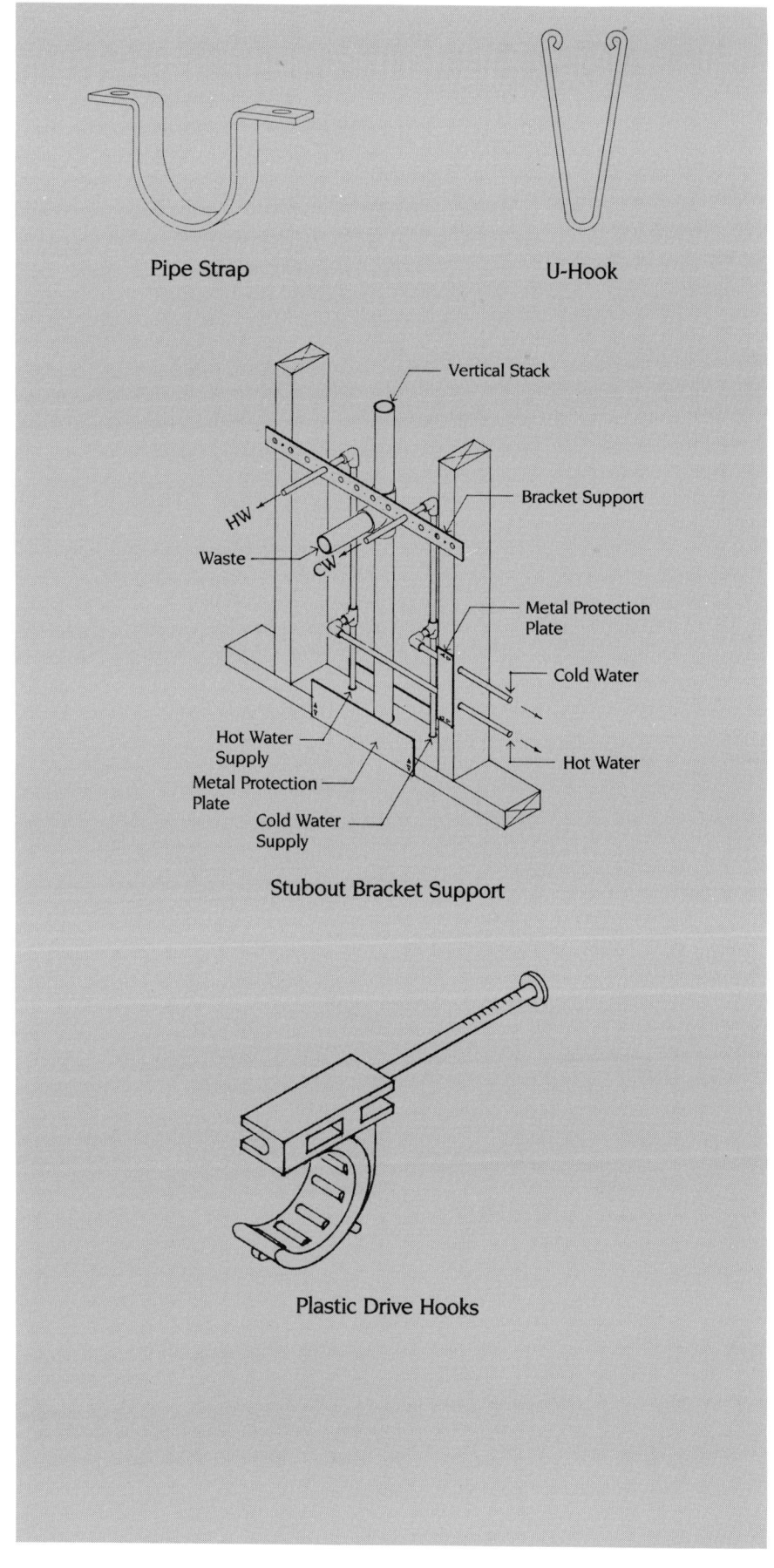

Figure 13.21

Saddles for cold water piping may be installed by the pipe coverer (installation instructions should first be read).

Piping support may be inconsequential or it may be critical to an estimate; do not overlook it.

Takeoff Procedure: From the piping takeoff totals, the number of hanger assemblies can be determined, by size and system, as shown in Figure 13.22.

Cast iron soil pipe should be supported every 5 feet. Thermoplastic piping may have to be supported as much as every 3 feet, depending on ambient temperatures. In many cases, it may be more cost effective to support plastic pipe using continuous angle, channel, or other rigid members between widely spaced hangers.

In addition to the table shown in Figure 13.22, additional support must be added for runouts, change of direction, large valves, headers, etc. A short cut estimating method for new construction is to use a 10 foot average span for all sizes and types of piping from the takeoff totals.

Cost Modifications: In place of individual hangers, trapeze or gang hangers, consisting of two rods and one horizontal member, may be used to support several lines at a labor and material savings. Field fabrication is typically required to make up these assemblies.

Support Spacing for Metal Pipe or Tubing			
Nominal Pipe Pipe Size Inches	Span Water Feet	Steam, Gas, Air Feet	Rod Size
1	7	9	↑
1½	9	12	3/8"
2	10	13	↓
2½	11	14	↑
3	12	15	½"
3½	13	16	↓
4	14	17	↑
5	16	19	5/8" ↓
6	17	21	3/4"
8	19	24	↑
10	20	26	7/8"
12	23	30	↓
14	25	32	↑
16	27	35	
18	28	37	1"
20	30	39	↓
24	32	42	↑
30	33	44	1¼" ↓

Figure 13.22

Chapter 14
PUMPS

A pump is a mechanical device used to convey, raise, or variate the pressure of fluids, such as water, oil, or sewage. In the building construction industry, pumps are employed in dewatering the site of excavation, increasing water pressure for potable or fire protection systems, circulating potable, cooling or heating water, transferring fuel oil, draining wet basements, or ejecting sewage from sub-basement pits. Representative pumps are shown in Figure 14.1.

The most commonly used pumps are centrifugal, driven by an electric motor. In some instances fire pumps may be driven by diesel engine as a precaution against electrical failure. Most pump bodies are constructed of cast iron. For potable water pumps, bronze is usually substituted for cast iron at an increase in cost.

To insure proper operation, a check valve should be installed on the discharge side of the pump for backflow protection. If a throttling or flow regulating valve is required, it should also be on the discharge side to prevent pump cavitation.

Units of Measure: Pumps or pumping units are taken off, recorded, and priced as each.

Material Units: As a general rule, the pump and its motor are obtained as a complete unit with no other exterior trim or fittings. Some exceptions to this rule are vacuum heating pumps, condensate return, or boiler feed systems. These pumps may be purchased mounted on a steel or cast iron tank or receiver, complete with interconnecting piping, valves, gauges, thermometers, controls, etc.

Fire pumps, sewage pumps or ejectors, and factory assembled or prepackaged water pressure booster systems are also available. These prepackaged assemblies usually include more than one pump, as well as the aforementioned piping and trim.

An estimator's library should include manufacturers' catalogues of pumps and pump packages. These will show pipe sizes, weights, horsepower, and trim, and aid in visualizing the installation of the specified equipment or its equivalent. Many pump installations also require in line vibration isolating flexible connections and inertia bases.

Labor Units: Installing in line pumps is comparable to a valve installation, depending on pipe size and weight. Base mounted pumps normally require the pre-installation of a masonry foundation and anchor bolts. For larger base mounted pumps, extraordinary vibration elimination bases may be specified and included in both the material and labor cost.

The size and weight of a pump or prepackaged assembly will dictate the crew size and the mechanical equipment or tools required to unload, store, move to installation area, and set in place on the base or foundation. When weight and physical size indicate a larger crew than would normally be available, or if the item is impossible to handle by a normal crew, a rigging service should be considered.

The piping connection labor is taken from the actual joint connections at the suction and discharge terminals.

Takeoff Procedure: Pumps, like other mechanical equipment or fixtures, should be taken off and colored in prior to the piping system takeoff. In some instances, an equipment schedule is listed on the plans. This schedule should be checked carefully against the drawings. A separate pricing list should be made for equipment. Taking off equipment first has some advantages, listed below:

- The estimator becomes familiar with both the plans and the proposed building geography.
- The pieces of equipment become terminal points for the piping system takeoff.
- Vendors can be contacted to prepare quotations prior to bid day, while the estimator is busy with the time consuming piping takeoff.

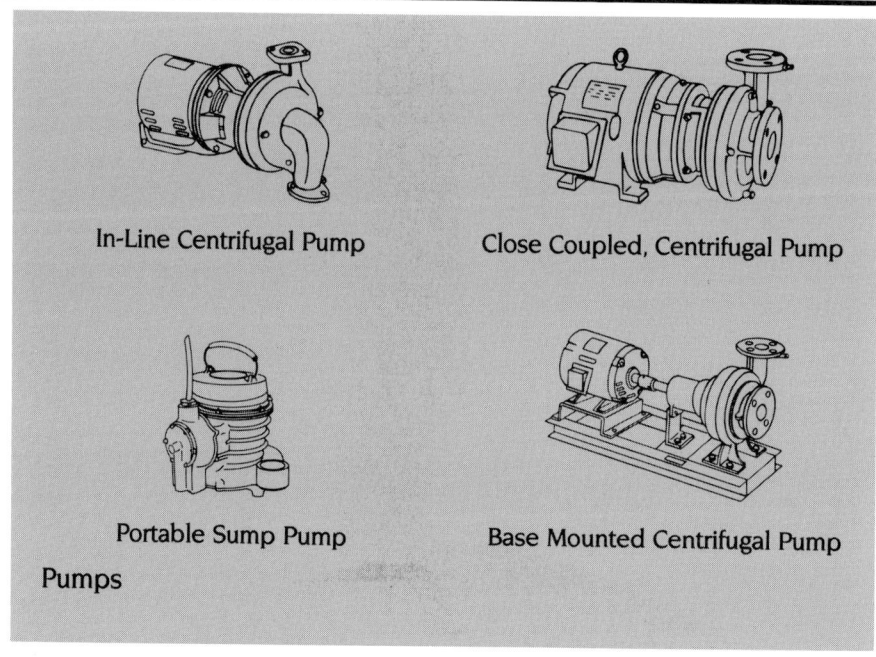

Pumps

Figure 14.1

Chapter 15
PLUMBING SYSTEMS

The plumbing portion of a mechanical estimate includes the piping and fixtures for the potable water, domestic hot water, storm, sanitary waste, and interior gas piping systems. In buildings such as hospitals, schools, or laboratories, medical gas and acid waste piping might be added to this list. Piping systems have already been discussed in detail in Chapter 13, "Piping". Plumbing fixtures, appliances, drains, carriers, and other plumbing accessories will be discussed in this chapter.

Knowledge of applicable local plumbing codes is a must. The estimator should never assume that the designer has complied with these codes. Fire standpipes, which are also the work of the plumber, will be covered in Chapter 16, "Fire Protection".

Fixtures and Appliances

Plumbing fixtures and appliances are the receptacles connected to and served by the water distribution, drainage, waste, and vent systems. Fixtures normally used in plumbing systems of commercial or residential buildings include bidets, bubblers or drinking fountains, emergency eyewashes, interceptors, lavatories, receptors, shower stalls, sinks, tubs, urinals, wash centers, and water closets. Appliances for these buildings may include electric water coolers, dishwashers, dryers, washing machines, water heaters, or water storage tanks.

Plumbing fixtures requiring cold water, waste, and vent connections are shown in Figure 15.1. The plumbing fixtures shown in Figure 15.2 require both hot and cold water supplies, plus waste and vent connections. Figure 15.3 shows plumbing appliances requiring only water piping connections, and no waste or vent. Both gas and oil water heaters require fuel piping and smoke pipe or flue connections. The water and relief piping connectors are all necessary, regardless of the type of fuel or electricity being used. The plumbing appliances shown in Figure 15.4 require cold water supply plus waste and vent connections. Figure 15.5 shows plumbing appliances requiring hot and cold water supplies, mixing valve, waste, and vent connections.

Electric water coolers may have remotely located chiller sections to serve more than one dispensing unit. In this case, additional insulated piping must be included in the estimate.

The plumbing contractor is usually responsible for the connection of piping appliances and systems furnished by others. These may include vending machines, kitchen and cafeteria items, heating and cooling systems, window washing apparatus, etc. The additional piping for these

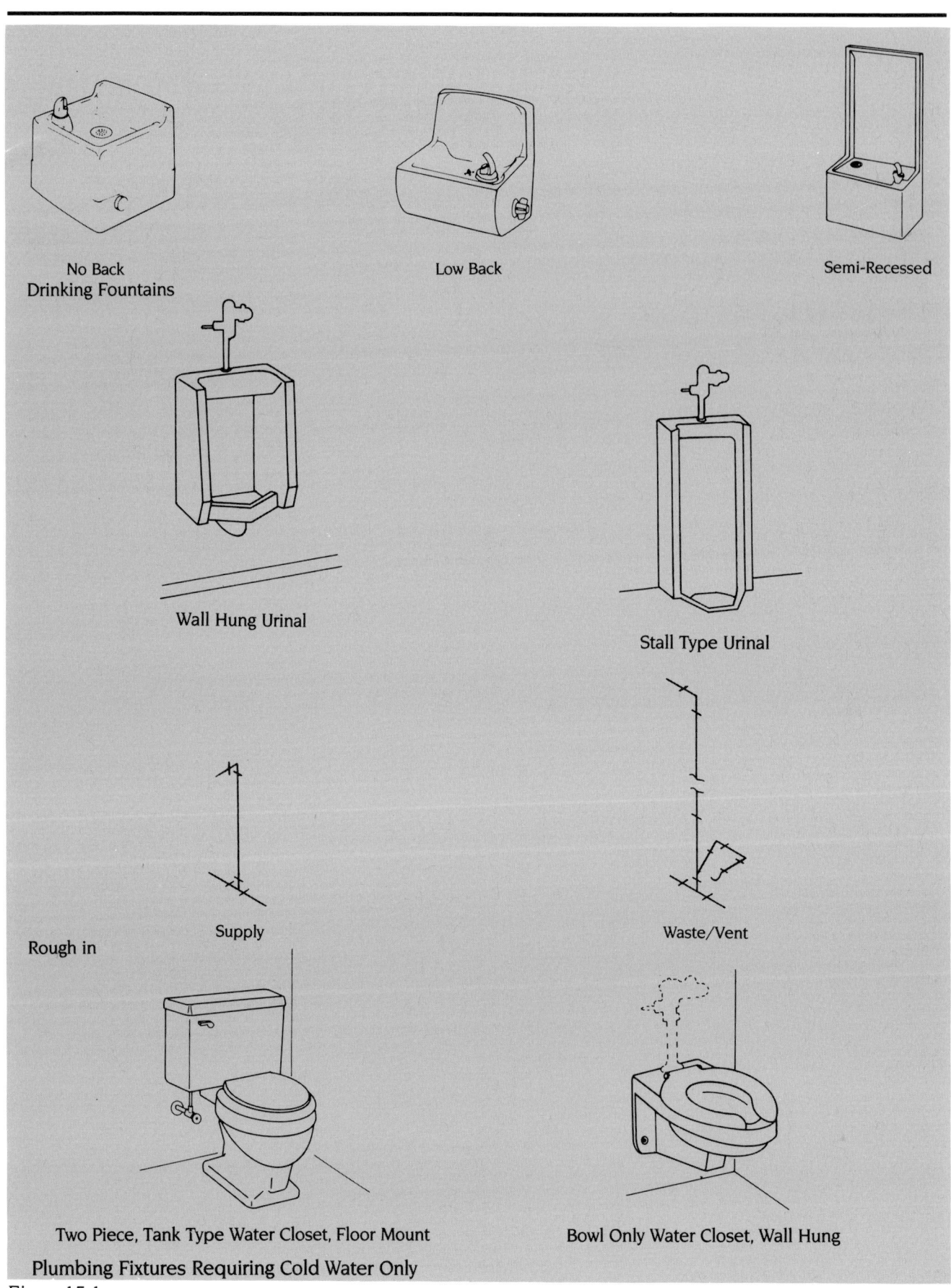

Plumbing Fixtures Requiring Cold Water Only

Figure 15.1

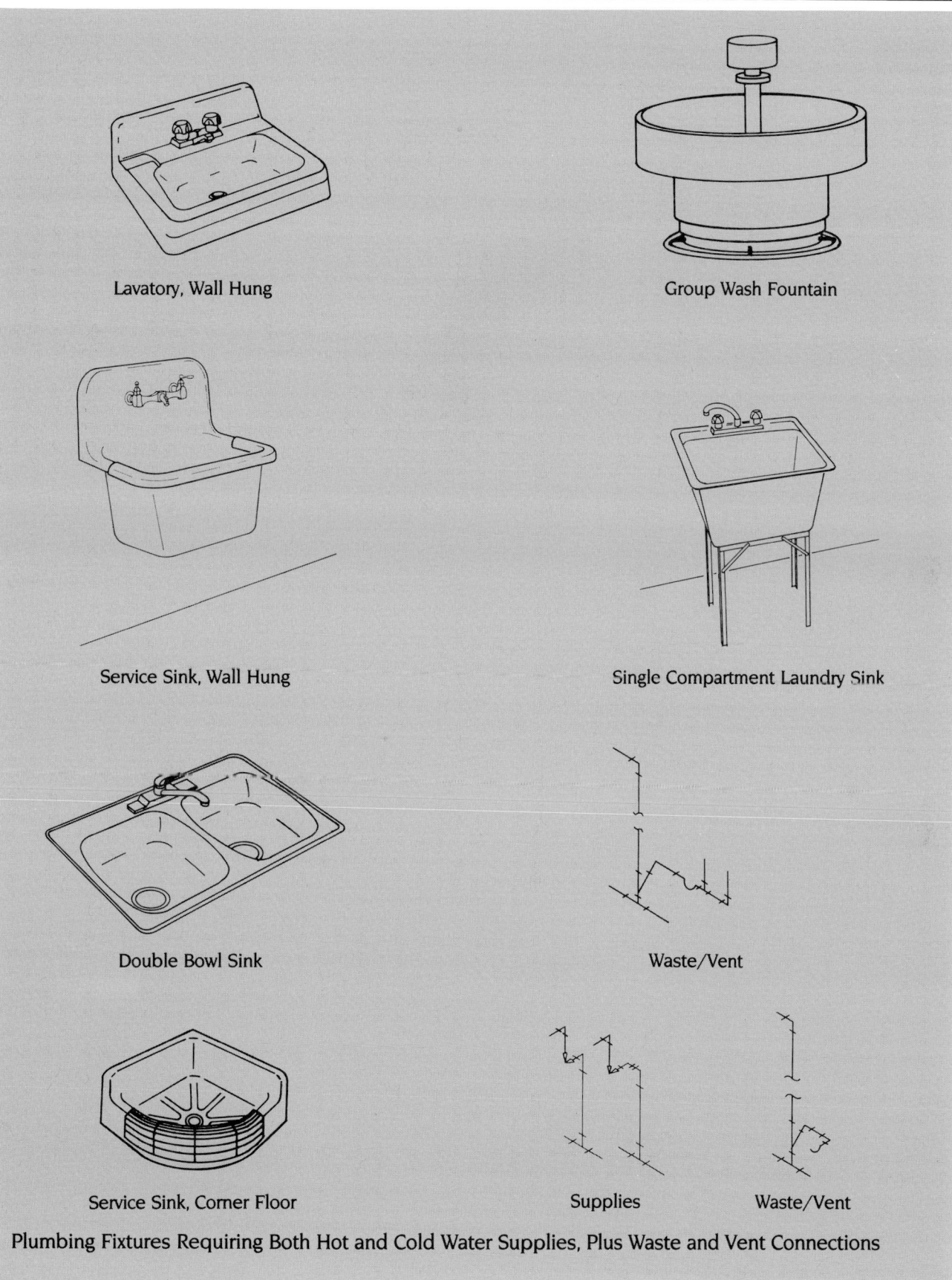

Figure 15.2

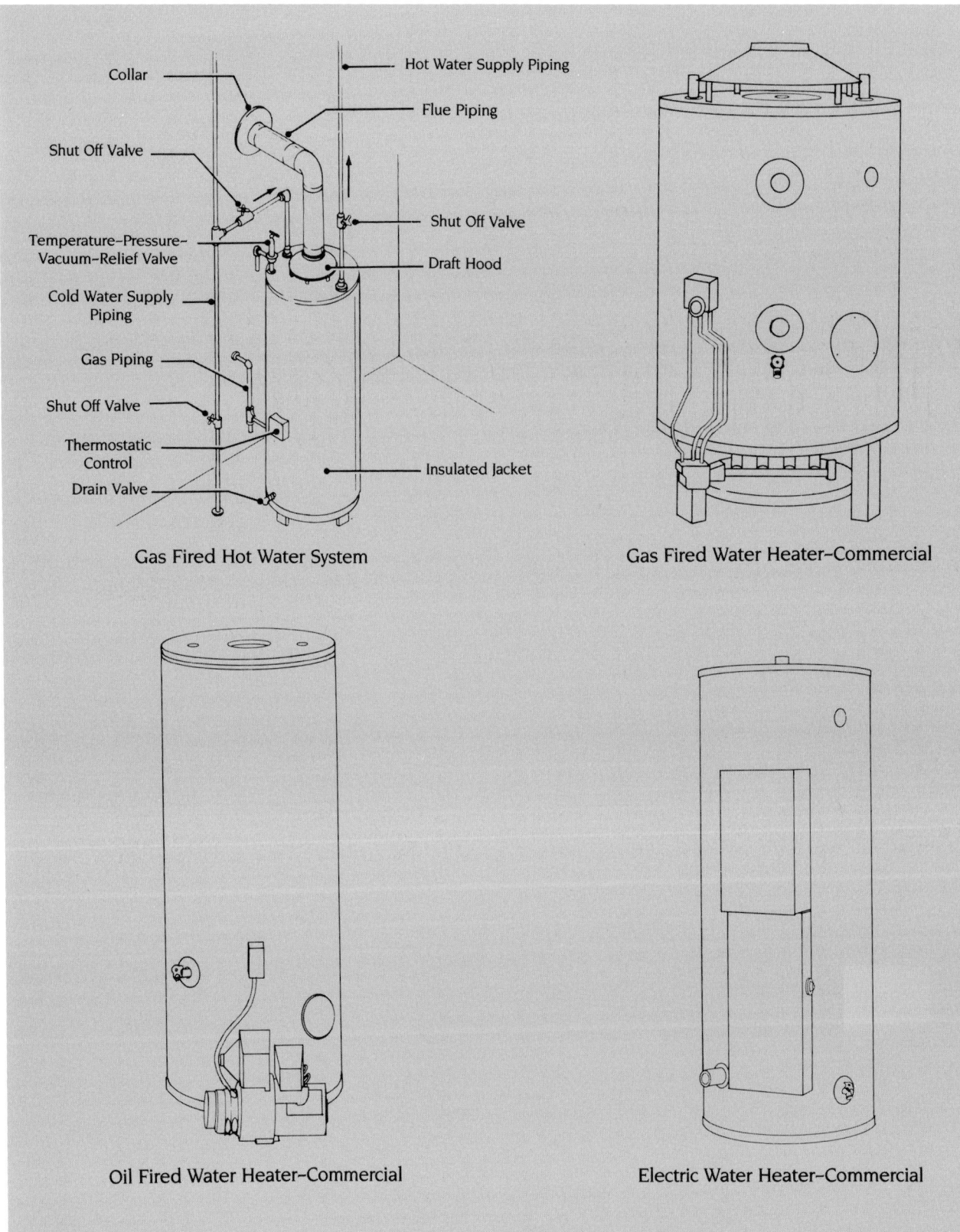

Figure 15.3

systems is not usually shown on the plumbing drawings, however, the estimator should read through all drawings for these plumbing connections.

Units of Measure: Plumbing fixtures are taken off and counted as each.

Material Units: Each plumbing fixture manufacturer offers various optional "trim" packages for each type of fixture. A trim package for a lavatory, for example, might include the faucets, pop-up waste, supply stops, flexible supplies, and a trap. Plating or other finishes and styles are optional. These trim packages should be included in the individual fixture prices.

Labor Units: Fixture installation is performed in two phases at two separate stages of construction. Initially, a fixture is *roughed in* prior to installation of the finished wall or floor. Roughing in is the installation of

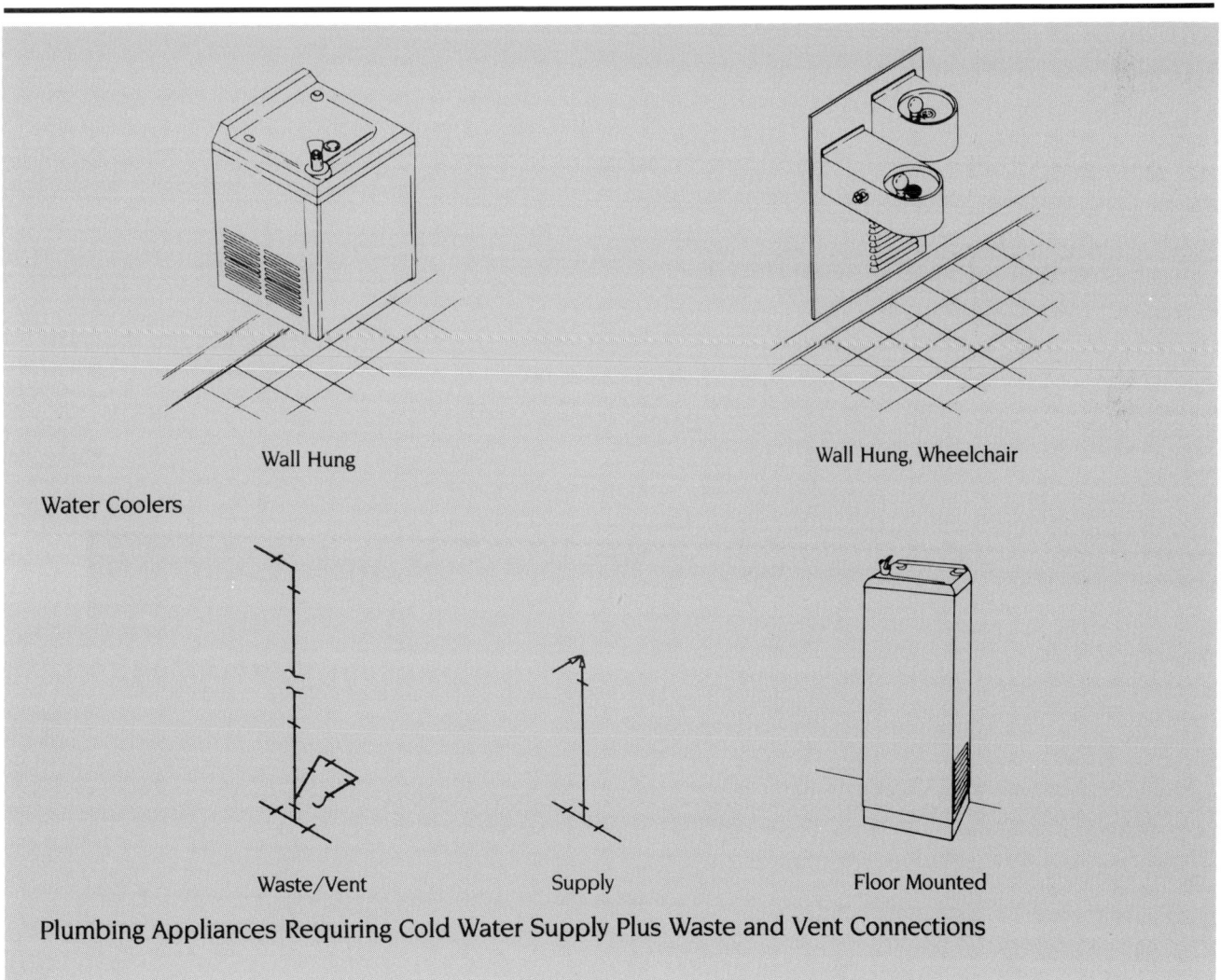

Figure 15.4

the pipes that are to be concealed in the wall or beneath the floor. After the wall or floor finish is completed, test nipples, drain connections, carrier arms, or studs will extend through the wall, ready for the actual setting and connecting of the fixture or appliance, the second phase of installation.

Takeoff Procedure: A separate fixture takeoff sheet should be utilized for each floor. After completion of the takeoff, the quantities from each floor should be totalled on a summary sheet. Similar fixtures should be listed separately if they have different trim packages or colors.

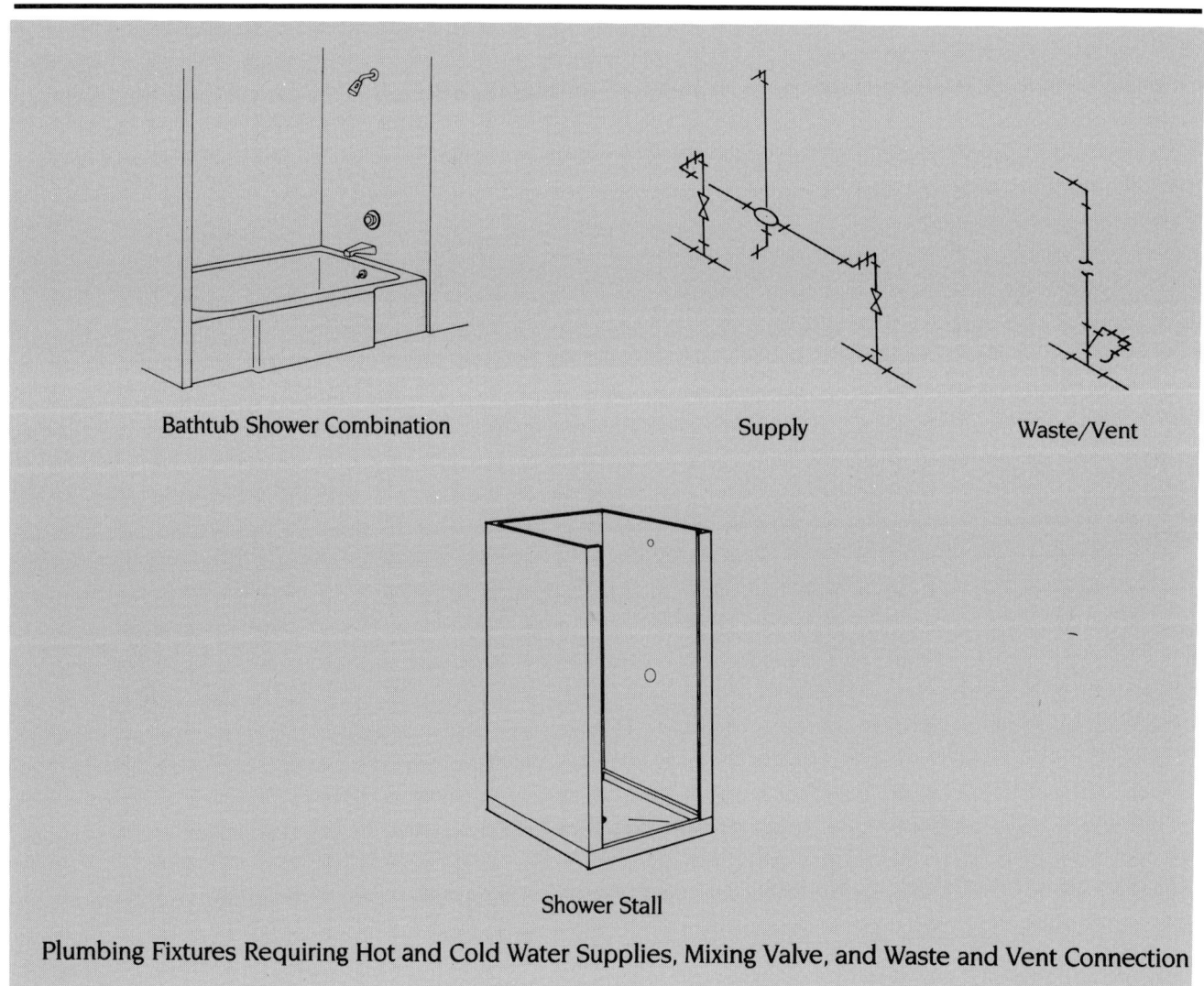

Plumbing Fixtures Requiring Hot and Cold Water Supplies, Mixing Valve, and Waste and Vent Connection

Figure 15.5

Drains and Carriers

Drains and carriers can have a significant impact on the plumbing estimate, for both material and labor. Drains are used extensively throughout a building, including roof, floor, parapet, area, trench, yard, deck, swimming pool drains, and floor sinks. Also used in buildings are wall and yard hydrants, backwater valves, cleanouts, and interceptors. The remainder of this labor-intensive group includes fixture carriers and over 100 special drain waste and vent fittings to be combined with the water closet carriers. All of the above-mentioned items are normally available from a single source of supply.

Storm drainage systems convey rainwater from roofs and the upper portions of structures where the water could cause damage to the building or constitute a hazard to the public. The water is discharged into a storm sewer or other approved location where it will not cause damage. The drainage system is not usually connected to a public sanitary sewage line.

Above ground piping for storm drains may be fabricated using brass, cast iron, copper, galvanized steel, lead, ABS, or PVC. Materials acceptable for underground use include cast iron, heavy wall copper, ABS, PVC, extra-strength vitrified clay pipe, and concrete pipe.

Stormwater is usually conducted away from the drainage area at the same rate that it collects. The discharge capacity is based on the size of the horizontal area to be drained, the design rate of the system, and the average rainfall rate. When the roof area abuts a sidewall, 50% of the wall area should be added to the roof area to determine the area to be drained. Local government departments, plumbing codes, and weather bureaus can usually supply climatological data on rainfall. This flow rate is the basis for sizing the system. The chart in Figure 23.19 is a guide for sizing. The chart is based on a maximum rate of rainfall of 4" per hour. Where maximum rates are more or less than 4" per hour, the figures for drainage should be adjusted. This can be done by multiplying by a factor of 4 and dividing by the local rate in inches per hour. The chart is applicable for round, square, or rectangular rainwater pipe. All of these shapes are considered equivalent if they can enclose a circle equal to the leader diameter.

Traps are not required for stormwater drains connected to storm sewers. Leaders and drains connected to a combined sewer and floor drain, which is then connected to a storm drain, need to be trapped. The size of the trap for an individual conductor should be the same size as the horizontal drain to which it is connected. The illustration in Figure 15.6 is a typical roof storm drainage system.

Floor and Area Drains provide adequate liquid drainage of floors and other surfaces in and around buildings. The local plumbing code will determine the minimum standards for size, location, and type. The code will also determine when and where interceptors and separators (for grease, oil, sand, hair, etc.) should be used. It also determines the when and where of traps, cleanouts, and backwater valves. Another major consideration in drain selection is the capability of the grate surface to be able to withstand the intended pedestrian or vehicular traffic.

Floor drains are manufactured in various matals, including bronze, alloy and cast iron tops, ductile iron, brass, and alloy bodies. The material used is governed by the drain location, (e.g., shower rooms, toilet rooms, hospitals, zoos, kitchens, etc.). A floor sink differs from a floor drain; a floor sink has rounded edges and an acid resistant finish, for improved sanitation and to allow thorough cleansing. Floor sinks, or receptors, should be used in food preparation areas, hospitals, and laboratories. Flushing rims for floor sinks and drains, can washers, and trap primers, all entail additional piping.

A plumbing estimator's library should contain complete drain and carrier manufacturers' catalogues. The estimator should be familiar with the accessories and various combinations which will affect the installation labor. Figure 15.7 shows common drains and castings. Carriers are shown in Figure 15.8.

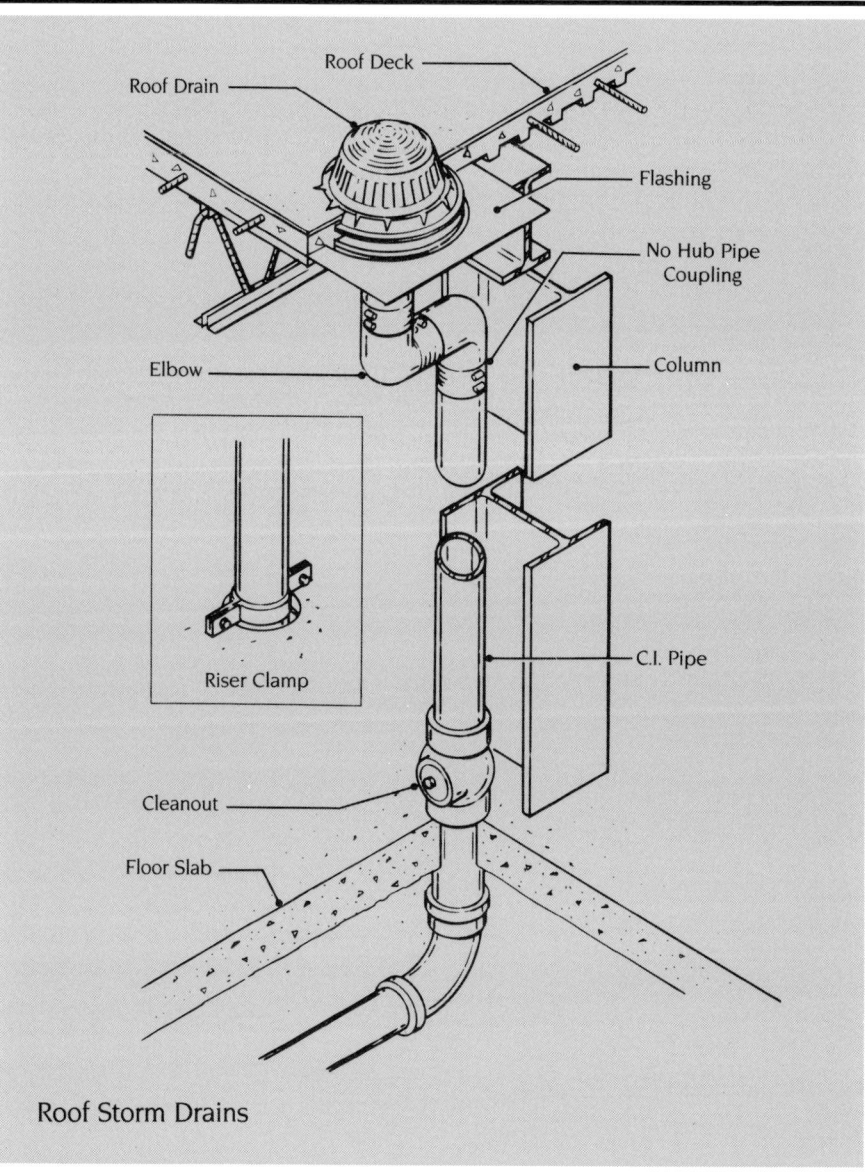

Figure 15.6

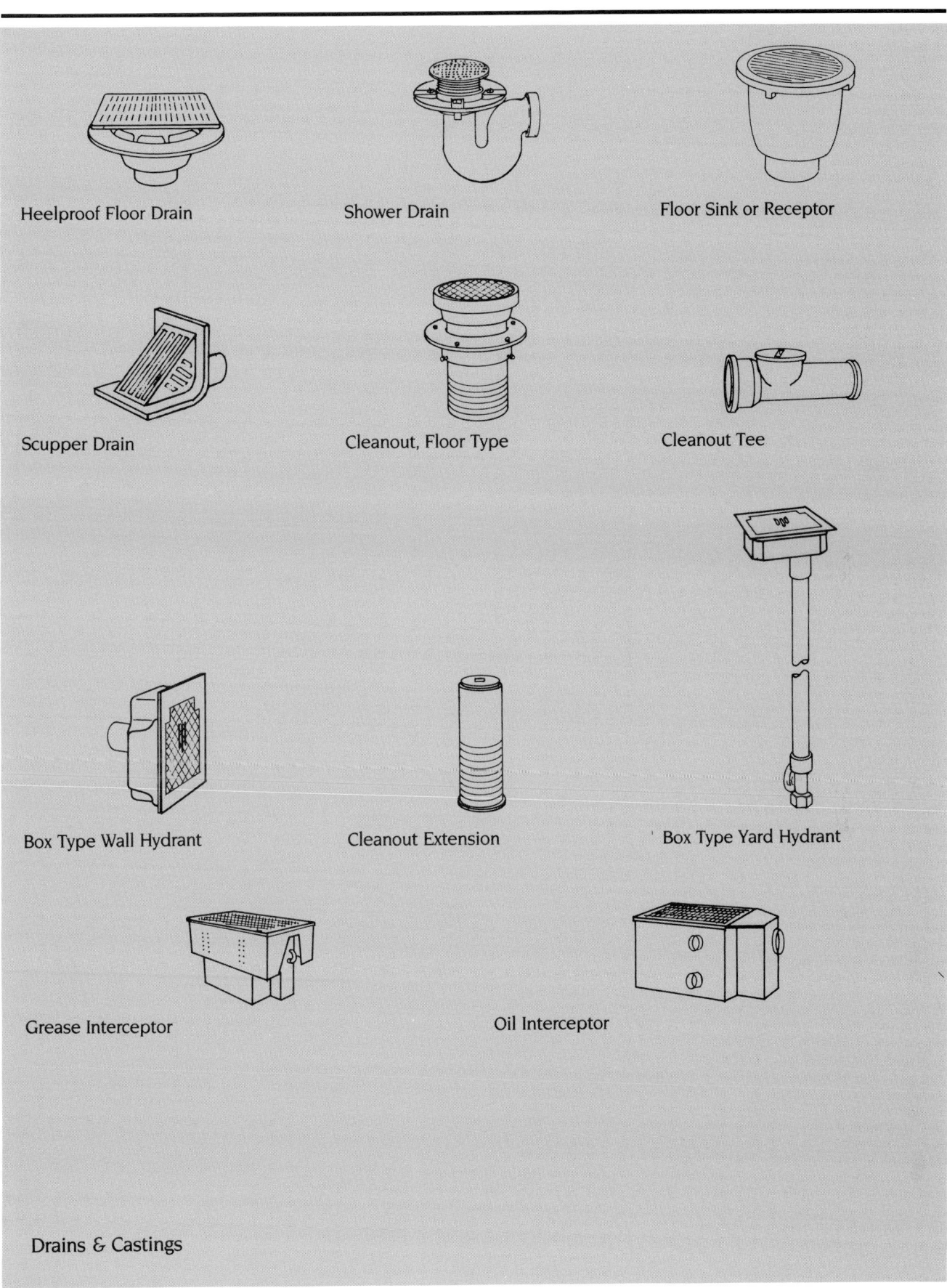

Drains & Castings

Figure 15.7

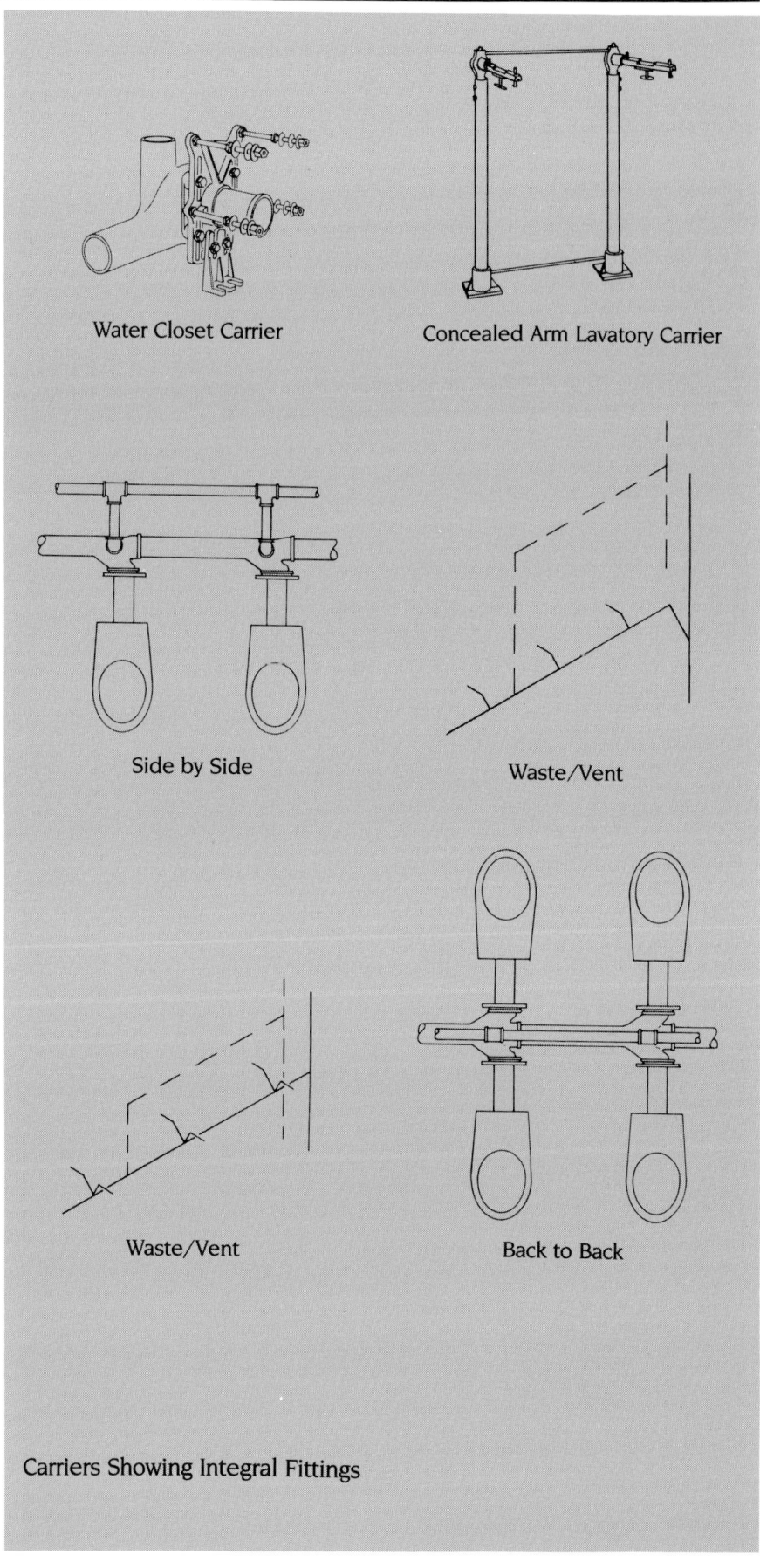

Figure 15.8

Cleanouts are an integral part of the drainage system, as they provide access for removal of accumulated solids. It is good practice, and often a code requirement, to include cleanouts and extensions at changes of direction greater than 45 degrees and at appropriate distances in straight piping.

Piping connections for drains and cleanouts are available in sizes of 1-1/2", 2", 3", 4", 5", 6", and 8" for threaded, gasketed, caulked, or no hub connections.

Hydrants supply potable water to building exteriors and to hose outlets for remote building areas. For exterior walls in climates subject to freezing temperatures non-freeze features are mandatory. Automatic backflow shutoff, vacuum breaker, and pressure relief options are also available. Yard and post hydrants are used outside the building structure, and offer safety features similar to those of wall hydrants.

Wall hydrants have 3/4" IPS female and 1" male connections on the inlet, and 3/4" male hose connections on the outlet. Yard hydrants, both post type and box type, are available with 3/4" through 2" IPS female inlets and 3/4" through 2" male hose connections. Depth of bury is a consideration when taking off yard hydrants.

Carriers have evolved from simple supports for water closets, lavatories, sinks, and perineal baths, to being integrated into the drainage system itself. These devices secure wall-mounted plumbing fixtures above the floor. Special carriers are available for drinking fountains, electric water coolers, wheelchair lavatories, and penal or institutional fixtures. Floor mounted back outlet water closets may also be integrated in the carrier support system. The use of carriers and wall hung fixtures not only provides a neater and more sanitary environment, but saves man-hours of roughing and installation time.

Carrier closet fittings have optional waste and vent locations to accommodate any code requirement. They are made in over 100 variations and are available with 2", 3", 4", 5", and 6" pipe connections.

Carriers will accommodate all types of metal and plastic pipe accepted by code, and can accept back to back or side by side multiple closet fixture installations. Figure 15.8 shows water carriers with integral waste fittings for back to back and side by side group installation.

Units of Measure: Drains, cleanouts, hydrants, interceptors, and carriers are all estimated as each.

Material Units: Using the specifications as a guide and a manufacturer's catalogue as a reference, the options and/or accessories for the various drains and carriers can be visualized and listed to comply with the designer's intent.

Labor Units: The following procedures must be considered and included in the labor estimate per drain for sleeving or cutting through each floor or roof:
- Setting the drain body and trim
- Returning at a later time to add the strainer, grate, funnel, or dome

Wall hydrants require the setting of the wall unit and making the inlet water connection. The installation of the wall unit must be coordinated with the construction of the wall. Setting may be accomplished by the wall construction contractor or by the plumber.

Interceptors are installed on the floor, partially recessed, flush with the floor, or in a pit or chamber beneath the floor. Size and weight must be considered in locating and setting an interceptor. Piping connections consist of vent, waste, inlet, and skimmer ports. When volatile gasses are to be piped, prevailing codes (be sure to check) may require a multiple venting arrangement. As with drains and wall units, installation should be coordinated with the general contractor.

Carrier labor considerations range from very simple installation of hangers and arms for sinks, lavatories, water fountains, or urinals, to complex integrated systems consisting of wall or floor-mounted carriers anchored to the building floor behind the finished wall, or in the case of substantial wall construction, bolted to or through the wall itself. More complex water closet carriers with integral waste and vent fittings are secured to the floor, piped into the waste and vent rough-in, all within the wall cavity or pipe chase. The fixture stubs and waste extension protrude through the finished wall for final fixture mounting.

Takeoff Procedure: Roof and area drains are taken off from the plumbing roof plans. The floor drains and interior specialty drains are taken off from the various floor plans and detail drawings, just as the backwater valves and interceptors might be.

Yard and wall hydrants are found on the ground floor plan and the site plan. Interior wall hydrants for special use might be found in swimming pool areas, kitchens, etc. These specialties can all be recorded on the same takeoff and summary sheet.

Carrier totals should be taken from the estimator's fixture takeoff sheet and listed with the drains for pricing.

Chapter 16
FIRE PROTECTION

Fire protection systems in the mechanical contracting industry include the use of standpipes, sprinklers, fire and smoke control dampers, fans, controllers, and alarms. This chapter contains descriptions of the materials and labor unique to installing standpipe systems, sprinkler systems, chemical systems, and combinations thereof. Piping, sheet metal, and other components have been or will be addressed in their own respective chapters.

Standpipe Systems

Standpipe systems, properly designed and maintained, are effective and valuable time-saving aids for extinguishing fires. This is especially true in the upper stories of tall buildings, the interior of large commercial or industrial malls, or other areas where construction features or access make the laying of temporary hose lines time consuming and/or hazardous. Adequate pressure is obtained by city water pressure, a reservoir at the roof or upper story of the building, or by booster pumps. Standpipes are frequently installed with automatic sprinkler systems for maximum fire protection.

Standpipe systems with fire hose valves, hose cabinets, fire department, and pumper connections are installed by plumbers. Sprinkler piping systems with spray heads, special valve arrangements, alarms, air compressors, and controls are installed by sprinkler fitting specialists.

The standard for standpipe system design is set by the National Fire Protection Association (NFPA) volume 14. However, the local authority having jurisdiction must be consulted for special conditions, local requirements, and approval.

There are three general classes of service for standpipe systems, listed below:
- **Class I** systems are used by fire departments and personnel with special training for heavy hose (2-1/2" hose connections).
- **Class II** systems are used by building occupants until the arrival of the fire department (1-1/2" hose connector with hose).
- **Class III** systems are used by either the fire department and trained personnel or by the building occupants (both 2-1/2" and 1-1/2" hose connections or one 2-1/2" hose valve with an easily removable 2-1/2" by 1-1/2" adapter).

Standpipe systems are also classified by the method that water is supplied to the system. The four basic types are listed below:

- **Type 1** denotes a wet standpipe system with its supply valve open and water pressure maintained at all times.
- **Type 2** is a standpipe system so arranged through the use of approved devices to admit water to the system automatically when a hose valve is opened.
- **Type 3** is a standpipe system arranged to admit water to the system through manual operation of approved remote control devices located at the hose stations.

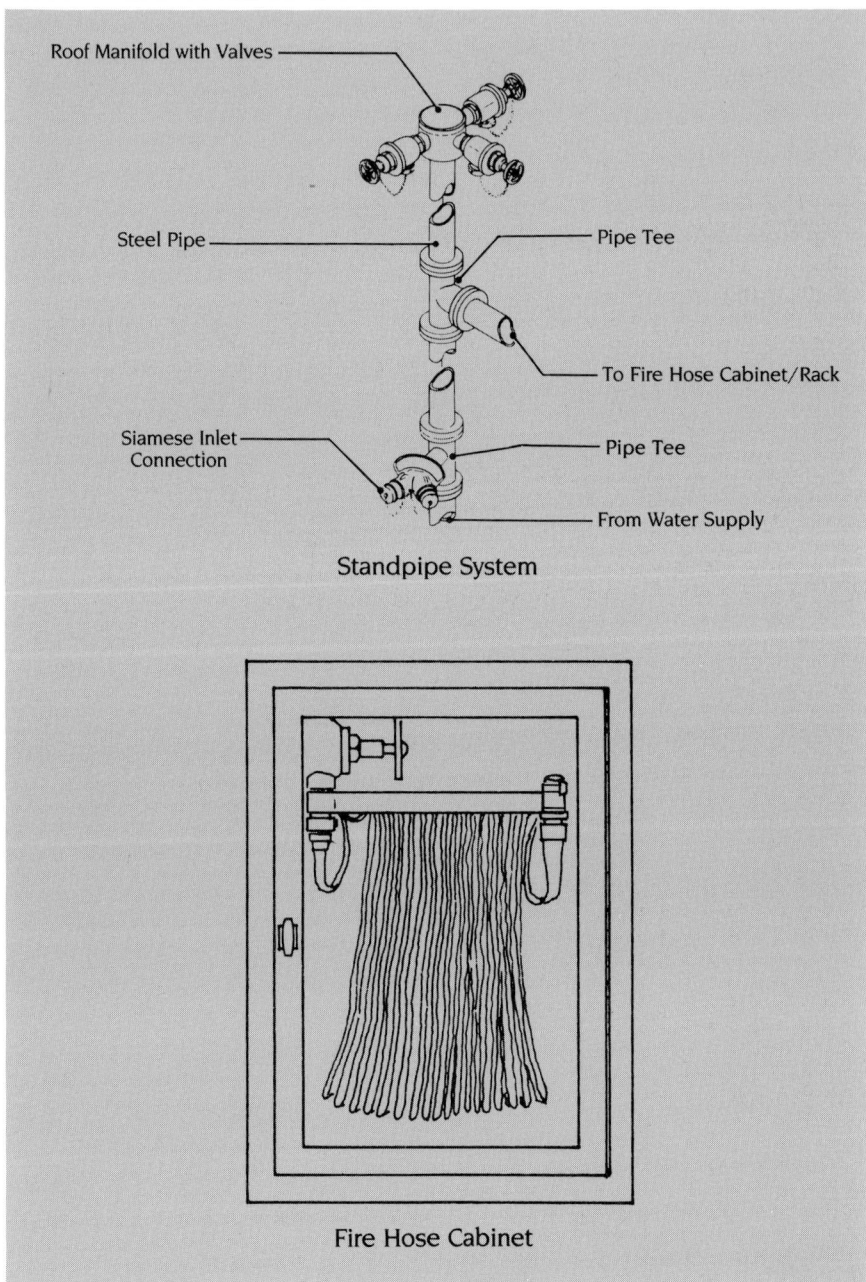

Figure 16.1

- **Type 4** is a dry standpipe having no fixed water supply.

Standpipe systems (shown in Figure 16.1) are usually made of steel pipe. Other forms of ferrous or copper piping may be used where acceptable to local authorities. Joints may be flanged, threaded, welded, soldered, grooved, or any method compatible with the approved piping materials.

Units of Measure: Fire hose standpipe systems are taken off according to their components. The pipe itself is taken off and recorded by the linear foot. Valves, fittings, siamese connections, cabinets, etc., are taken off as each. Hose racks and hose are priced as each assembly, which includes selections of hose length in 50, 75, and 100 foot increments.

Material Units: Material units for pipe and fittings are described in Chapter 13, "Piping". Care should be taken when pricing the required valves, adapters, couplings, finishes, type of hose, and type of nozzle. Hose threads in all standpipe systems must comply with the requirements of the local fire department. Local authorities may require that fire hose cabinets contain enough space, or a separate compartment, for a portable fire extinguisher. U.L. approval or listing may also be required for some of the component items. Materials and components that are U.L. listed are usually more and sometimes much more expensive than non-listed items.

Labor Units: Labor units for the piping portions of standpipe protection systems are described in Chapter 13, "Piping". Standpipe risers are required by law or code for high rise buildings. They are installed piecemeal during construction as each floor is finished to ensure that there is always a live hose station within one or two floors of the top. Normally closed valves on this riser should not be left open, as this will inhibit flow to upper floors during a fire. These valves should, instead, be wired closed and tagged.

Glass fronts should not be installed on recessed hose cabinets until construction is in the finishing stages. These items will add man-hours to the overall installation labor. Any mechanical system that is to perform temporary duties during construction will also require additional labor, for installation, final cleaning, and possible replacement or repair before final acceptance. Labor units for standpipe systems should be given special consideration during the estimating process.

Takeoff Procedure: Piping portions of these systems are taken off by the linear foot. All other components and accessories are taken off and recorded as each. Each standpipe system should be listed on separate summary sheets, totalled, and priced. The totals from each summary sheet should then be listed separately on the plumbing estimate sheet.

Sprinkler Systems

The *wet pipe sprinkler system* (shown in Figure 16.2) is the most popular type of automatic fire-suppression system in use today. This type of piping system is filled with water under pressure and connected to a municipal or private supply. Water is released immediately through one or more sprinkler heads when opened by fire or heat. A fusible element melts, opening the sprinkler head when a specified temperature is reached. Once opened, the head or heads must be replaced before the system can be reactivated. Quick response and low first cost are the main reasons for the popularity of the wet pipe system.

In areas subject to freezing, *dry pipe sprinkler systems* (shown in Figure 16.3) are used. The dry pipe system is similar to the wet pipe system, except that the dry pipe system is filled with compressed air. The compressed air is maintained under a constant pressure higher than the available water pressure. Water is kept out of the system by means of a valve which contains the opposing pressures. When one or more sprinkler heads open, after sensing fire or heat, the air pressure drops. Water enters the piping system and is released through the sprinkler heads. The dry pipe system does not respond as quickly as the wet pipe system and costs more to install because of the required compressed air and special dry pendant heads. Further, dry pendent heads and the piping must be installed without low points or pockets that cannot be drained. The dry type valve itself must be enclosed in an area not subject to freezing.

For areas where water damage must be kept to a minimum during fire suppression, a *firecycle system* is employed (shown in Figure 16.4). The firecycle system is a dry system with preaction fire-detection devices and a control valve. Electrical controls in this system have the capability to close the flow control valve after fire detectors sense that the fire is out. A time-delay feature built into the system allows water to flow through the open type sprinkler head for a predetermined period before closing the flow valve. Should the fire re-ignite, the valve opens and the cycle begins again. Battery backup is a requirement for this type of system.

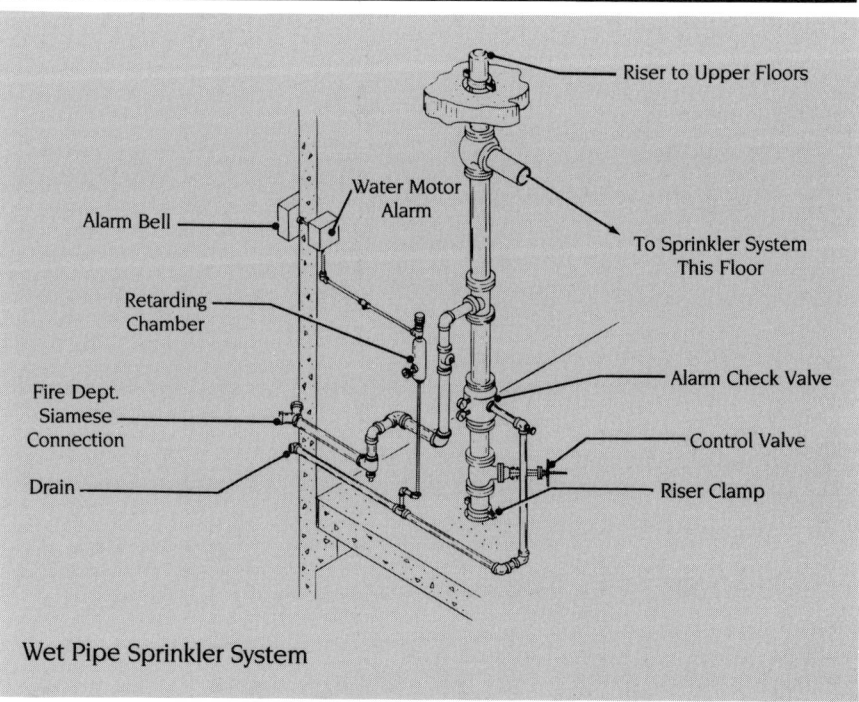

Figure 16.2

The firecycle system is always available for subsequent fires since it does not have to be shut down for head replacement, as do systems employing sprinkler heads with fusible links. This type of system also eliminates the need for manual shut-off valves and position indicators.

Preaction systems (shown in Figure 16.5) are used in areas subject to freezing. They are also used where accidental damage to sprinkler heads or piping, and subsequent water leakage, is unacceptable. Because the preaction system is more sensitive to fire than a sprinkler head, it provides a quicker response than the dry pipe system.

The preaction system is filled with air that does not have to be pressurized. Conventional closed type sprinkler heads are used, however, the initial detection of heat or fire occurs via heat-activated devices that are more sensitive than the sprinkler heads. These devices detect a rapid temperature rise and open the preaction sprinkler valve. This valve fills the system with water and activates the alarm. With a further rise in temperature, one or more sprinkler heads in the fire area will open and begin the extinguishing process. This type of system has two important advantages: early warning, which allows occupants to evacuate; and quick notification of fire-fighting personnel. Like the dry pipe system, if the preaction system is installed in an area subject to freezing, pendent heads (if used) must be of the dry type and the other drainage provisions must be adhered to. The preaction valve must be installed in a non-freeze area.

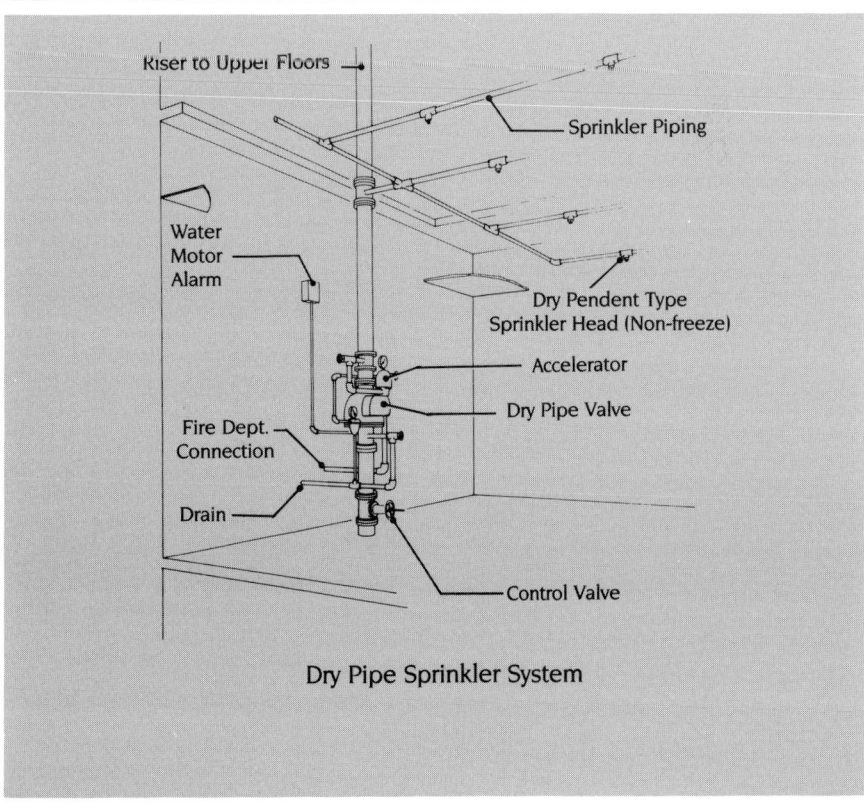

Dry Pipe Sprinkler System

Figure 16.3

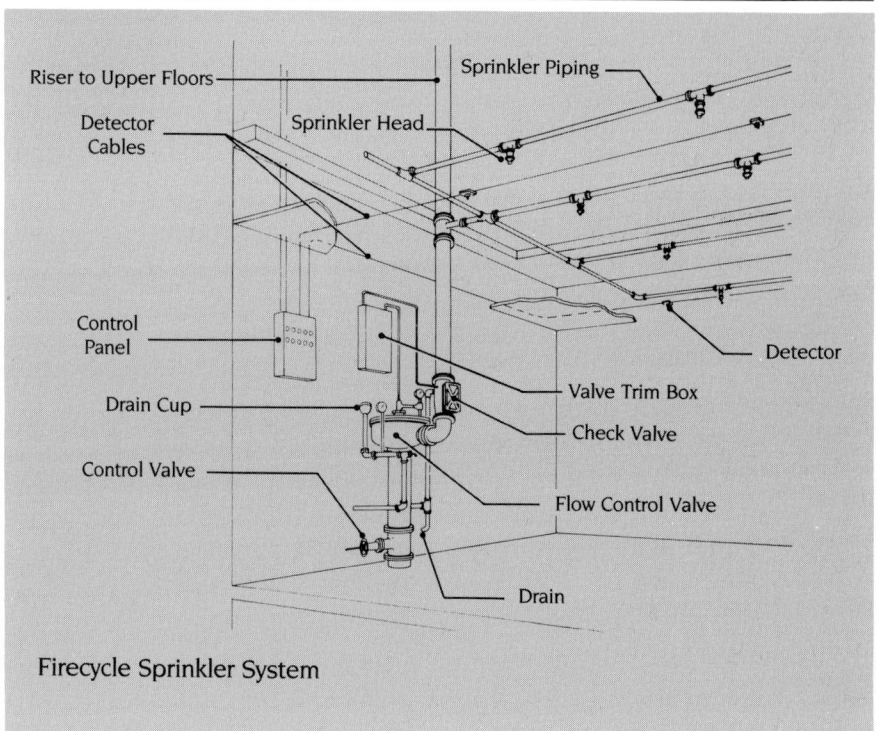

Figure 16.4

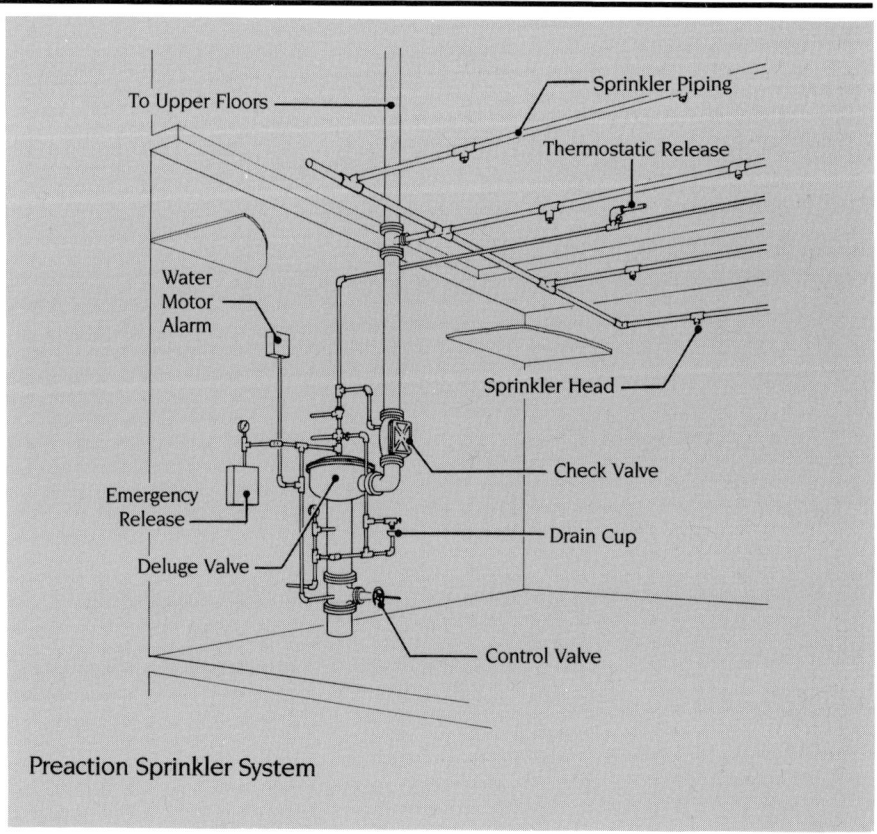

Figure 16.5

A *deluge system* (shown in Figure 16.6) is a dry system connected to a water supply. This water supply is contained by a preaction deluge valve which is controlled by heat-activated devices. This type of system employs open type nozzles rather than fusible heads. When the deluge valve opens, it admits water to the entire system, or valved zone. The valve then discharges water from all the nozzles in the protected area. By using preaction fire-detection devices and by wetting down the entire area instantly, the deluge system prevents the spread of the fire. For this reason, the deluge system is particularly suitable for storage or work areas involving flammable liquids.

All of the aforementioned types of sprinkler systems can be combined with fire standpipe systems. Sprinkler systems should be designed and installed in accordance with the current provisions of the National Fire Protection Agency standard number 13. Traditional piping methods and materials, black steel threaded and flanged piping, are often replaced with thin wall steel pipe and grip type fittings. The abundance and frequency of sprinkler heads, each requiring a tee, almost mandates the use of threaded pipe and fittings or the grip type system. In some situations, other acceptable materials include copper tubing and several types of plastic tubing. Local building codes govern the acceptability of these materials.

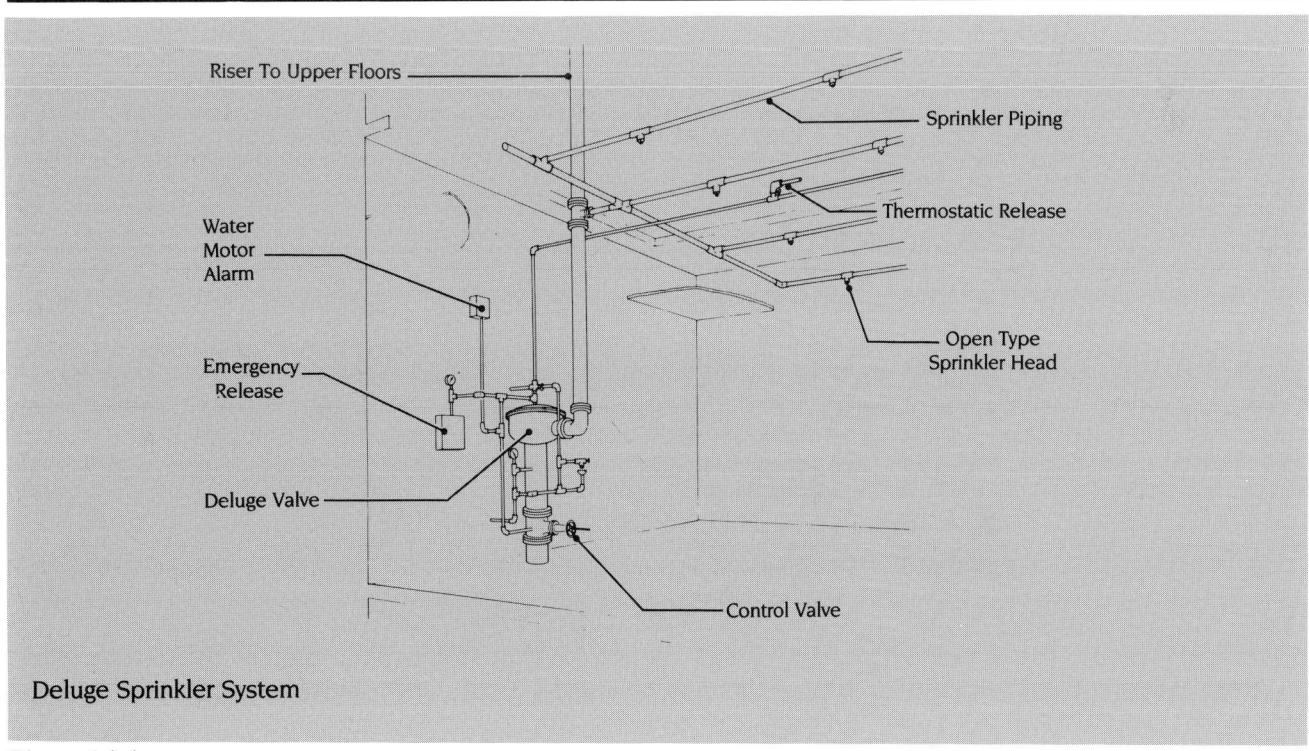

Deluge Sprinkler System

Figure 16.6

System Classification

Rules for installation of sprinkler systems vary depending on the classification of building occupancy falling into one of the following three categories.

Light Hazard Occupancy: The protection area allotted per sprinkler should not exceed 200 S.F. with the maximum distance between lines and sprinklers on lines being 15'. The sprinklers do not need to be staggered. Branch lines should not exceed eight sprinklers on either side of a cross main. Each large area requiring more than 100 sprinklers and without a sub-dividing partition should be supplied by feed mains or risers for ordinary hazard occupancy. Building types included in this group are:

- Auditoriums
- Churches
- Clubs
- Educational
- Hospitals
- Institutional
- Libraries
  (except large stack rooms)
- Museums
- Nursing Homes
- Offices
- Residential
- Restaurants
- Schools
- Theaters

Ordinary Hazard Occupancy: The protection area allotted per sprinkler shall not exceed 130 S.F. of noncombustible ceiling and 120 S.F. of combustible ceiling. The maximum allowable distance between sprinkler lines and sprinklers on line is 15'. Sprinklers shall be staggered if the distance between heads exceeds 12'. Branch lines should not exceed eight sprinklers on either side of a cross main. Building types included in this group are:

- Automotive garages
- Bakeries
- Beverage manufacturing
- Bleacheries
- Boiler houses
- Canneries
- Cement plants
- Clothing factories
- Cold storage warehouses
- Dairy products manufacturing
- Distilleries
- Dry cleaning
- Electric generating stations
- Feed mills
- Grain elevators
- Ice manufacturing
- Laundries
- Machine stops
- Mercantiles
- Paper mills
- Printing and publishing
- Shoe factories
- Warehouses
- Wood product assembly

Extra Hazard Occupancy: The protection area allotted per sprinkler shall not exceed 90 S.F. of noncumbustible ceiling and 80 S.F. of combustible ceiling. The maximum allowable distance between lines and between sprinklers on lines is 12'. Sprinklers on alternate lines shall be staggered if the distance between sprinklers on lines exceeds 8'. Branch lines should not exceed six sprinklers on either side of a cross main. Included in this group are:

- Aircraft hangars
- Chemical works
- Explosives manufacturing
- Linoleum manufacturing
- Linseed oil mills
- Paint shops
- Shade cloth manufacturing
- Solvent extracting
- Varnish works
- Oil refineries
- Volatile or flammable liquid manufacturing and use

Units of Measure: For bidding purposes, sprinkler systems are often estimated by the square foot of floor area to be protected. Square foot prices are developed from the contractor's own records for each type of system and material specified. While it is preferable to use prices from completed projects, the estimator may also wish or need to consult cost publications such as *Means Mechanical Cost Data*. A detailed unit price estimate may be required if the square foot cost data is not sufficient or relevant to the project. An item by item takeoff and pricing exercise is required for a unit price estimate. The pipe is taken off by the linear foot, according to the specified material and joining methods. Fittings, valves, and accessories are recorded as each.

Material Units: Material units for steel, copper, and plastic piping have been described in Chapter 13, "Piping". Alarm valves, controllers, and sprinkler heads should be recorded on the takeoff sheets by type and class. From these totals, a list of required accessories (e.g., escutcheons or guards) can be determined. A chrome finish may be specified for heads in various areas. Wax coated heads are utilized in corrosive atmospheres. In climates subject to freezing, special non-freeze pendent heads must be used for drop ceiling installations. These non-freeze pendent heads are complex and expensive to purchase and to install because the exact measure of the drop to a finished ceiling must be determined prior to installation. Final adjustment of the telescoping escutcheon allows up to 1/2" play for field adjustment. Not shown on the drawings but usually required are spare heads, cabinets, wrenches, and other miscellaneous devices; check the specifications for these requirements.

Labor Units: Labor units for piping are covered in Chapter 13, "Piping", under the appropriate piping material classification. Fabrication and installation of sprinkler piping, however, are unique. Most sprinkler piping is cut and fabricated at the shop or at a jobsite pipe shop. Fabricated piping assemblies are bundled together and marked or coded by the location of final installation. The pipe sizes, measures, and coding schedule are listed on a layout plan prepared by the sprinkler contractor from the architectural plans. These plans are coordinated with the electrical and sheet metal contractors to avoid scheduling conflicts. In some instances, shop welding is used in the pipe fabrication, however, welding is not permitted for modification or repair of an existing system. An installation layout plan does not have to be prepared by the contractor until construction is to begin. At this time, a complete installation plan is required. If a detailed unit price estimate is made, some of the installation plans may be required in advance. A square foot estimate normally requires little advance planning.

Since sprinkler heads are precisely located by code and the piping has to maintain a required pitch, other trades must defer to the sprinkler piping route.

Takeoff Procedure: For a unit price estimate, the sprinkler heads are taken off first and summarized by type for pricing. The main valves, alarms, and controllers should be listed accordingly. The pipe, fittings, control or zone valves are taken off as described in Chapter 13, "Piping". From these totals the necessary amounts of hangers and other appurtenances are estimated.

For a square foot estimate, building area is taken off in square feet. This is used along with the system type and materials to determine cost.

Chemical, Foam, and Gas Fire Suppression

Chemicals, foams, and gasses of various types delivered through portable or fixed spray systems are used to extinguish fires in special applications. Fixed spray systems include foams and gasses, primarily carbon dioxide or halon, which extinguish fires by suffocation. Portable spray systems include foams and gasses, as well as a variety of dry chemicals and pressurized water.

Halon Fire Suppression

Halon fire suppression systems are used because they are fast, effective, and clean (shown in Figure 16.7). These systems are called "clean" because they leave no residue that must be cleaned up or that could contaminate valuable items (i.e., records or electronic equipment). Because halon gas is a non-conductor of electricity and is several times as heavy as air, it permeates the working area and penetrates cabinets or other electric or electronic enclosures where chemical powders cannot.

Depending on its concentration, halon gas ranges from non-toxic to low toxicity. Halon 1301 is normally used in a concentration of 5 to 7 percent. This concentration has no effect on personnel in the area.

Seven to ten percent concentration requires evacuation within one minute of exposure; concentrations above 10 percent require evacuation prior to discharge. Halon 1301 is also colorless and has minimal visual impedance to hamper evacuation.

Halon may be applied to fires by portable extinguishers, local application of strategically placed cylinders (or prepackaged systems), or by a centrally located cylinder or battery of storage cylinders connected to a piping

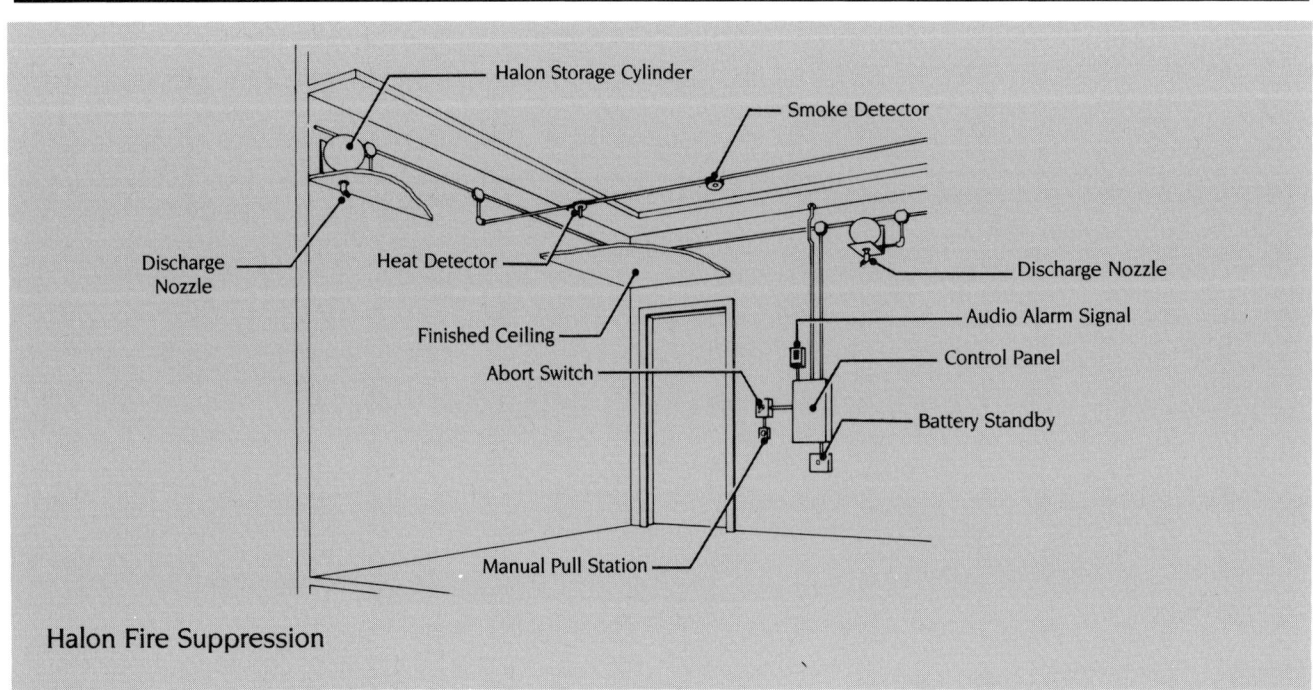

Halon Fire Suppression

Figure 16.7

distribution system and discharge nozzles. The local application or modular system is the most cost effective type of system, because it eliminates the piping installation costs encountered in a centrally located system.

Detection and actuation are critical requirements for fast extinguishing, to eliminate not only fire damage but also the accompanying risks of smoke, heat, carbon monoxide, and oxygen depletion. Fire and smoke detectors are wired to a control panel that activates the alarm systems, vertifies or proves the existence of combustion and releases the extinguishing agent, all in a matter of seconds.

The halon fire suppression system is most effective in an enclosed area. The control system may also have the capability of closing doors and shutting off exhaust fans. These systems are in popular use in the following places:

- Aircraft (both cargo and passenger)
- Libraries and museums
- Bank and security vaults
- Electronic data processing
- Transformer and switchgear rooms
- Tape and data storage vaults (rooms)
- Telephone exchanges
- Laboratories
- Radio and television studies
- Flammable liquid storage areas

Units of Measure: Self-contained or modular fire suppression systems are taken off and priced as each. For larger "pre-engineered" or "engineered" systems (central storage configuration) components should be taken off individually as each. Pipe should be taken off by the linear foot (see Chapter 13 for piping and components).

Material Units: A halon piping system (or other fixed spray system) is similar to a sprinkler piping system in material costs; hangers and supports are determined from the piping totals and cylinder locations. Dispersion nozzles (rather than heads) are utilized, although not as frequently as heads are used in sprinkler systems. Other materials to be taken off and priced are release stations, detectors, deflectors, alarms, annunciators, abort switches, cylinders (containers), and controllers.

Labor Units: Before labor units can be estimated, discharge nozzles and detectors should be located and a piping route laid out from the nozzles back to the storage cylinders. The quantities should also be taken off to determine fabrication labor. Steel T&C (threaded and coupled) piping and grip type connectors are most often used in these systems.

Takeoff Procedures: The takeoff should begin with the nozzles and other piping components, and proceed to the pipe and fitting takeoff. Any valves used in these systems will be found at the storage cylinders or manifolds. Piping takeoff should proceed as described in Chapter 13, "Piping".

Other Fire Extinguishing Systems

In addition to the methods of fire protection already covered in this chapter, the plumbing contractor may be required to furnish portable extinguishers, and the sheet metal contractor may be required to install or furnish kitchen hoods with built-in fire dampers, spray assemblies, and fan disconnect apparatus.

Other spray systems similar to halon are used in the form of portable extinguishers or fixed pipe installations. These spray systems include foam used primarily for fuel fires and other flammable liquids, and carbon dioxide. This type of system extinguishes fires by smothering. Because carbon dioxide is slightly toxic it is limited in use to fires in classifications B or C (listed below). It should also be noted that carbon dioxide dissipates and may allow reignition.

A third type of extinguishers covers a variety of dry chemicals and powders which are available for the range of fire classifications. Care should be taken to match the proper system with the expected hazard. To facilitate proper use of extinguishers on different types of fires, the NFPA Extinguisher Standard has classified fires into the following four types.

- **Class A** fires involve ordinary combustible materials (such as wood, cloth, paper, rubber, and many plastics) requiring the heat-absorbing (cooling) effects of water, water solutions, or the coating effects of certain dry chemicals which retard combustion.
- **Class B** fires involve flammable or combustible liquids, flammable gasses, greases, and similar materials where extinguishment is most readily secured by excluding air (oxygen), inhibiting the release of combustible vapors, or interrupting the combustion chain reaction.
- **Class C** fires involve live electrical equipment where safety to the operator requires the use of electrically non-conductive extinguishing agents. (Note: when electrical equipment is de-energized, the use of Class A or B extinguishers may be indicated).
- **Class D** fires involve certain combustible metals (such as magnesium, titanium, zirconium, sodium, potassium, etc.) requiring a heat-absorbing extinguishing medium not reactive with the burning metals.

Some portable fire extinguishers are of primary value on only one class of fire; some are suitable on two or three classes of fire; none is suitable for all four classes of fire.

Most currently manufactured extinguishers are labeled with a classification system so that users may quickly identify the class of fire for which a particular extinguisher may be used. The classification system is contained in the NFPA Extinguisher Standard which gives the applicable class symbol (or symbols) with supplementary words to recall the meaning of the letters. Color coding is also used. UL and ULC require the classification system on their labels. Numerals are used with the identifying letters for extinguishers labeled for Class A and Class B fires. Color coding is part of the identification system, and the triangle (Class A) is colored green, the square (Class B) red, the circle (Class C) blue, and the five-pointed star (Class D) yellow. These symbols are shown in Figure 16.8.

Units of Measure: Portable fire extinguishers and related accessories are taken off and priced as each. Fixed pipe systems, such as halon systems, are taken off as each or by component, depending on system configuration.

Material Units: Material units for portable fire extinguishers are relatively simple. Individual wall hung extinguishers, when specified, will be furnished with mounting brackets. These same brackets may be used for mounting within certain types of extinguisher cabinets as well.

Specifications for cabinets may require one or more extinguishers or hose and valve combinations.

The cabinets would have the same material units as previously outlined under standpipe systems. Cabinet doors and trim may be painted steel, aluminum, or stainless steel. The cabinet front may be solid panel, glass, wire glass, plexiglass, or a combination of solid panel and glass. Decals, blankets, spanner wrenches, axes, and alarms are among the safety and operating options available.

Fixed or permanent type systems will have piping and heads or nozzles similar to the apparatus in the halon systems. Kitchen exhaust hoods would probably arrive on the job with the piping assembly in place. Unique fuel loading areas might require a piping system and components (covered in Chapter 13, "Piping").

Labor Units: Labor considerations for portable extinguishers include the following:
- Anchoring the wall bracket to the building structure
- Securing the extinguisher in its bracket

Pressurized water type extinguishers are shipped empty and must be charged in the field. Any extinguishers that have been set in place for standby use during construction will have to be removed, cleaned, and inspected or recharged before final acceptance.

Recessed cabinets are built into the walls by the General Contractor. The finished frames and doors are installed by the plumber as the extinguishers are placed. This is similar to the procedures previously outlined for fire hose cabinets.

The labor units for permanent installations are similar to the units for halon systems. Labor units for piping components will be found in Chapter 13, "Piping".

Takeoff Procedure: Portable fire extinguishers and special built in permanent spray systems are often not shown on the mechanical drawings, however, the specifications will indicate the types and locations

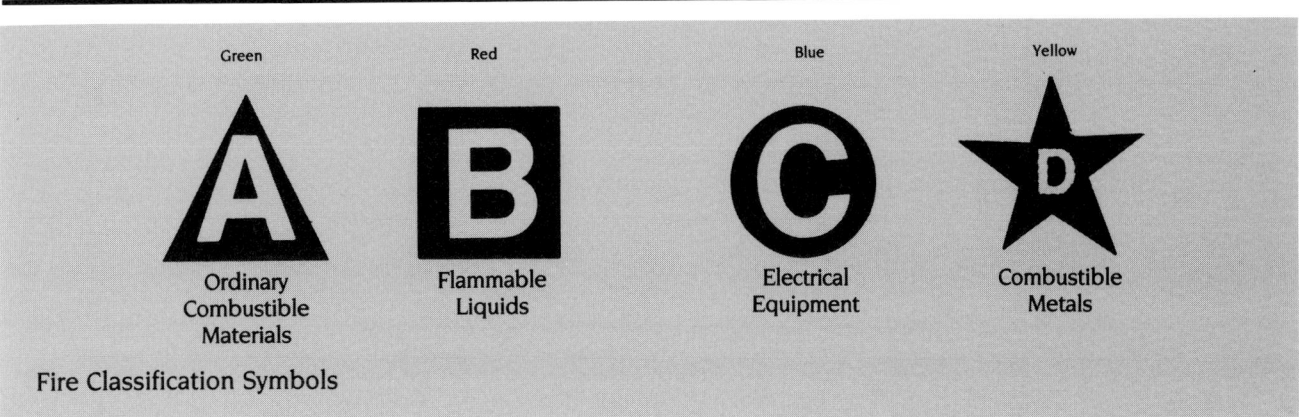

Figure 16.8

required. The architectural floor plans and elevations may indicate each extinguisher station. The plumbing drawings will indicate whether or not extinguishers will be housed with fire hoses (in the same enclosure). Spray systems for kitchen exhaust hoods should be found either on the kitchen or HVAC equipment drawings or both. Piping arrangements for fuel storage and piping systems will be indicated on the drawings.

The estimator, in beginning the takeoff, should separate the extinguishers by type and floor. Permanent spray installations should have a pipe and fittings takeoff as described in Chapter 13, "Piping". The spray heads and any valves, pumps, or storage cylinders should be recorded and priced separately from any other piping system.

The project documents should be carefully read to determine who is responsible for supply and/or installation of these special items.

Chapter 17
HEATING, VENTILATING, AND AIR CONDITIONING

The word air conditioning is often used to mean air cooling. Actually, the field includes much more then just cooling; heating, cleaning, humidifying, dehumidifying, and replacement of the air within a structure or contained space are other functions performed by mechanical air conditioning systems. The heating or cooling medium might be air, water, or steam, and the energy source, coal, electricity, gas, or oil. The conveyance of these liquids and gasses (piping and ductwork) is covered under other sections of this book. This chapter contains descriptions of equipment (both unitary and packaged), and hydronic and air handling systems.

Heating Boilers

Heating boilers are designed to produce steam or hot water. The water in the boilers is heated by coal, oil, gas, wood, electricity, or a combination of these fuels. Boiler materials include cast iron, steel, and copper. Several types of boilers are available to meet the hot water and heating needs of both residential (see Figure 17.1) and commercial (see Figures 17.2 and 17.3) buildings.

Cast iron sectional boilers (see Figure 17.1, 17.2, and 17.3) may be assembled in place or shipped to the site as a completely assembled combustion package. These boilers can be made larger on site by adding intermediate sections.

Steel boilers (see Figure 17.2 and 17.3) are usually shipped to the site completely assembled. Large steel boilers may be shipped in segments for field assembly and testing. The components of a steel boiler consist of tubes within a shell, plus a combustion chamber. If the water being heated is inside the tubes, the unit is called a *water tube boiler*. If the water is contained in the shell and the products of combustion pass through tubes surrounded by this water, the unit is called a *fire tube boiler*. Water tube boilers are manufactured with copper tubes or coils. Electric boilers have elements immersed in the water and do not fall into either category of tubular boilers.

Heating boilers are rated by their hourly output expressed in "British Thermal Units" (B.T.U.). The output available at the boiler supply nozzle is referred to as the gross output. The gross output in B.T.U. per hour divided by 33,475 indicates the boiler horsepower rating. The net rating of a boiler is the gross output less allowances for the piping tax and the pickup load. The net load should match the building heat load.

Due to the high cost of fuels, efficiency of operation is a prime consideration when selecting boiler units, and because of this, recent innovations in the manufacturing field have provided the opportunity to install efficient and compact boiler systems.

The Department of Energy has established test procedures to compare the "Annual Fuel Utilization Efficiency" (AFUE) of comparably sized boilers. Better insulation, heat extractors, intermittent ignition, induced draft, and automatic draft dampers contribute to the near 90 percent efficiencies claimed by manufacturers today.

In the search for higher efficiency, a new concept in gas-fired water boilers has recently been introduced. This innovation is the pulse condensing type boiler, which relies on a sealed combustion system rather than on a conventional burner. The AFUE ratings for pulse-type boilers are in the low to mid 90 percent range. Pulse-type boilers cost more initially than conventional types, but savings in other areas help to offset this added cost. Because these units vent through a plastic pipe to a side wall, no chimney is required. The pulse type boiler also takes up less floor space, and its high efficiency saves on fuel costs.

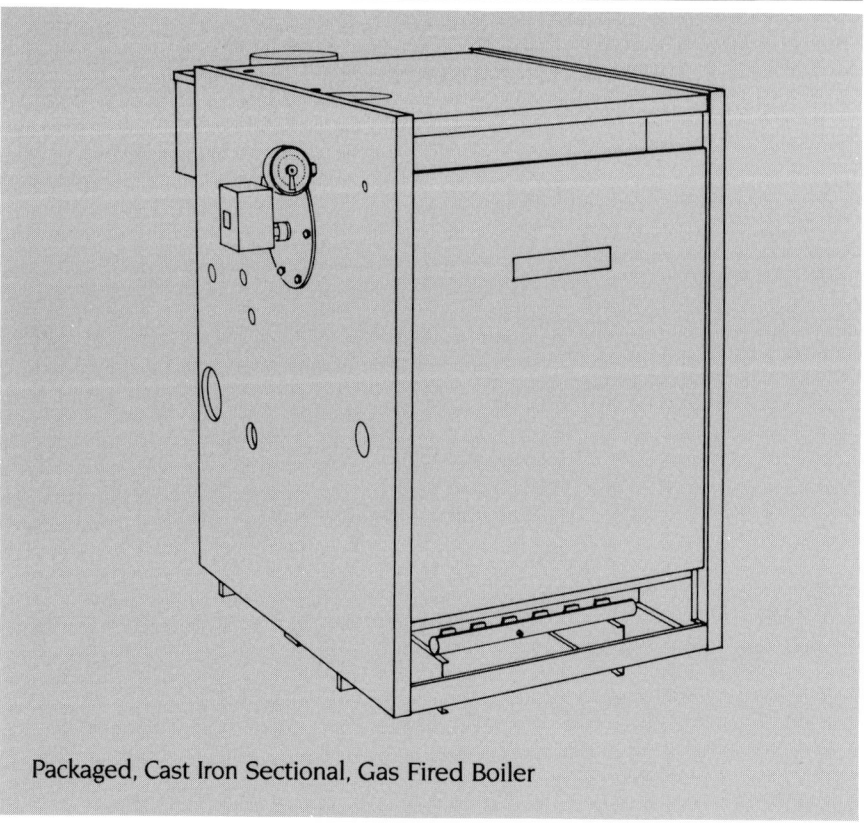

Packaged, Cast Iron Sectional, Gas Fired Boiler

Figure 17.1

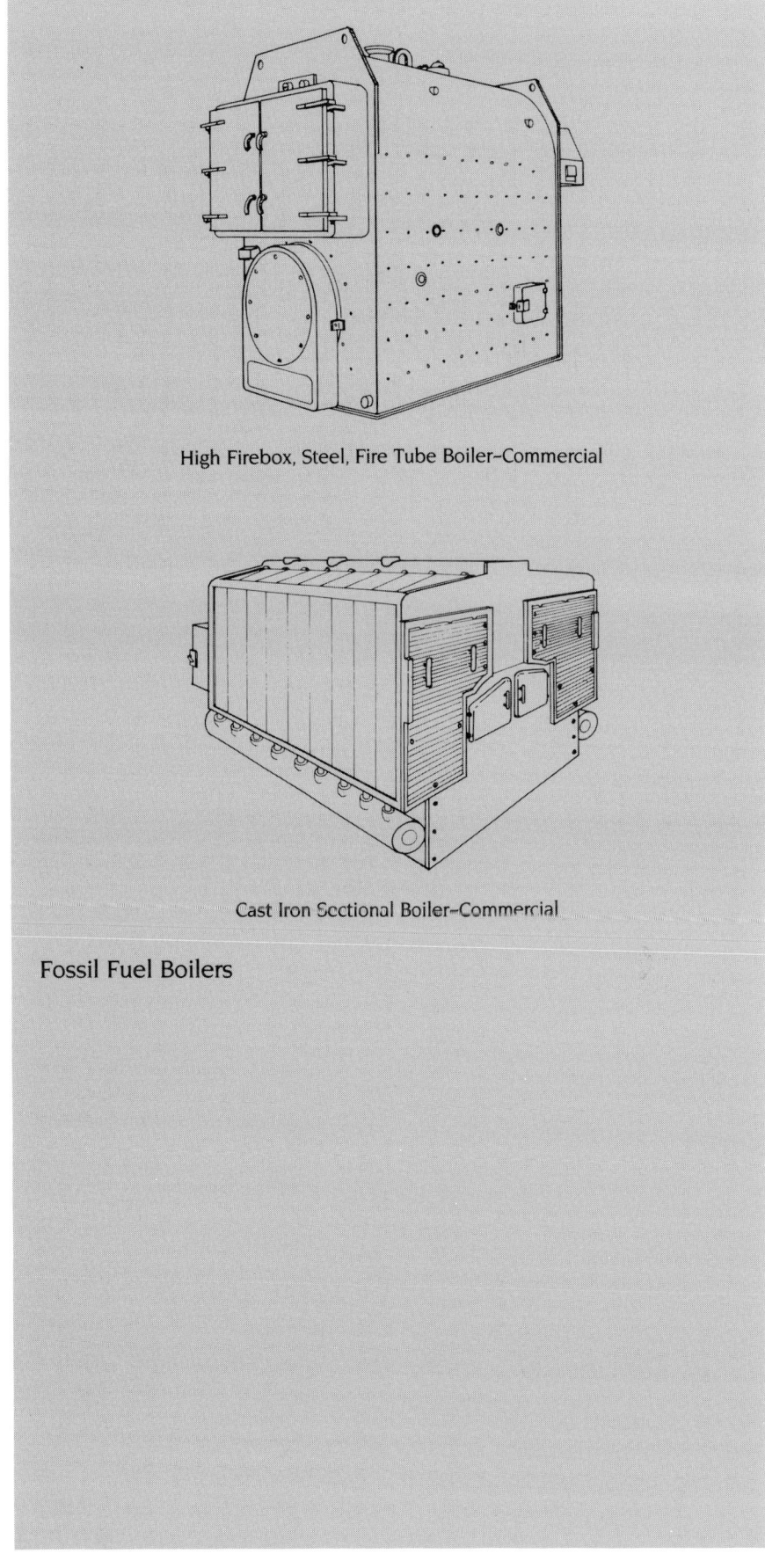

High Firebox, Steel, Fire Tube Boiler–Commercial

Cast Iron Sectional Boiler–Commercial

Fossil Fuel Boilers

Figure 17.2

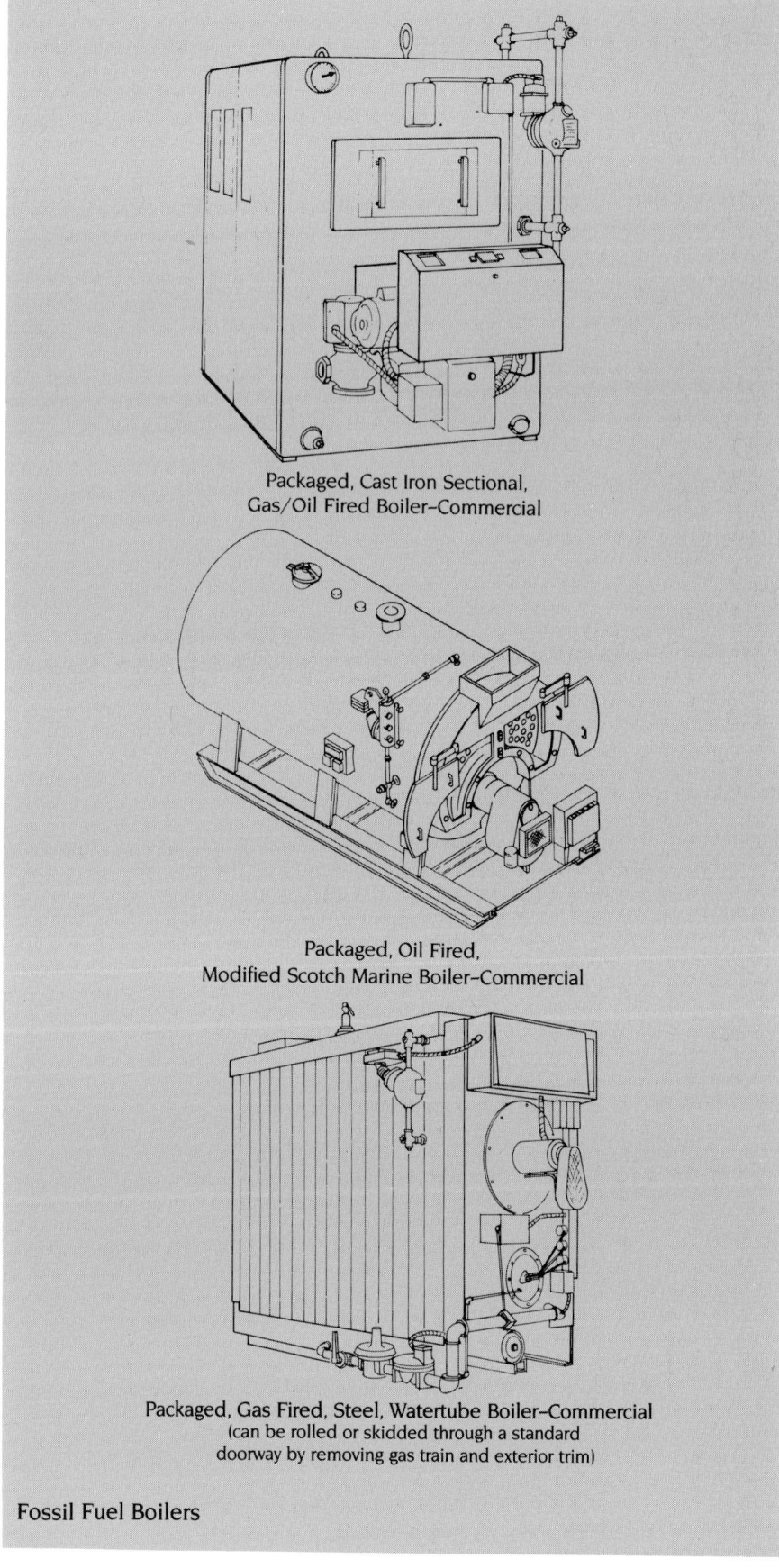

Fossil Fuel Boilers

Figure 17.3

Another innovation in the boiler field is the introduction from Europe of wall-hung, residential-size boilers. These gas-fired, compact, efficient (up to 80 percent AFUE) boilers may be directly vented through a wall or to a conventional flue. Combustion make-up air is directed to a sealed combustion chamber similar to that of the pulse-type boiler. Storage capacity is not needed in these boilers because the water is heated instantaneously as it flows from the boiler to the heating system. The boiler material consists mostly of steel. Heat-exchanger water tubes are made of copper or stainless steel, with some manufacturers using a cast iron heat exchanger. Governing conditions in boiler selection include the following:

- **Accessibility**: Both for installation and for future replacement, cast iron boiler sections can be delivered through standard door or window openings. Some steel water-tube boiler sections can be delivered through standard door or window openings. Some steel water-tube boilers are made long and narrow to fit through standard door openings.
- **Economy of installation**: Packaged boilers that have been factory fired require minimal piping, flue, and electrical field connections.
- **Economy of operation**: In addition to the AFUE ratings, boiler output should be matched as closely as possible to the building heat loss. Installation of two or more boilers (modular), piped and controlled so as to step-fire to match varying load conditions, should be carefully evaluated. Only on maximum design load would all boilers be firing at once. This method of installation not only increases boiler life, but also provides for continued heating capacity in the event that one boiler should fail.

The two boilers shown in Figure 17.2 may be trimmed out to fire any fossil fuel and generate either steam or hot water. The cast iron boiler may be shipped completely assembled, or the nine "pork chop" sections (18 castings) of the boiler may be assembled in place by two men.

Truly packaged boilers are factory fired and tested. They arrive on the job ready to be rigged or manhandled into place. Connections required are minimal; fuel, water, electricity, supply and return piping, plus a flue connection to a stack or breeching. Figure 17.3 shows three different packaged burner units.

If steam or hot water service is available to a building from a remote source of supply, the need for a boiler is eliminated. If the proposed system is to be forced hot water and the remote source of supply is to be steam or high temperature hot water, a shell and tube type heat exchanger (converter) must be provided. A heat exchanger would also be required if the building is served by a low temperature chilled water service or brine for building cooling.

Units of Measure: Boilers and heat exchanges are taken off and recorded as each.

Material Units: Manufacturers' catalogues should be consulted to determine options available for the specified boiler or its equal. Steel boilers are usually less expensive than cast iron boilers, however, it is not up to the estimator to substitute on so grand a scale. A cast iron boiler shipped "knocked down" in sections may appear to be less expensive than a packaged unit, but labor and material costs to assemble the unit should be considered before a choice between the two options is made.

Quotations should be carefully examined to ensure that the proper steam or water trim, insulated jacket, operating controls, burner package, and delivery to the site or a staging area (i.e., the contractor's or rigger's yard) are included in the price. In some instances the burner gas, oil, or combination must be purchased from a separate source. Allowance must be made for a proper mounting front plate, refractory, and possible electrical work. An oil-fired unit will require an oil storage tank, and a pumping and piping system to the boiler. Grates and stokers should be added to the boiler cost for coal-fired burners. The job specifications should be thoroughly read to ensure that all requirements are met.

Labor Units: Residential boilers are normally placed by two men while still in the shipping crate. Larger boilers, whether packaged or knocked down, require mechanical aids for unloading and positioning. These tools may range from rollers assisted by a come-a-long, rope or chain fall, or a rigging crew, to a crane or even a helicopter, depending on location. There is often a piece of excavating or hoisting equipment on the job which can be used to set boilers, oil tanks, chillers, cooling towers, air handling units, etc., at a considerable cost savings. This advantage, however, cannot be reliably preplanned, therefore, the estimator should not consider it in the estimating procedure.

Takeoff Procedure: Boilers or boiler systems can be directly entered on the estimate or summary sheet as they will usually be taken off from the same plan. If the estimate is for tract housing or multi-family units with many boilers, a separate takeoff sheet is necessary to list all the boilers. Fuel oil storage and piping systems should also be handled on a separate sheet.

Hydronic Terminal Units

Hot water or steam systems transfer heat to desired locations via radiators, convectors, coils, heat exchangers, non-ducted fan coil units, humidifiers, water heaters, condensate meters, and fuel oil heaters, as well as to hospital and kitchen equipment. Figure 17.4 illustrates typical layouts for hydronic systems.

Radiation is broken down into two classes, direct and indirect. Direct radiation includes free standing or wall hung cast iron sectional tubular radiators, cast iron baseboard, fin and tube bare elements or with expanded metal grilles. Also available (manufactured in Europe) are stamped steel panels containing channels or flattened tubes for water or steam circulation, classed as direct radiation.

Indirect radiation, which heats by air circulating through its enclosure by convection, includes cast iron or copper heating elements enclosed in sheet metal casings called convectors. Convectors may be free standing, wall hung, semi- or fully-recessed. Residential copper tube aluminum fin baseboard radiation encased in a metal enclosure as well as commercial fin tube (both copper and steel) are methods of indirect radiation. Two types of hydronic heating terminal units are shown in Figure 17.5.

Duct coils are suspended in the supply ductwork. Suspended, floor, or wall mounted unit heaters are heating coils and a fan within the same enclosure.

Units of Measure: Free standing and wall hung radiators are taken off as each and listed according to type, size, or output. Unit heaters and duct coils are recorded in the same manner. Baseboard radiation, fin tube, and panel must be taken off as each, but also recorded by linear foot of

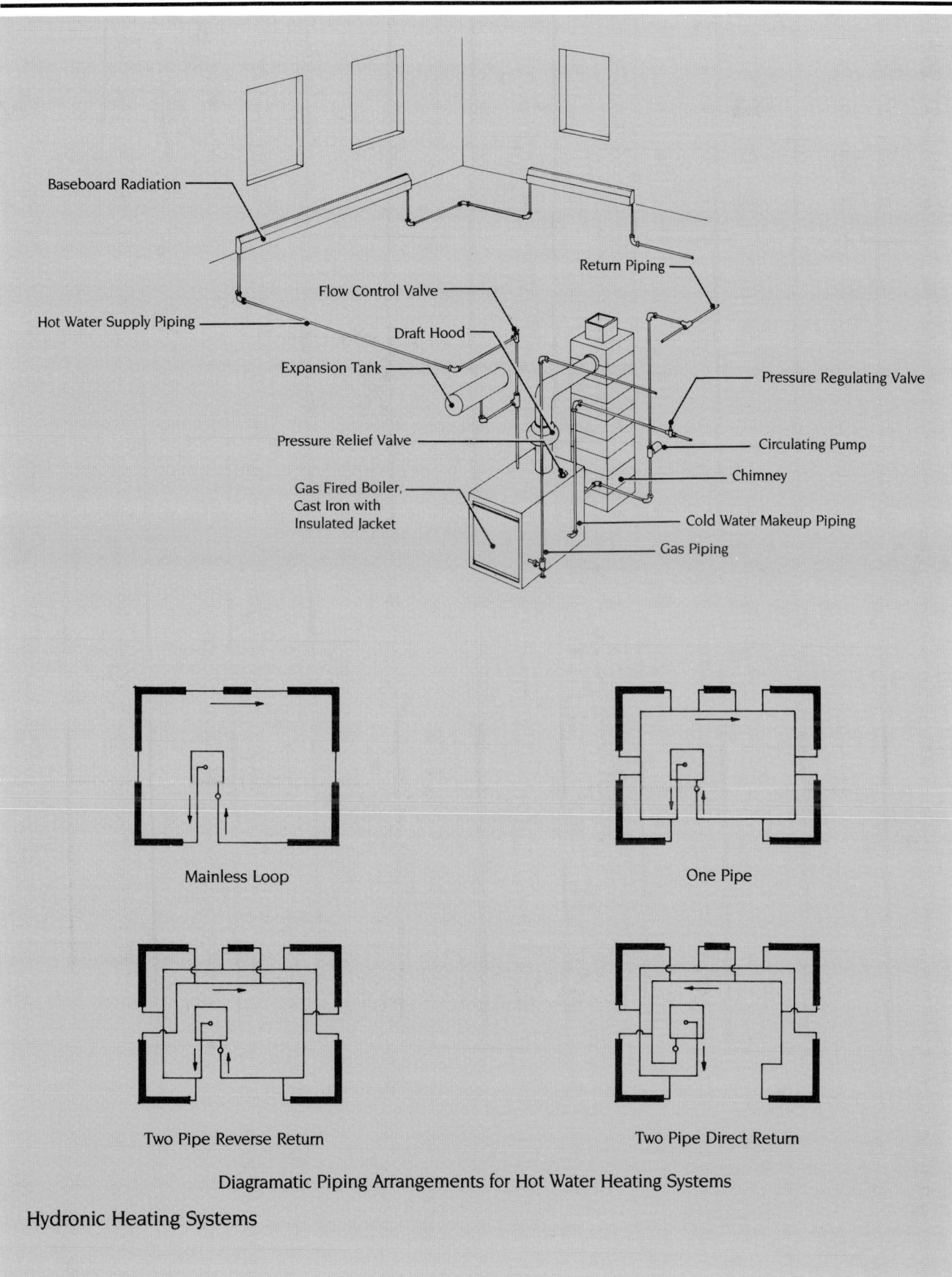

Figure 17.4

enclosure. In the case of wall to wall, radiation is recorded by the linear foot of heating element (and the number of rows). All other terminal units should be recorded as each and kept with their specialized system (e.g., hospital equipment, fuel oil heaters, etc.). Cast iron radiators are priced by output in square feet of radiation. Fin tube radiation is priced by the linear foot plus the various trim items. Unit heaters, coils, and panel radiators are priced as each.

Material Units: Wall hung radiation of all types require supports. Indirect radiation requires enclosures, knob or chain dampers, access doors, special grilles, and finishes. All radiators usually require a supply control valve (manual or automatic), a balancing valve, on the return, or a radiator trap for steam radiation. Both water and steam radiation require automatic or manual air venting capability. If the radiator is down feed hot water, it should have a drain cock. Unit heaters, duct coils, and all terminal units would have almost identical trim (the size will vary) with the radiation. Manufacturers' catalogues showing all the options and accessories should be referred to while preparing the estimate.

Labor Units: The labor to install and pipe hydronic units should be based on the unit count (i.e., a certain number of each type per day). For fin tube and all baseboard, the linear footage of enclosure must also be considered as hanging so many feet per day of both back and front panels. For long runs of dummy enclosure in wall to wall installations, be sure to capture both the material and labor costs of the bare pipe concealed within the enclosure. Duct coils are installed by the tin knocker and piped by the pipefitter.

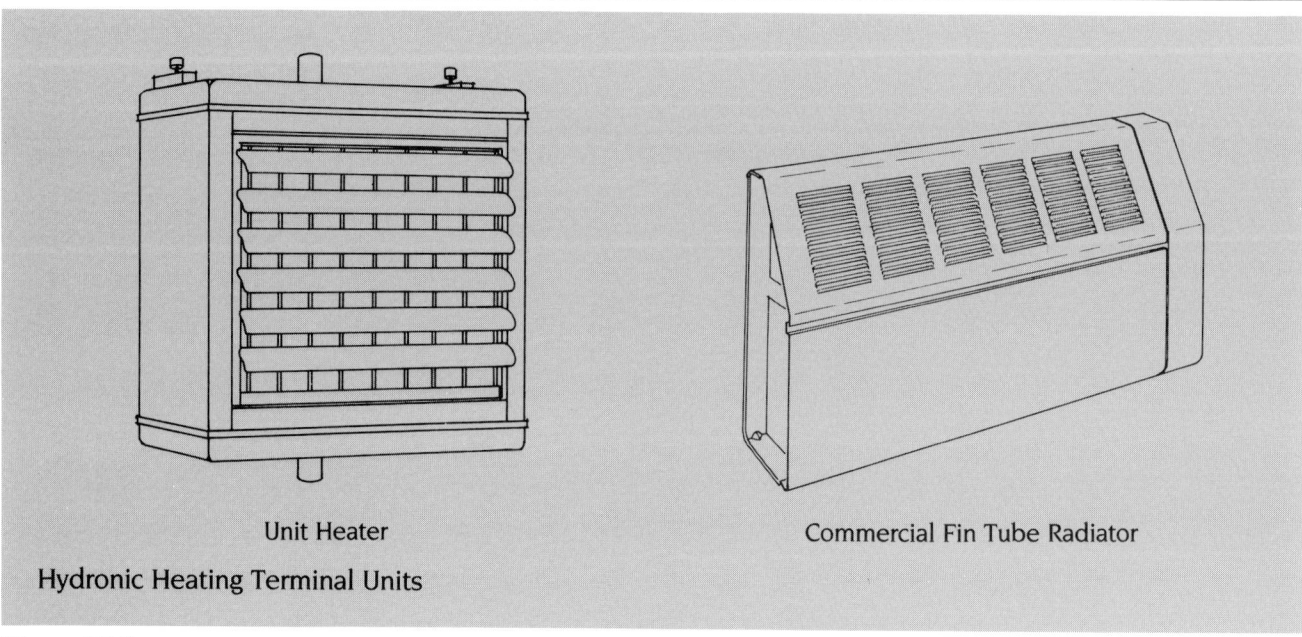

Unit Heater   Commercial Fin Tube Radiator

Hydronic Heating Terminal Units

Figure 17.5

Takeoff Procedure: On a hydronic radiation job the takeoff should begin at the radiation or heating units and proceed floor by floor, starting at the top. This procedure helps to familiarize the estimator with the building. The heating units can also be used later as targets for the piping takeoff. The units should be color coded to correspond with the piping color codes. All like types of radiation can be totalled together for pricing and labor estimating. For example, if 42 cast iron wall hung radiators are specified for 1260 S.F., this means 42 supply valves, 42 air vents, 84 wall brackets, and 42 traps are needed. (The radiation itself is priced by the S.F. of radiation (E.D.R.).) If 42 fin tube radiators, 1 row of slope top, 600 feet of enclosure, and 504 linear feet of element were specified, then 42 supply valves, 42 air vents, 84 element supports, 36 intermediate enclosure supports, and 84 access doors or valve enclosures are needed. If inside or outside corners are required, they should be listed separately.

Unit heaters, coils, and the related accessories are taken off and totalled on the summary sheets. These items may be priced in with fans and air handling units by some manufacturers (or manufacturers' representatives) in order to gain an advantage over competitors who cannot quote a complete line. A well prepared estimator will have several sources of pricing and supply to compare individual prices with lump sum quotes. An example of this is shown in Figure 17.6.

Chillers and Hydronic Cooling

The central component of any chilled water air conditioning system is the water chiller. Packaged water chillers (shown in Figure 17.7) are available in three basic designs: the reciprocating compressor, direct-expansion type; the centrifugal compressor, direct-expansion type; and the absorption type. Chillers and other air conditioning apparatus are sized by the ton; that is, a ton of cooling which equals the melting rate of one ton of ice in a 24-hour period, or 12,000 B.T.U. per hour. The three types of chillers vary significantly in their cooling power, as well as in their operation. The reciprocating compressor chiller, which generates cooling capacities in the range of 10 to 200 tons, is usually powered by an electric motor. The centrifugal compressor, which generates cooling capacities ranging from 100 to several thousand tons, is also commonly powered by an electric motor, but it may be designed for a steam-turbine drive as well. In some instances both of these types of chillers may be powered by internal combustion engines.

Absorption-type chillers provide cooling capacities ranging from 3 to 1600 tons. Because it uses water as a refrigerant and lithium bromide or other salts as an absorbant, this system consumes about 10 percent of the electrical power required to operate the conventional reciprocating and centrifugal direct-expansion chillers. This low consumption advantage is particularly desirable in buildings where an electrical power failure triggers an emergency backup system, as in hospitals, data processing centers, electronic switching systems locations, and other buildings that must continue to function on auxiliary power. Absorption-type chillers are economically advantageous in areas where electric power is scarce or costly, where gas rates are low, or where waste or process steam or hot water is available during the cooling season. Solar power may also be used in some areas to generate the heat required for the absorption process.

The chillers themselves may be air or water cooled. Very small chillers are available with an air-cooled condenser built into the package. For larger

Means Forms

COST ANALYSIS

SHEET NO. CH-8

PROJECT: Office Building

ESTIMATE NO.

ARCHITECT

DATE

TAKE OFF BY: QUANTITIES BY: JJM   PRICES BY: FPW   EXTENSIONS BY: JJM   CHECKED BY: AHF

Item	Qty	ABC Co.	USA Inc.	XYZ Assoc.	Individual Misc. Quotes
Chiller	1	65000	60500 *	88000	no
Start & Service		YES	no	YES	1000
AH Units	6	25000	no	YES	23000
Vibr. Bases	6	no	no	YES	1000
Starters	1	YES	YES	YES	
Reheat Coil	1	200 *	no	YES	280
Roof Fan	1	no	no	1200 *	1220
FOB Job		YES	YES	YES	YES
Total		90200 —	60500 *—	89200 —	

60500 —	Chiller
1000 —	Service
23000 —	Units
1000 —	Bases
200 —	Coil
1200 —	Fan

* Best Prices   # 86900

Estimate Summary Sheet Used for Cost Comparison

Figure 17.6

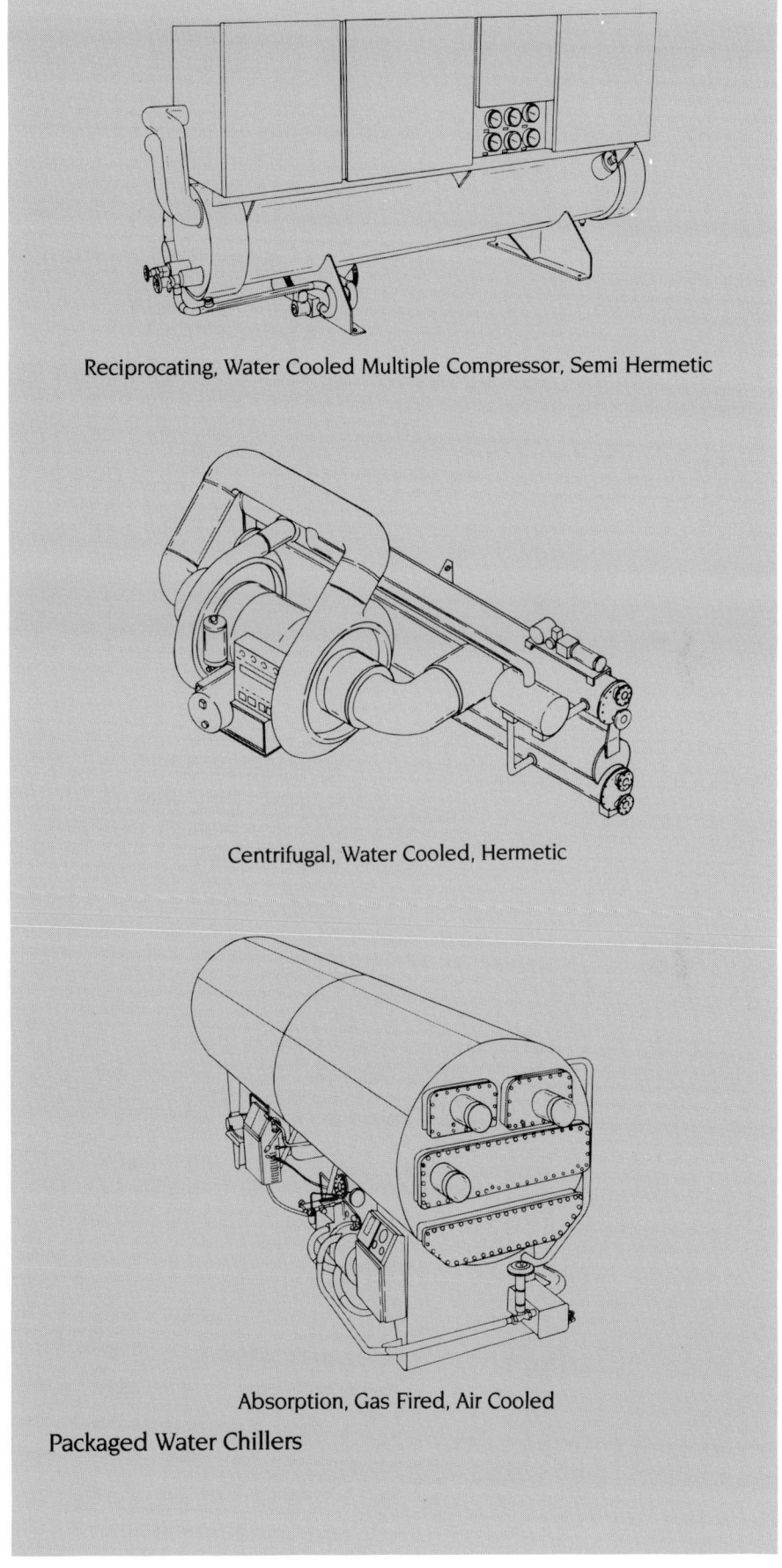

Figure 17.7

systems and for systems that may create too much noise for their location, the air-cooled condenser is installed at a distance from the chiller, and the two units are connected with refrigerant piping.

The basic process for air conditioning systems using refrigerants as the cooling medium is the cooling and condensing back to liquid form of the refrigerant gas that was heated during the evaporation stage of the cycle. This condensation is achieved by cooling the gas with air, water, or a combination of both. Air-cooled condensers cool the refrigerant by blowing air directly across the refrigerant coil; evaporative condensers use the same method of cooling, with the addition of a spray of water over the coil to expedite the process.

The condenser for water cooled chillers is piped into a remote water source, such as a cooling tower, pond, or river, via the "condenser water system". Completely packaged chiller systems may include built-in chilled water pumps, and all interconnecting piping, wiring, and controls. All of the components of the completely packaged unit are factory installed and tested prior to shipment to the installation site for connection to the chilled water and condenser water systems.

When water is used as the condensing medium and is abundant enough that recycling is not required, it may be piped to a drain after performing its cooling function and returned to its source. The source may be a river, a pond, or the ocean. If water supply is limited, expensive, or regulated by environmental restrictions, a water conserving or recycling system must be employed. Several types of systems may be installed to conform to these limitations. For example, a water regulating valve, a spray pond, a natural draft cooling tower, or a mechanical draft cooling tower (shown in Figure 17.8) may be used.

In very small cooling systems a temperature-controlled, water-regulating valve may be used, provided that such a system is permitted by local environmental and/or building codes. The regulating valve system functions by allowing cooling water to flow when the condenser temperature rises, and, conversely, by stopping the flow as the temperature falls. The problem with this system, however, is that the heated condenser water cannot be recycled and is therefore wasted during the flow cycle.

The *spray pond* and *natural draft cooling tower systems*, although they are viable and available methods of cooling, are not commonly used for building air conditioning. For reasons of water loss caused by excessive drift and the large amount of space required for their installation and operation, they are less desirable than the mechanical draft cooling tower method.

Mechanical draft cooling tower systems are classified in two basic designs: *induced draft* and *forced draft*. In an induced draft tower (as shown in Figure 17.8), a fan positioned at the top of the structure draws air upwards through the tower as the warm condenser water spills down. A cross-flow induced draft tower operates on the same principle, except that the air is drawn horizontally through the spill area from one side. The air is then discharged through a fan located on the opposite side. In a forced draft tower the fan is located at the bottom or the side of the structure. Air is forced by the fan into the water spill area, through the water, and then discharged at the top. All designs of mechanical draft towers are rated in tons of refrigeration; three gallons of condenser water per minute per ton is an approximate tower sizing method.

After the water has been cooled in the tower it passes through a heat exchanger, or condenser, in the refrigeration unit. Here, it again picks up heat and is pumped back to the cooling tower. The piping system is called the condenser water system. Figure 17.9 shows the layout of a typical condenser water system.

The actual process of cooling within the mechanical draft cooling tower takes place when air is moved across or counter to a stream of water falling through a system of baffles or "fill" to the tower basin. After the cooled water reaches the basin it is piped back to the condenser. Some of the droplets created by the fill are carried away by the moving air as "drift" and some of the droplets evaporate. This limited loss of water is to be expected as part of the operation of the tower system. Because of the loss of water by drift, evaporation, and bleed off, replenishment water must be added to the tower basin to maintain a predetermined level and to assure continuous operation of the system. To prevent scale buildup, algae, bacterial growth, or corrosion, tower water should be treated with chemicals or ozone applications. The materials used in constructing mechanical draft cooling towers include redwood (which is the most commonly employed material), other treated woods, asbestos, various metals, plastics, concrete, or ceramic materials. The fill, which is the most important element in the tower's operation, may be manufactured from the same wide variety of materials used in the tower structure. Factory assembled, prepackaged towers are available and are usually preferable to built in place units, with multiple tower installations now being used for large systems.

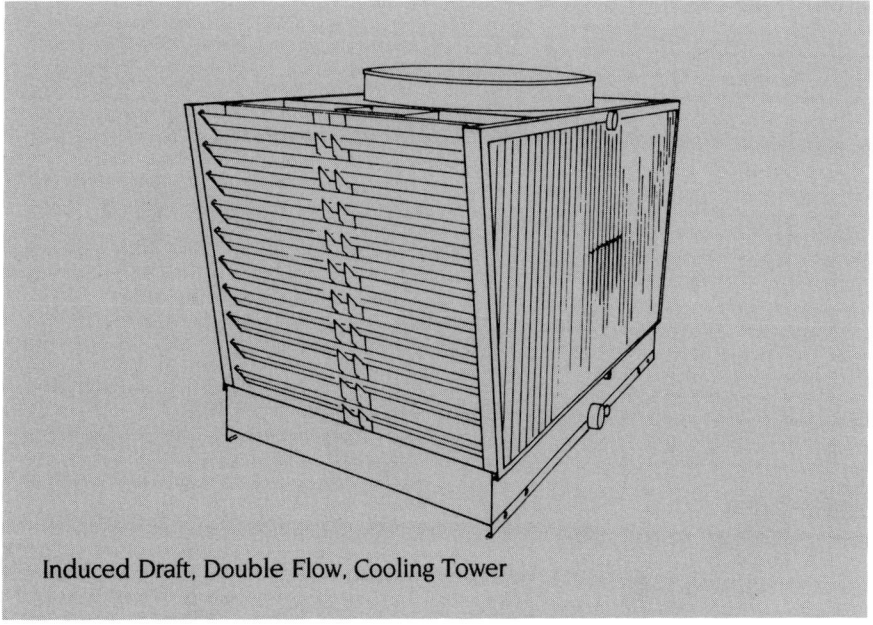

Induced Draft, Double Flow, Cooling Tower

Figure 17.8

The location of cooling towers is an important consideration for both practical and aesthetic reasons. They may be located outside of the building: on its roof or on the ground. If space permits, a cooling tower may be installed indoors by substituting centrifugal fans for the conventional noisy propeller type, adding air intake and exhaust ductwork. The tower discharge should not be directed into the prevailing wind, or towards doors, windows, and building air intakes. In general, common sense should be used when determining tower placement so that the noise, heat, and humidity the system creates do not interfere with building operation and comfort. The manufacturer's guidelines for installation should be strictly followed, especially those sections that address clearances for maximum air flow, maintenance, and future unit replacement.

Economical operation of a cooling tower system may be achieved through effective control and management of several critical aspects of its operation, including: careful monitoring of water treatment, selecting and maintaining the most efficient condensing temperature, and controlling water temperature with fan cycling.

A recent novel development in tower water system operation allows the tower to substitute for the water chiller under certain favorable climactic conditions. This new method cannot be implemented in all cases, but during periods when it can be employed, substantial savings result in the reduced cost of chiller operation.

Units of Measure: Chillers, cooling towers, air cooled condensers, and any other major components of hydronic water chilling systems are taken off and recorded as each.

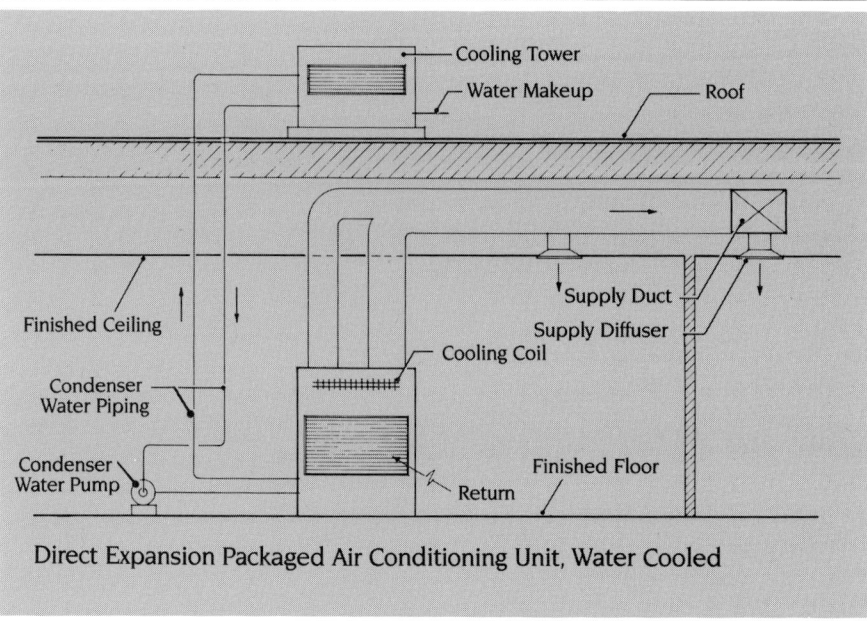

Figure 17.9

Material Units: Manufacturers' catalogues and quotations should be consulted for each estimate to confirm that all required components are included and how each chiller or tower will be shipped (in one piece or broken down). Packaged chillers have built-in motor control panels and starters, whereas motor starters for cooling towers are purchased and shipped separately. If motor starting devices are to be furnished by the mechanical contractor and are not duplicated in the electrical specifications, the tower supplier's quotation should be checked and the starter cost added, if it has not been included. A successful bidder who is required to furnish motor starting equipment can save money by culling the starters from all of the equipment quotes and purchasing them as a package from an electrical outlet dealer. Packaged equipment, however, must be purchased with these controls built in.

The refrigerant charge and oil for centrifugal chillers will be quoted and shipped separately. The charging, startup, and warranty service must be estimated by an experienced technician, either in house, by the manufacturer, or by a qualified subcontractor. This also applies to absorption units and their charges of lithium bromide, etc.

Labor Units: Because chillers, towers, or air cooled condensers are large and cumbersome, they will always require mechanical assistance to be set into place. Placement can vary greatly, ranging from simple manhandling with skids or rollers (for smaller units) to the use of hoists and cranes (for larger sizes). Equipment location, regardless of size, will also dictate special hoisting apparatus. Cooling towers and air cooled condensers may be installed on the roof of a building, indoors in one of the upper stories of a high rise building, in mechanical equipment rooms where the chillers and boilers are no longer in the traditional basement, on the roof, or on intermediate floors. Size and location will dictate whether chillers and towers should be shipped in one piece or knocked down for assembly in place. The manufacturer's quote usually includes assembly and installation.

The supplier's quotes for equipment, particularly for large shipments, should include job site delivery. Delivery may be F.O.B. the job, or F.O.B. the factory with freight allowed to the job or staging area.

Takeoff Procedure: All materials for air cooling should be taken off and recorded on a summary sheet where several manufacturers' quotations can be compared, as shown in Figure 17.6. The estimator can then compare choices, and can better determine the best price.

Air Handling Equipment

Air handling units consist of a filter section, a fan section, and a coil section on a common base. They are used to distribute clean, cooled, or heated air to the occupied building spaces. These units (some of which are shown in Figures 17.10, 17.11 and 17.12) are available in a wide range of capacities, from 200 cubic feet per minute to tens of thousands of cubic feet per minute. The units also vary in complexity of design and versatility of operation. Small units tend to have relatively simple coil and filter arrangements and modestly sized fan motor. Larger, more sophisticated units usually require remote placement. Because of the need to overcome losses caused by intake and supply ductwork, and by complex coil, filter, and damper configurations, the fan motor horsepower must be dramatically increased.

Small air handling units may be located and mounted in a variety of settings and by different methods (as shown in Figure 17.10). They may be mounted on the floor or hung from walls or ceilings with no discharge ductwork required in the room they service. Small air handling units require supply and return piping for heating and/or cooling. If these units are used for cooling, a drain connection is also required.

Determining the proper size, number, capacity, type, and configuration of coils in the unit is a prime consideration when selecting and/or designing an air handling unit. As a general rule, the amount of air (in cubic feet per minute) to be handled by the unit determines the size and number of the various coils. Electric or hydronic coils are used for heating; chilled water or direct expansion coils are used for cooling. As the units increase in size and complexity, the coil configurations and arrangements become limitless. A simple heating and cooling unit, for example, may use the same coil for either hot or chilled water. A large unit usually demands different types of coils to perform many separate functions. In humid conditions, the air temperature may be intentionally lowered to remove

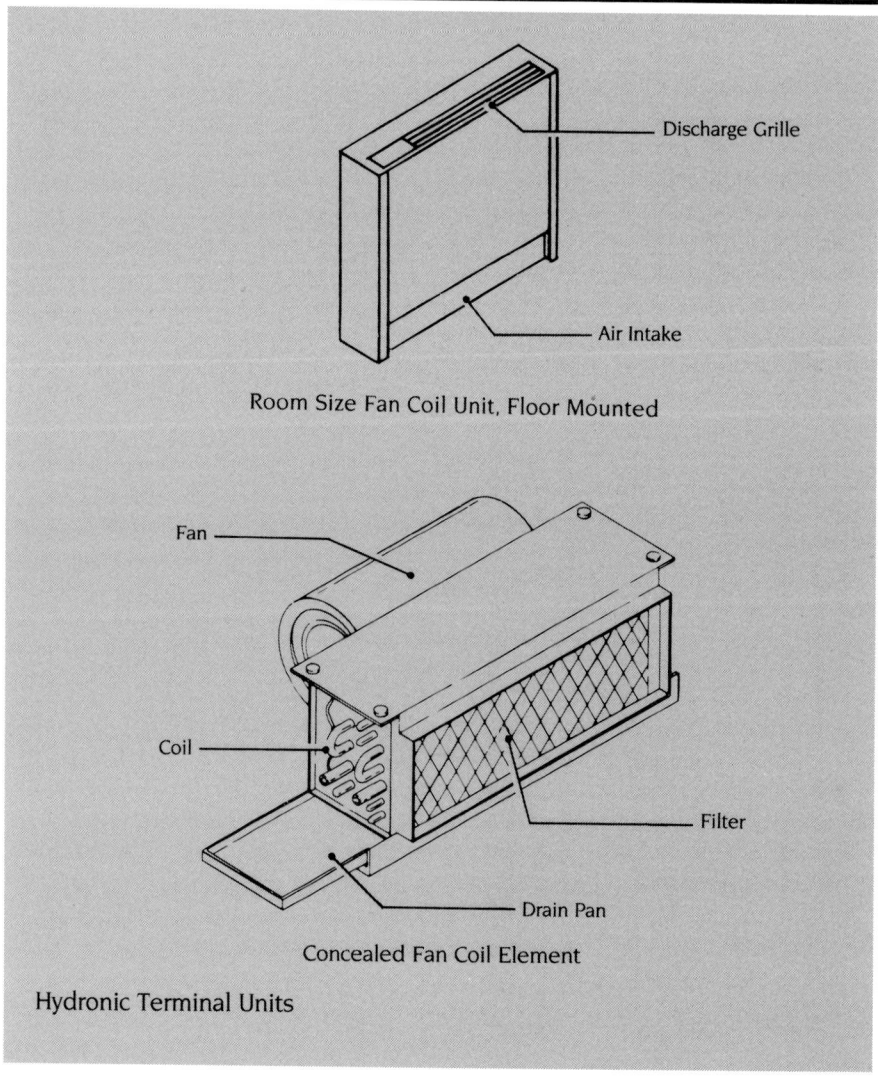

Figure 17.10

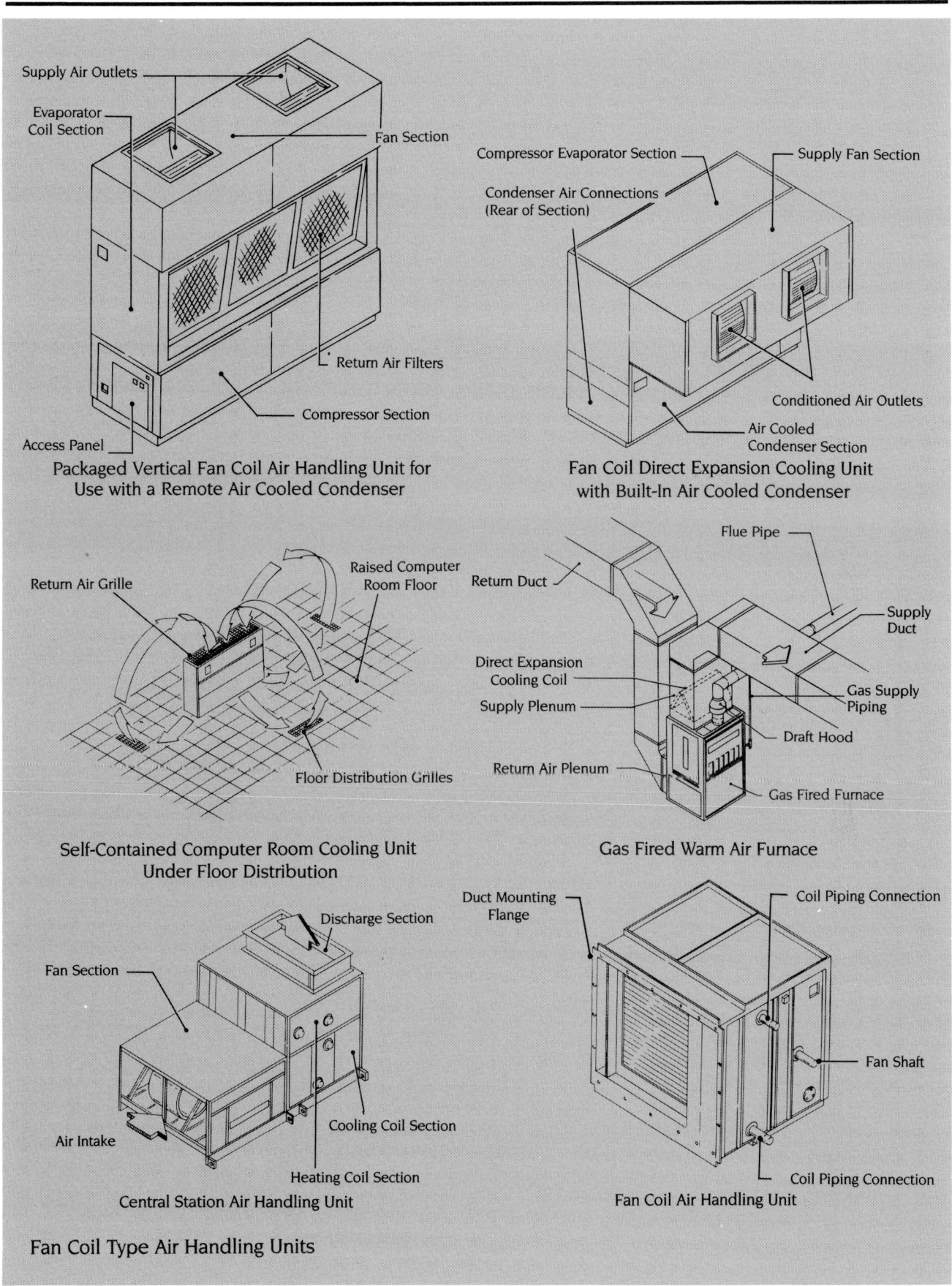

Figure 17.11

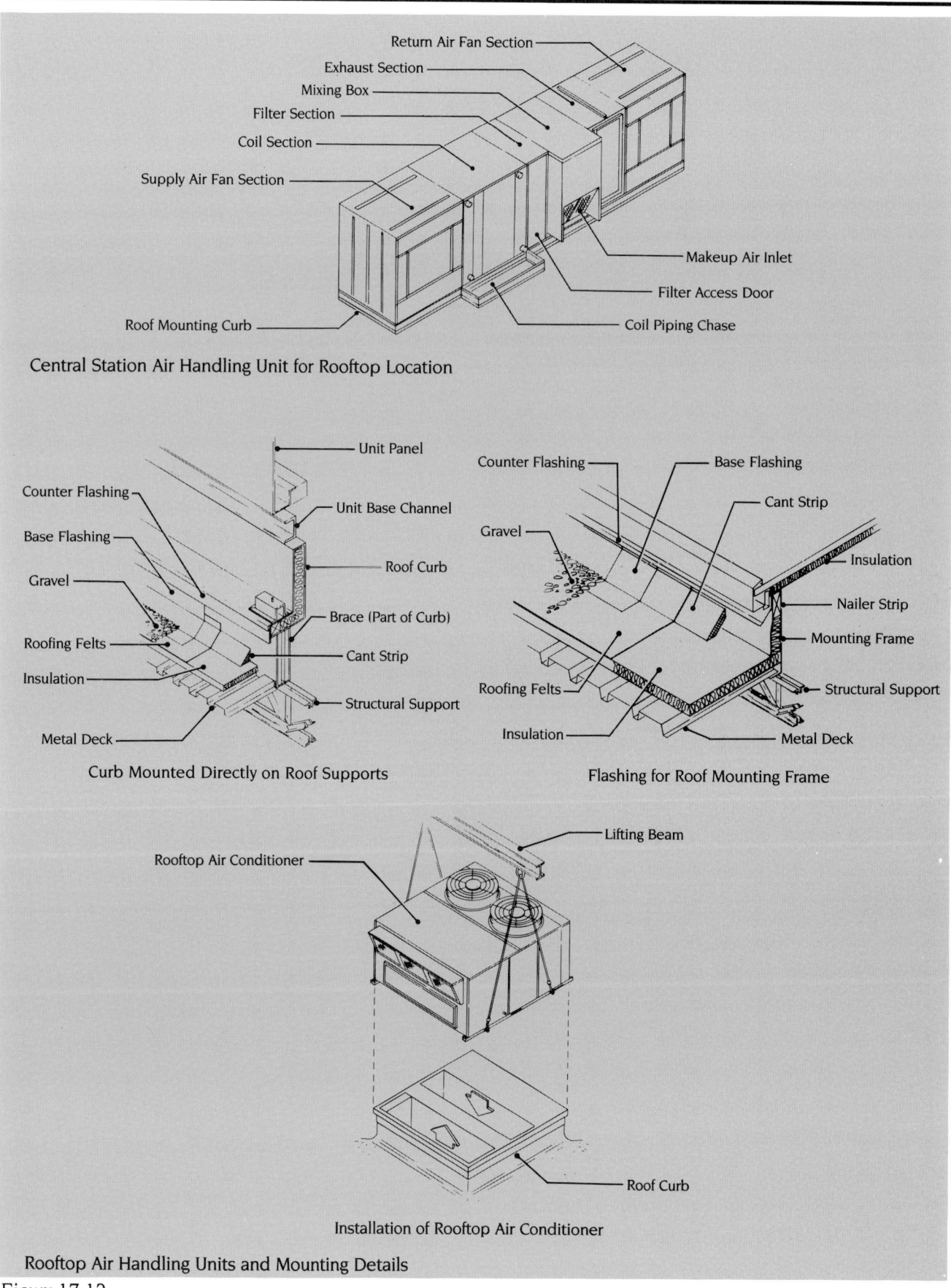

Rooftop Air Handling Units and Mounting Details

Figure 17.12

moisture. In this case, a reheat coil is added to return the temperature to its desired level. Conversely, in dry conditions, a humidifier component is built into the unit. If outside air is introduced to the unit at subfreezing temperatures, then a preheat coil is placed in the outside air intake duct.

Certain precautions should be taken to prevent damage and to assure the efficiency of the unit's coils and other components. To protect the coil surfaces from accumulating dust and other airborne impurities, a filter section is a necessary addition to the unit. If a unit is designed to cool air, a drain pan must be included beneath the coiling section. This pan is then piped to an indirect drain to dispose of the unwanted condensation.

Another protection precaution involves the insulation of the fan coil casing internally. If this precaution is not taken to protect the cooling coil section and all other sections "down-stream", corrosive or rust-causing condensation will damage the casing and discharge ductwork. Insulating this casing also helps deaden the noise of the fan. Noise can also be controlled by the installation of flexible connections between the unit and its ductwork, mounting the unit on vibration-absorbing bases (if located on the floor), or suspending the unit from vibration absorbing hangers (if secured to the ceiling or wall).

Units of Measure: All air handling units are taken off and recorded as each, separated by type for pricing and labor estimating.

Material Units: Material considerations in addition to the units themselves include curbs for roof mounted units (shown in Figure 17.12), vibration mounts for suspended units, large base mounted equipment, and filters. Complete self-contained packaged units are available with built-in limit and operating controls. Central station fan coil units are not supplied with these controls or starters which must be supplied by the mechanical or electrical contractor. Job specifications and manufacturer's instruction should be carefully read for additional material required.

Labor Units: Air handling equipment must be set in its designated place either manually or with mechanized equipment. A composite crew consisting of sheet metal workers and pipefitters usually handle large equipment on union jobs. After each unit has been assembled and set on its base, piping and duct connection must be made.

Takeoff Procedure: Air handling equipment is taken off from the drawings and designated by equipment room or system. Like equipment should be listed together in one quote from various suppliers.

Fans and Gravity Ventilators

Fans are used to supply, circulate, or exhaust air for human comfort, safety, and health reasons. Fans may be exposed in the area being served or in a remote location, connected by ductwork to the served area. Fans may also be located outside of the building, on the roof or on side walls. In general, a fan consists of an electric motor and drive, blades, and a wheel or propeller. All of these parts are contained within an enclosure. The drive assembly may operate via belts and pulleys or may be directly connected to the motor. While the direct-drive type fan is less expensive, objectionable noise can result as the size or speed of the fan increases. Belt drive affords greater flexibility in speed and performance. Proper fan selection is important not only because of noise but also to avoid the feeling of air movement or drafts due to excess velocity. Fans are sized according to the cubic feet of air they can handle in one minute (C.F.M.).

Fans are classified in two general groups, *centrifugal* and *axial-flow*. Centrifugal fans are further classified by the position of the blades on the fan wheel, either forward-curved or backward-curved. Axial-flow fans, where the air flows around the axis of the blade and through the impeller, are classified as propeller, vane-axial, and tube-axial. Some typical fans are shown in Figure 17.13.

Self-contained air handling or air conditioning units depend on a centrifugal fan for reasons of adaptability to duct configurations, and quiet and efficient operation. Air filters are used in air supply systems to protect the heating or cooling coils from dust or other particles picked up by the air flow.

The air handling capacity or volume delivered by a fan may be varied by a motor speed control, outlet damper control, inlet vane control, or fan drive change. The most efficient method is a variable-speed motor, but this is also the most costly type of fan control.

Roof-mounted ventilators (shown in Figure 17.14) are designed to remove air from a building without the use of motor-driven fans. In some cases, this process is achieved by a rising of warm air and its displacement by denser or heavier cold air. Some ventilators, however, use the action of the wind to syphon air through the ventilator. Relief hoods may be used for exhaust air, as well as for makeup air intakes, through the use of dampers or self-acting shutters. Hoods and ventilators are usually constructed of galvanized steel or aluminum.

Gravity roof ventilators, which are not very efficient, are used in situations where rapid removal of stale air is not a factor. The cost of a motor-driven roof fan is two to three times that of a gravity ventilator.

Units of Measure: Fans and ventilators are taken off and priced as each.

Material Units: Roof mounted fans and ventilators may be supplied (when specified) with self-flashing and sound attenuating curbs. Fans will require motor control switches to be furnished by the mechanical or electrical contractor (check both mechanical and electrical specification for duplication). Dampers, shutters, and special finishes or linings may be required on a job by job basis. Large fans, whether supply or exhaust, will require vibration mounts or bases.

Labor Units: Fans and ventilators are installed by sheet metal workers, even though the motor driven exhaust fan and the intake/exhaust hoods may be furnished by the mechanical contractor in accordance with local custom. Fans should be rigged or hoisted into place when other mechanical or roof mounted equipment is being placed, to take advantage of hoisting equipment availability.

Takeoff Procedure: Fans are taken off and recorded by type, size, or system. If they are being taken off by the trade that will also be purchasing the ventilators, the fans should be totalled on the same summary sheet. If not, there should be individual takeoff, summary, and pricing procedures. The sheet metal subcontractor may be recording these fans for labor only, and the mechanical contractor for material cost only.

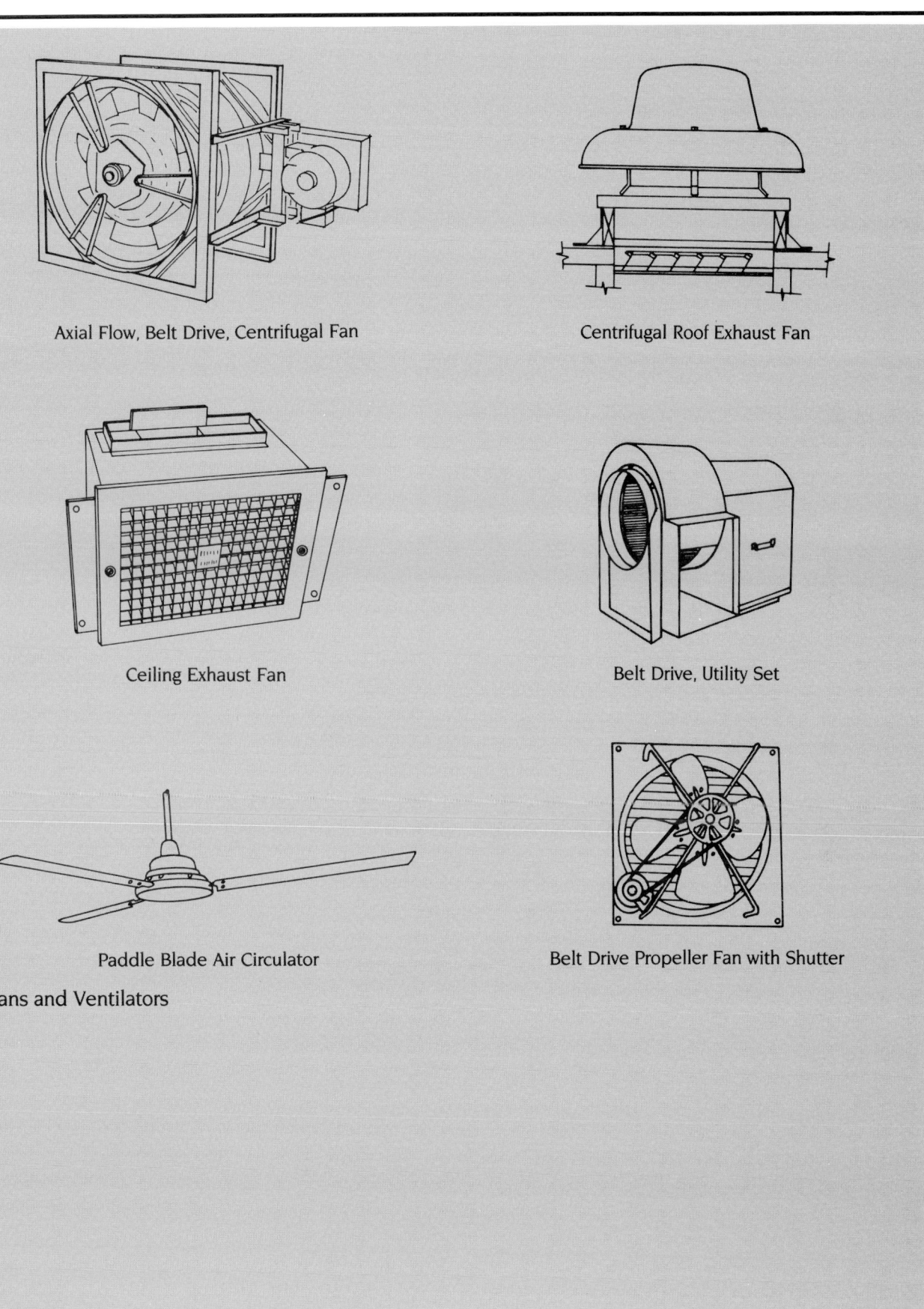

Fans and Ventilators

Figure 17.13

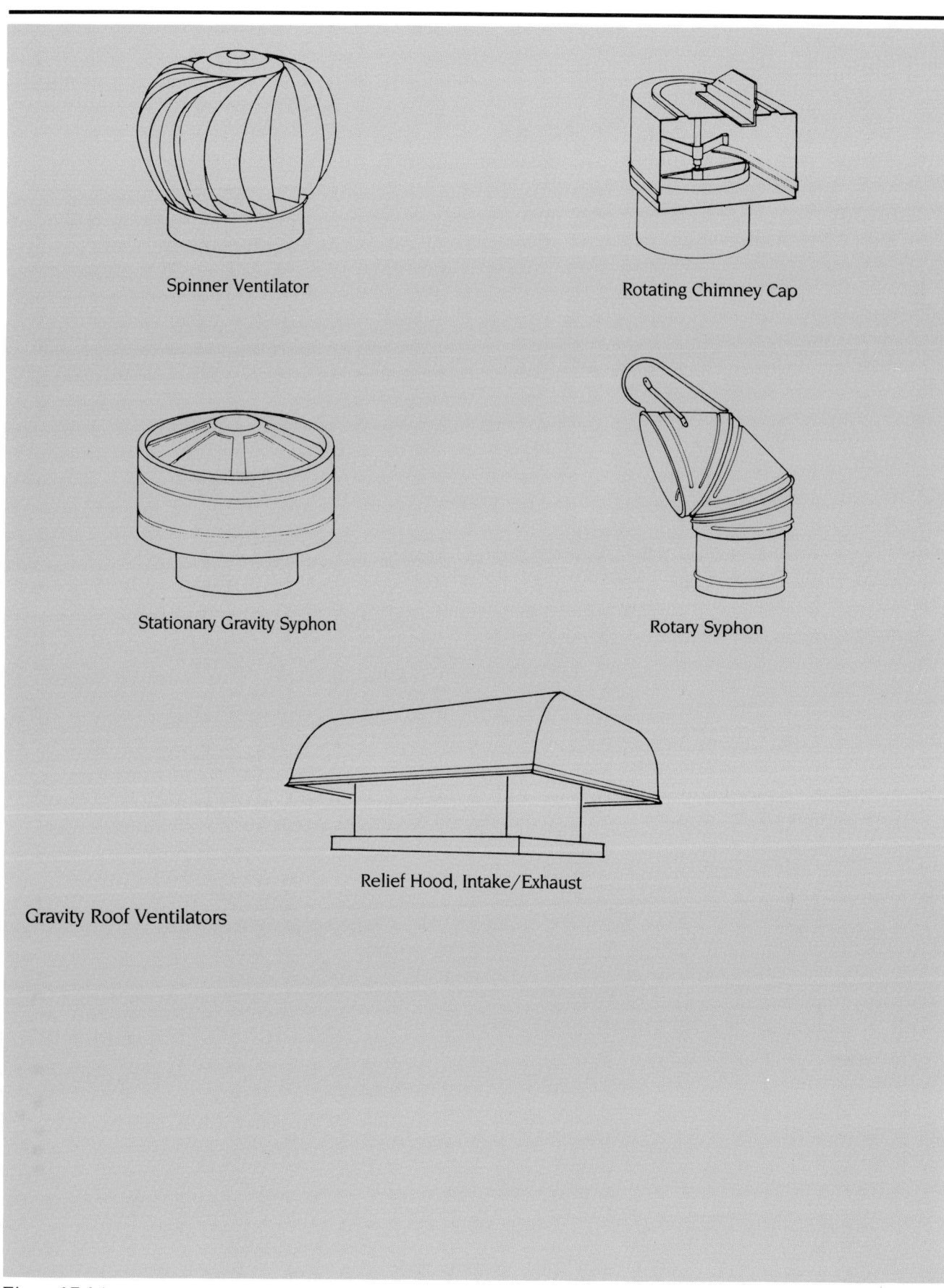

Figure 17.14

Chapter 18
DUCTWORK

Ductwork is the conduit or pipe which transports air through heating, ventilating, or air conditioning systems. Although ductwork has traditionally been fabricated from sheets of galvanized steel, it is also made of aluminum sheets and fiberglass board. Stainless steel or one of several plastics are preferred materials for use in corrosive atmospheres.

Ductwork may be rectangular, round, or oval in shape, depending on requirements of size and application. Flexible or spiral-wound ductwork is also available for special installations. Fittings are available to accommodate changes in the direction, shape, and size of ductwork. Elbows, tees, transitions, reducers, and increasers come under this category. Typical ductwork systems depicting many of these items are shown in Figures 18.1 and 18.2.

Industry standards regulate the thickness or gauge of metal duct material. The basis for these regulations is the size of the ductwork and the pressure of the air contained. The appropriate weight for ductwork material is determined using these same standards. Based on this weight, the fabrication labor and installation labor may be calculated.

Unit of Measure: Ductwork is taken off and recorded by the linear foot for each size. For metal ducts, the footage is then converted into pounds of metal and priced accordingly. The calculation for this conversion takes into account the gauge and weight per square foot of the metal being used. The table shown in Figure 18.3 can be used to obtain weight and area for any size and gauge of galvanized steel duct. The chart shown in Figure 18.4 can also be used to obtain weight per foot for various types of sheet metal. Both the table and the chart shown in these Figures can be found in the reference section of *Means Mechanical Cost Data*. Smaller sized ducts and lighter gauges of metal will cost more per pound than longer and heavier ducts. Fiberglass and plastic ductwork is converted to square feet for pricing. Ductwork specialties (e.g., grilles, registers, diffusers, dampers, access doors, mixing boxes etc., several of which are shown in Figure 18.5) are taken off and priced as each. Sound lining for ductwork is estimated by the square foot. Flexible connectors used at equipment connections and angle iron used for supports and special bracing are taken off and priced by the linear foot.

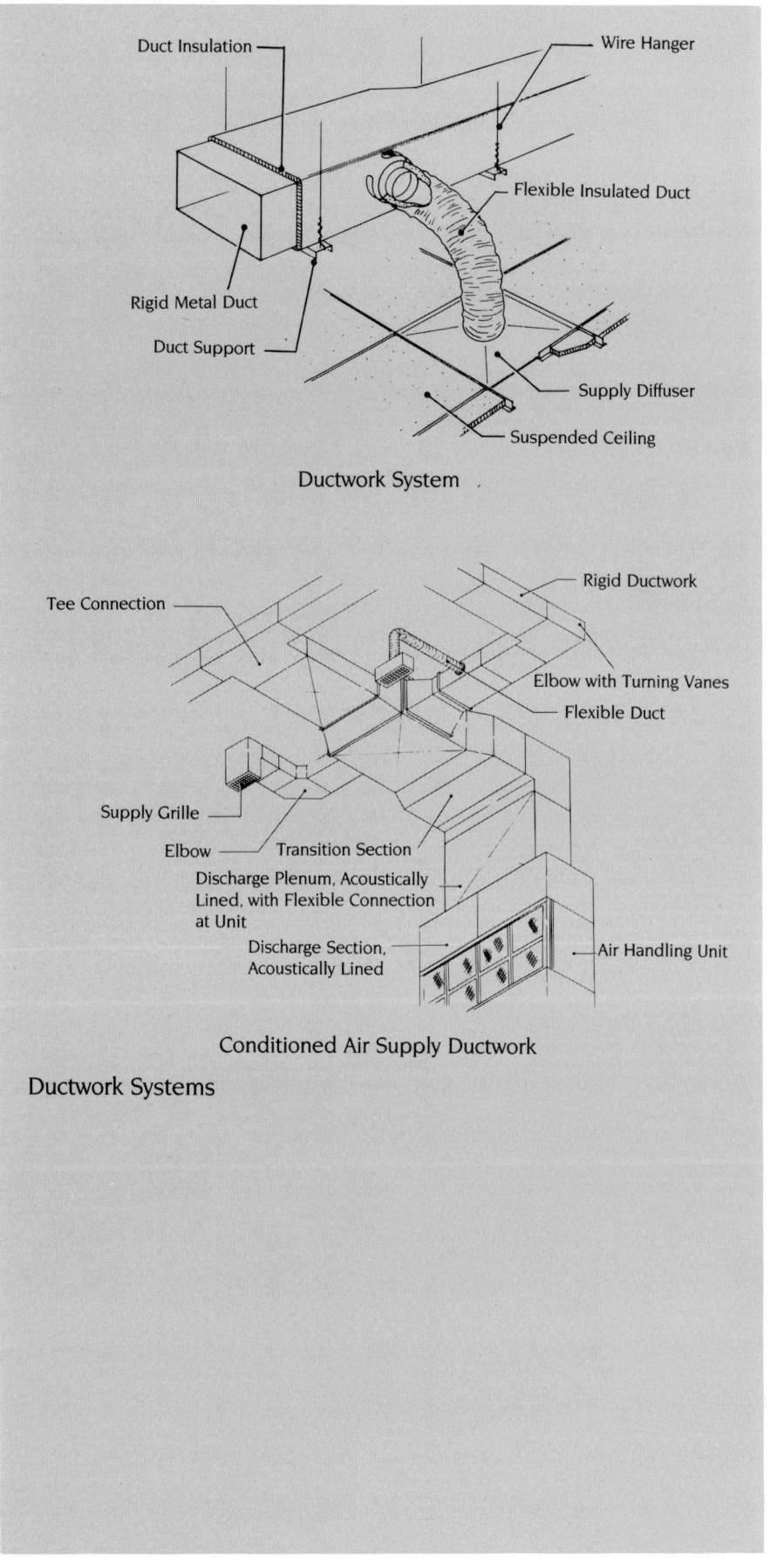

Ductwork Systems

Figure 18.1

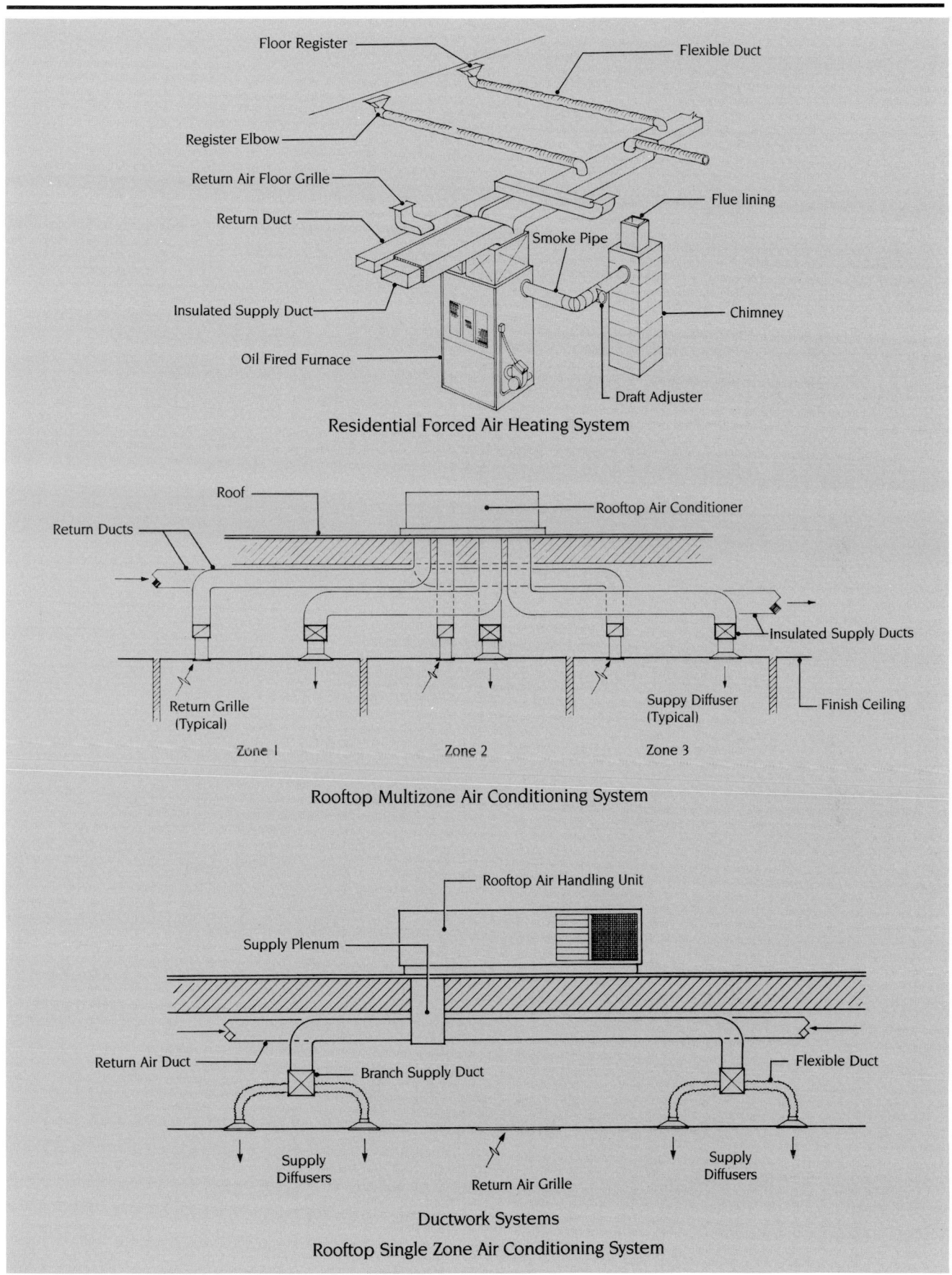

Figure 18.2

AIR CONDITIONING | **C8.4-000** | **Air Distribution**

Table 8.4-009 Sheet Metal Calculator (Weight in Lb./Ft. of Length)

Gauge	26	24	22	20	18	16	Gauge	26	24	22	20	18	16
Wt.-Lb./S.F.	.906	1.156	1.406	1.656	2.156	2.656	Wt.-Lb./S.F.	.906	1.156	1.406	1.656	2.156	2.656
SMACNA Max. Dimension - Long Side		30"	54"	84"	85" Up		SMACNA Max. Dimension - Long Side		30"	54"	84"	85" Up	
Sum-2 Sides							Sum-2 Sides						
2	.3	.40	.50	.60	.80	.90	56	9.3	12.0	14.0	16.2	21.3	25.2
3	.5	.65	.80	.90	1.1	1.4	57	9.5	12.3	14.3	16.5	21.7	25.7
4	.7	.85	1.0	1.2	1.5	1.8	58	9.7	12.5	14.5	16.8	22.0	26.1
5	.8	1.1	1.3	1.5	1.9	2.3	59	9.8	12.7	14.8	17.1	22.4	26.6
6	1.	1.3	1.5	1.7	2.3	2.7	60	10.0	12.9	15.0	17.4	22.8	27.0
7	1.2	1.5	1.8	2.0	2.7	3.2	61	10.2	13.1	15.3	17.7	23.2	27.5
8	1.3	1.7	2.0	2.3	3.0	3.6	62	10.3	13.3	15.5	18.0	23.6	27.9
9	1.5	1.9	2.3	2.6	3.4	4.1	63	10.5	13.5	15.8	18.3	24.0	28.4
10	1.7	2.2	2.5	2.9	3.8	4.5	64	10.7	13.7	16.0	18.6	24.3	28.8
11	1.8	2.4	2.8	3.2	4.2	5.0	65	10.8	13.9	16.3	18.9	24.7	29.3
12	2.0	2.6	3.0	3.5	4.6	5.4	66	11.0	14.1	16.5	19.1	25.1	29.7
13	2.2	2.8	3.3	3.8	4.9	5.9	67	11.2	14.3	16.8	19.4	25.5	30.2
14	2.3	3.0	3.5	4.1	5.3	6.3	68	11.3	14.6	17.0	19.7	25.8	30.6
15	2.5	3.2	3.8	4.4	5.7	6.8	69	11.5	14.8	17.3	20.0	26.2	31.1
16	2.7	3.4	4.0	4.6	6.1	7.2	70	11.7	15.0	17.5	20.3	26.6	31.5
17	2.8	3.7	4.3	4.9	6.5	7.7	71	11.8	15.2	17.8	20.6	27.0	32.0
18	3.0	3.9	4.5	5.2	6.8	8.1	72	12.0	15.4	18.0	20.9	27.4	32.4
19	3.2	4.1	4.8	5.5	7.2	8.6	73	12.2	15.6	18.3	21.2	27.7	32.9
20	3.3	4.3	5.0	5.8	7.6	9.0	74	12.3	15.8	18.5	21.5	28.1	33.3
21	3.5	4.5	5.3	6.1	8.0	9.5	75	12.5	16.1	18.8	21.8	28.5	33.8
22	3.7	4.7	5.5	6.4	8.4	9.9	76	12.7	16.3	19.0	22.0	28.9	34.2
23	3.8	5.0	5.8	6.7	8.7	10.4	77	12.8	16.5	19.3	22.3	29.3	34.7
24	4.0	5.2	6.0	7.0	9.1	10.8	78	13.0	16.7	19.5	22.6	29.6	35.1
25	4.2	5.4	6.3	7.3	9.5	11.3	79	13.2	16.9	19.8	22.9	30.0	35.6
26	4.3	5.6	6.5	7.5	9.9	11.7	80	13.3	17.1	20.0	23.2	30.4	36.0
27	4.5	5.8	6.8	7.8	10.3	12.2	81	13.5	17.3	20.3	23.5	30.8	36.5
28	4.7	6.0	7.0	8.1	10.6	12.6	82	13.7	17.5	20.5	23.8	31.2	36.9
29	4.8	6.2	7.3	8.4	11.0	13.1	83	13.8	17.8	20.8	24.1	31.5	37.4
30	5.0	6.5	7.5	8.7	11.4	13.5	84	14.0	18.0	21.0	24.4	31.9	37.8
31	5.2	6.7	7.8	9.0	11.8	14.0	85	14.2	18.2	21.3	24.7	32.3	38.3
32	5.3	6.9	8.0	9.3	12.2	14.4	86	14.3	18.4	21.5	24.9	32.7	38.7
33	5.5	7.1	8.3	9.6	12.5	14.9	87	14.5	18.6	21.8	25.2	33.1	39.2
34	5.7	7.3	8.5	9.9	12.9	15.3	88	14.7	18.8	22.0	25.5	33.4	39.6
35	5.8	7.5	8.8	10.2	13.3	15.8	89	14.8	19.0	22.3	25.8	33.8	40.1
36	6.0	7.8	9.0	10.4	13.7	16.2	90	15.0	19.3	22.5	26.1	34.2	40.5
37	6.2	8.0	9.3	10.7	14.1	16.7	91	15.2	19.5	22.8	26.4	34.6	41.0
38	6.3	8.2	9.5	11.0	14.4	17.1	92	15.3	19.7	23.0	26.7	35.0	41.4
39	6.5	8.4	9.8	11.3	14.8	17.6	93	15.5	19.9	23.3	27.0	35.3	41.9
40	6.7	8.6	10.0	11.6	15.2	18.0	94	15.7	20.1	23.5	27.3	35.7	42.3
41	6.8	8.8	10.3	11.9	15.6	18.5	95	15.8	20.3	23.8	27.6	36.1	42.8
42	7.0	9.0	10.5	12.2	16.0	18.9	96	16.0	20.5	24.0	27.8	36.5	43.2
43	7.2	9.2	10.8	12.5	16.3	19.4	97	16.2	20.8	24.3	28.1	36.9	43.7
44	7.3	9.5	11.0	12.8	16.7	19.8	98	16.3	21.0	24.5	28.4	37.2	44.1
45	7.5	9.7	11.3	13.1	17.1	20.3	99	16.5	21.2	24.8	28.7	37.6	44.6
46	7.7	9.9	11.5	13.3	17.5	20.7	100	16.7	21.4	25.0	29.0	38.0	45.0
47	7.8	10.1	11.8	13.6	17.9	21.2	101	16.8	21.6	25.3	29.3	38.4	45.5
48	8.0	10.3	12.0	13.9	18.2	21.6	102	17.0	21.8	25.5	29.6	38.8	45.9
49	8.2	10.5	12.3	14.2	18.6	22.1	103	17.2	22.0	25.8	29.9	39.1	46.4
50	8.3	10.7	12.5	14.5	19.0	22.5	104	17.3	22.3	26.0	30.2	39.5	46.8
51	8.5	11.0	12.8	14.8	19.4	23.0	105	17.5	22.5	26.3	30.5	39.9	47.3
52	8.7	11.2	13.0	15.1	19.8	23.4	106	17.7	22.7	26.5	30.7	40.3	47.7
53	8.8	11.4	13.3	15.4	20.1	23.9	107	17.8	22.9	26.8	31.0	40.7	48.2
54	9.0	11.6	13.5	15.7	20.5	24.3	108	18.0	23.1	27.0	31.3	41.0	48.6
55	9.2	11.8	13.8	16.0	20.9	24.8	109	18.2	23.3	27.3	31.6	41.4	49.1
							110	18.3	23.5	27.5	31.9	41.8	49.5

Example: If duct is 34" x 20" x 15' long, 34" is greater than 30" maximum, for 24 ga. so must be 22 ga. 34" + 20" = 54" going across from 54" find 13.5 lb. per foot. 13.5 x 15' = 202.5 lbs. For S.F. of surface area 202.5 ÷ 1.406 = 144 S.F.

Note: figures include an allowance for scrap.

Figure 18.3

Material Units: The contractor's price per pound of metal ductwork should include a 10 to 15 percent allowance for waste, slips, and hangers. Fiberglass and plastic duct supports must be taken off and priced on an individual (as each) basis. Many other accessories are needed to complete a ductwork system. Supply and return "faces" include diffusers, registers, and grilles. Control devices include dampers and turning vanes.

Labor Units: Labor units for ductwork are unique in comparison to other mechanical trades in that the duct and fittings have to be fabricated as well as installed in place. Metal duct fabrication is normally done at the sheet metal worker's shop and transported to the site for installation. Shop or fabrication labor may be estimated in one of two ways: by the average number of pounds of metal fabricated per man, per hour; or by the number of sheets converted to ductwork per man, per day. Installation labor can also be estimated in one of two ways: by the number of pounds installed per man, per day; or by the linear feet of duct hung or installed per man, per day.

Factors affecting fabrication and installation labor are listed below:
- Job conditions
- Height
- Methods of reinforcement and support
- Duct size and gauge
- The use of sound lining
- Sealing or taping the joint to prevent air leakage

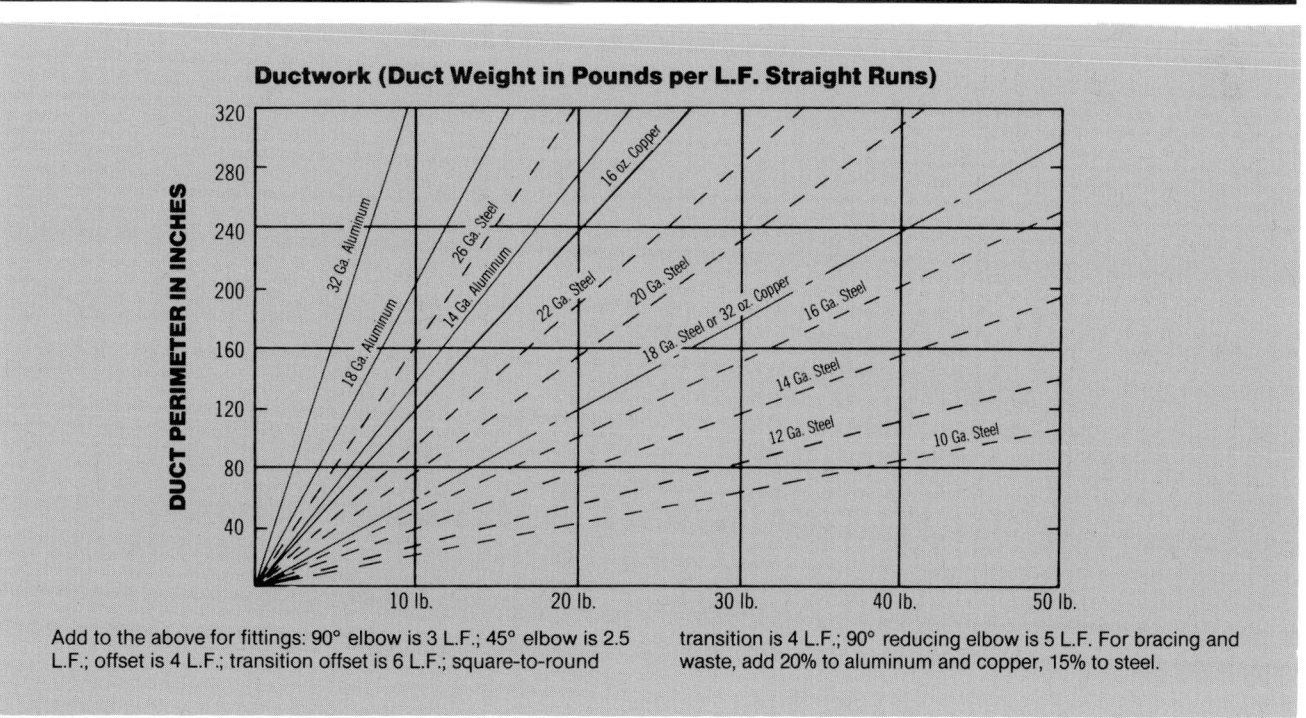

Figure 18.4

For duct installation over 10 feet high, the percentage modifiers shown in Figure 18.6 (for field installation labor only) should be used.

The temperature control contractor is usually responsible for supplying the motorized dampers used in temperature control systems. These dampers must be installed, however, by the sheet metal contractor. A schedule of dampers is supplied by the control company, identifying damper location and function, but final sizing to match the duct dimensions is the sheet metal contractor's responsibility. Other labor concerns may include equipment usually furnished by the prime mechanical contractor but installed by the sheet metal contractor (e.g., duct coils, induction boxes, mixing and terminal boxes, humidifiers, filters, fans, kitchen hoods, etc.).

Local labor practices may also require that a split crew of tin knockers and pipefitters install supply air handling equipment if coils are involved. The designer may not be aware of such local practices, especially if the design firm is from another locality. The sheet metal estimator should become familiar with all of the mechanical specifications to ensure that any "gray areas" of responsibility are properly defined, so as to avoid jurisdictional work stoppages during construction. Electrical and ceiling plans should be read in case ceiling troffer or lighting fixtures are to be integrated with air handling systems. Architectural plans and specifications might contain information about louvers and wall or roof openings, which could affect the sheet metal estimate as well. Again, be sure to check the job specifications carefully.

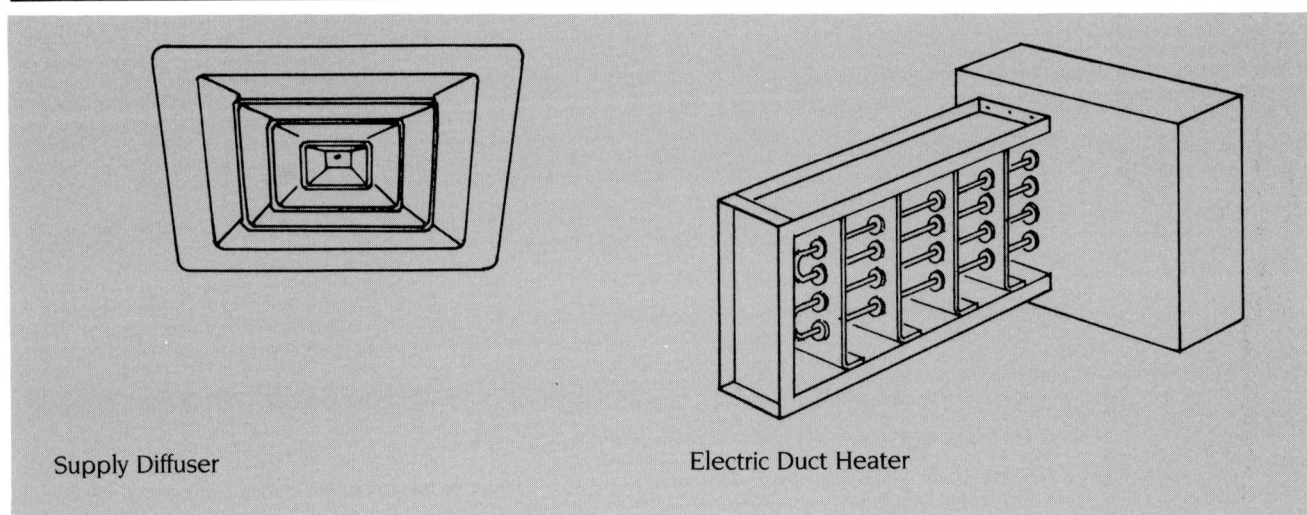

Supply Diffuser　　　　Electric Duct Heater

Figure 18.5

Takeoff Procedure: The takeoff should begin at the source, the mechanical room, or wherever the supply units or exhaust fans are located. This will help the estimator to visualize from the start the number of systems and zones to be contended with. The number and location of ducts to be lined should be noted. Ductwork should be listed on a separate takeoff sheet, recorded by size and length. Each system should be listed individually, recording lined duct, insulated duct, and bare duct separately. A ductwork takeoff form, designed for this purpose, was shown in Figure 4.2.

Each system should be marked on the plans in a different colored pencil. Different materials should be listed separately, if, for example, galvanized and aluminum are both used in the same system. Ductwork should be measured straight through, rather than taking off each fitting. A 10 to 50 percent allowance for fittings must then be added to the ductwork totals. This percentage should be based on the fitting frequency indicated, the system complexity, and the estimator's own experience. The estimator with no company experience factor to refer to should take off and record each fitting, and then use a standard conversion factor for each. This data should be kept until the estimator's experience factors have been established.

After the ductwork has been taken off, the specialties (e.g., supply and return devices or faces, dampers, access doors, turning vanes, extractors, etc.) should be taken off. The footage totals can be divided by 4' or 8' lengths to obtain the number of field joints to be made up. The drawings should be reread at this point, with attention paid to the building perimeter. Items without duct connections, but which are the sheet metal workers' responsibility to furnish or install, should be identified. For example, if louvers are built in to a prefabricated panel wall, the sheet metal contractor should check the architectural details and specifications for his involvement, as these items may have been ignored in the mechanical specifications. Other specialty items might be propeller or wall type fans, wall boxes, sleeves, drain pans, or radiator and convector casings.

| Height Modifications for Ductwork Installation ||
Height Above Floor Level	Modification to Labor
10 to 15 feet	10 percent
15 to 20 feet	20 percent
20 to 25 feet	25 percent
25 to 30 feet	35 percent
31 feet and up	50 percent

Figure 18.6

Chapter 19
INSULATION

Thermal insulation is used in mechanical systems to prevent heat loss or gain and to provide a vapor barrier for piping and ductwork systems, boilers, tanks, chillers, heat exchangers, air handling equipment casings, and the like. Most boilers, water heaters, chillers, and air handling units are provided with insulated metal jackets and require little, if any, field insulation.

Insulation is available in rigid or flexible form and is manufactured from fiberglass, cellular glass, rock wool, polyurethane foam, closed cell polyethylene, flexible elastomeric, rigid calcium silicate, phenolic foam, or rigid urethane.

Insulation is produced in a variety of wall thicknesses and may be applied in layers for extreme temperatures. Rigid board or blocks, or flexible blanket are used for ductwork and equipment. Preformed sections are available for use with pipe.

The standard length for pipe insulation has always been 3'. This is still true for fiberglass and calcium silicate, but foam insulations (e.g., polyurethane, polyethylene, and urethane) are produced in 4' lengths. This means that 1/3 less butt joints will have to be made in straight runs of pipe. Flexible elastomeric insulation is shipped in 6' lengths.

Units of Measure: Pipe insulation is measured by the linear foot. Fittings, joints, etc., are measured as each. Ductwork, breeching, and equipment insulation, whether block, board, or blanket, are measured by the square foot.

Material Units: The materials required to apply and seal insulation will vary, depending on the type of insulation and exterior finish specified, (e.g., fire retardant, vapor barrier, weatherproof, etc.). For some insulation, these special finishes are applied at the factory, while others require field application.

Required materials may also include preformed fitting, valve, and flange covers unless the insulation is to be mitered or cut in the field to allow for fittings and valves. In addition, butt joining strips, tape, wire, adhesives, stick clips, or welding pins, for adhering the insulation and protective jacketing, are required depending on the specified materials.

Labor Units: The simplest pipe covering installation method is slipping elastomeric insulation over straight sections of pipe or tubing, prior to pipe installation. This insulation can be formed around bends and elbows.

When the piping is already installed, the insulation must be slit lengthwise, placed, and both the butt joints and seams (at the slit) must be joined with a contact adhesive. This process requires more labor than does the simple slipping on method described above. Fittings, in this case, are covered by mitering or cutting a hole for a tee or a valve bonnet. Fiberglass insulation is already slit for placement around the pipe, and fittings are mitered. From straight sections, holes are cut for tees and valve bonnets using a knife. Elastomeric insulation usually does not require any additional finish.

A variety of jackets and fittings are available for all types of insulation, including roofing felt wired in place and preformed metal jackets used as closures where exposed to weather. Pre-molded fitting covers are available to give fittings and valves a finished appearance. Factory applied self sealing jackets are an advantage of fiberglass insulation. Flanges, which are removed frequently, often require unique insulated metal jackets or boxes which are fabricated by the insulation contractor.

Calcium silicate insulation is often specified for higher temperatures (above 850 degrees Fahrenheit). Installation of calcium silicate is more labor intensive than installation of fiberglass or elastomeric insulation. This rigid insulation is made in half sections for pipe and in 3' lengths. Multiple segments must be used in layers to achieve larger outside diameters. The segments are wired in place 9" on center, normally requiring two workers. Fittings for calcium silicate insulation are made by mitering sections using a saw. Flanges and valves require oversized sections cut to fit. These valve and fitting covers are wired in place and finished with a troweled coat of insulating cement. This type of insulation has a high waste factor, due to crumbling and breaking during cutting and ordinary use.

Special thicknesses of any insulation must be obtained by adding multiple layers of oversized insulation, and sealing and securing using the same method as was used for the base layer. This additional work should be included in the labor estimate.

Hangers and supports require special treatment for insulated piping systems. The pipe hanger or support must be placed outside the insulation, to allow for expansion and contraction in steam or hot water systems, and to maintain the vapor barrier in cold water systems. As a result, oversized supports should be used to fit the O.D. of the insulation, rather than the pipe.

For heated piping, preformed steel segments are welded to the bottom of the pipe at each point of support. These saddles are sized according to the insulation thickness plus the pipe O.D. For cold water (anti-sweat) systems, where no metal to metal contact between pipe and support can be tolerated, the hanger is oversized to allow for placement of the insulation. The insulation is then protected by a sheet metal shield which matches the outer radius of the insulation.

For flexible insulation, which cannot support the weight of the pipe, a rigid insert is substituted for the lower section of insulation at each point of support. This insert may be formed from calcium silicate insulation, cork, or even wood. The insulation jacket must enclose this insert within the vapor proof envelope. The labor for insulation around hangers should be estimated carefully. The insert, for example, might be installed by the pipe coverer, while the metal shield or protector is furnished by the pipefitter or plumber.

Duct and equipment insulation may be a wraparound blanket type insulation, wired in place and sealed with adhesive or tape. Rigid board is secured by the use of pins, secured to the duct exterior with mastic or spot welded into place. The insulation sheets are pressed onto these pins and secured with self locking washers.

The butt joints and seams are sealed with adhesive and tape. Block and segment insulation, when installed on round or irregular shapes, is wired on, and then covered with a chicken wire mesh and coated with a troweled application of insulation cement.

Labor units for installing and finishing insulation will vary from the simplest procedure to more complex and time consuming processes. The plans and specifications should be read carefully to determine the type of insulation required. This will help assure that the method and sequence of installation labor is properly estimated.

Takeoff Procedure: Pipe insulation is recorded by size on a specialized takeoff sheet which is also used to record fittings and valves similar to the takeoff method for the pipe itself. A separate takeoff sheet should be used for each system, broken down by thickness and exterior finishes required. Fittings are converted to linear feet of pipe for pricing purposes, as shown in Figure 19.1.

Ductwork insulation is taken off in the same manner as the duct itself, then converted to square feet for pricing. Separate forms should be used to keep the ductwork insulation takeoff separate from the equipment insulation takeoff. The ductwork takeoff form (shown in Chapter 4, Figure 4.2), the piping schedule (shown in Chapter 4, Figure 4.1), or the Quantity Sheet (shown in Chapter 8, Figure 8.1) might be employed for this purpose. In lieu of performing the takeoff, or to double check the takeoff, the estimator will often contact the sheet metal and/or piping contractors to obtain or verify quantities for ductwork insulation.

Conversion of Fittings into Equivalent Feet of Straight Pipe for Estimating Insulation:	
Fitting	Equivalent Pipe Footage
Elbows or Bends	Add 3 feet
Pair of Flanges	Add 3 feet
Reducing Couplings	Add 2 feet

Figure 19.1

Chapter 20
AUTOMATIC TEMPERATURE CONTROLS

Temperature control systems range from simple thermostats and boiler or furnace operating and limit controls to complex electronic HVAC controls integrated with energy management systems which are a part of the "intelligent building concept". To regulate indoor temperatures, the control system responds to indoor/outdoor temperature changes by opening, closing, and modulating valves and dampers (shown in Figure 20.1) in heating and cooling piping or ductwork. This includes, when necessary, resetting the temperature of the heating or cooling medium. Control systems also provide automatic temperature adjustments to correspond to occupied and unoccupied periods of building use. The temperature control system itself may be pneumatically, electrically, or electronically operated, or any combination thereof.

Units of Measure: Distances for wiring or tubing are recorded and priced by the linear foot. Control devices, such as controllers or sensors, are recorded as each. Complete temperature control systems are quoted as each, on either an "installed" or "supervised" basis.

Material Units: The experienced estimator should know which pieces of HVAC equipment to be furnished by the mechanical contractor will include temperature controls. Self-contained control valves will be furnished complete with built in thermostats and sensing elements. A packaged air conditioner, heat pump, warm air furnace, or boiler will arrive with operating and limit controls installed and pre-wired. Remote thermostats, programmable controllers, or outdoor sensors should be handled as optional equipment for field installation and wiring, and should be ordered as each if not part of a package.

Depending on the project specifications, the temperature control system could be bid by a manufacturer as a turn-key installation. In this case, the temperature control bidder, or his representative, would assume complete responsibility for the installation.

Alternatively, the specifications may call for the temperature control manufacturer to bid the material only. Installation might then be performed by the mechanical or electrical subcontractor, or by a subcontractor specializing in control work. In this case, the installer must provide the necessary material to connect and hook up the control equipment. The control manufacturer will generally provide supervision for installation of his equipment.

Servicing is one important factor that should be considered with all temperature control systems, regardless of installation method. Most of the callbacks on an HVAC system are for adjustment or calibration of the temperature controls. For this reason, the proposed service agreement with the controls manufacturer should be carefully read.

Labor Units: Labor units for valve installation and sensing element wells, which are installed by the pipefitter, have already been covered in Chapter 13, "Piping". Control dampers and ductmounted controllers are installed by the sheet metal worker (see Chapter 18, "Ductwork"). Motorized operators for these dampers and valves are field installed by the control contractor. These operators may be either electric or pneumatic. If the system is pneumatic, the tubing will be installed by a pipefitter employed by the control contractor. If the system is electric or electronic, it should be wired by an electrician under the supervision of the control contractor.

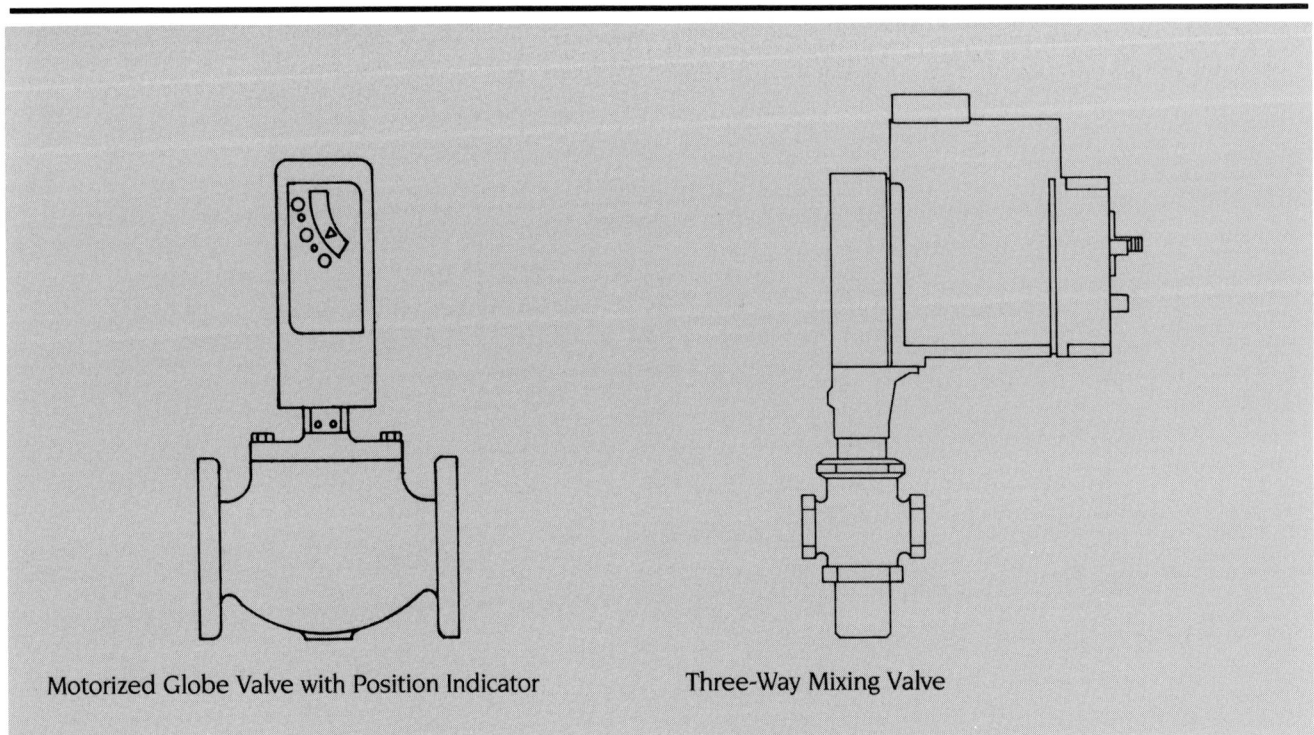

Motorized Globe Valve with Position Indicator   Three-Way Mixing Valve

Figure 20.1

Takeoff Procedure: Plans and specifications should be reviewed to determine placement and frequency of controls. The control specifications will dictate what has to be controlled and how. The drawings will indicate how many control points there should be and where they should be located.

After the specifications and the design sequence of operation have been read, a list should be made of control points. An average of forty or fifty feet of tubing or wiring per control device should be anticipated. For shorter runs it may be better to scale off the actual distances. Some estimators may even prefer to make an actual tubing or wiring layout, rather than rely on averages. A wiring layout or control diagram is essential if the estimator has to obtain pricing from an electrician.

Temperature control systems can be so complex and specialized that the majority of mechanical contractors will subcontract out each job to a company that specializes in the particular installation and service of the type of system specified. If a subcontractor or several subcontractors are used, the mechanical contractor will be responsible for their installation. For this reason, supervision time should be included to handle this requirement.

Chapter 21
SUPPORTS, ANCHORS, AND GUIDES

Supports, anchors, and guides must be fabricated on site for the many occasions when there is no stock item available to meet the job requirement. Heat exchangers, tanks, coils, fan coil units, and piping may, at some time or another, need job fabricated supports.

Some relatively simple supports can be made from perforated channel metal framework and the accompanying fittings and brackets which bolt together. More substantial supports, however, must be fabricated from steel structural shapes (i.e., angles, tees, channels, and beams). Figure 21.1 shows the properties of the various steel shapes available.

The steel shapes are priced by the pound and measured or taken off by the linear foot. To convert linear feet into pounds, the estimator should refer to a handbook or stock list readily available from any steel distributor or warehouse. A table for estimating angle lengths is shown in Figure 21.2. If a handbook is not on hand, an approximation can be made from the fact that 12 cubic inches of steel weighs 3.4 lbs. This means that a piece of steel 1" x 1" and 12" long, a 1/4" plate 4" wide by 12" long, or a 2" x 2" x 1/4" angle 12" long each weigh approximately 3.4 lbs. The following example demonstrates the method of estimating a field fabricated stand for a vertical plate type heat exchanger. The estimator first prepares a sketch indicating stock sizes and material lengths. The dimensions are shown in Figure 21.3.

Angle is stocked in 20' lengths but can be purchased in 5' increments. Therefore the material for this stand, using 4" x 4" x 1/4" angle shape steel, would be the following:

Material: 30 L.F. @ 6.6 lbs./ft. = 198 lbs. @ $.56	= $110.88
Fabricate: Cut to length, 8 pieces @ 1/4 hr. = 2 hrs. @ $23.30	= $46.60
Weld: 4 feet @ 4 L.F./hr. = 1 hour @ $23.30	= $23.30
Oxygen & Acetylene: 2 hours @ $5.80	= $11.60
Rental 300 AMP Arc Welder: 1-hour @ $4.20	= $4.20
TOTAL	$196.58

Hot rolled structural steel is available in the following shapes.

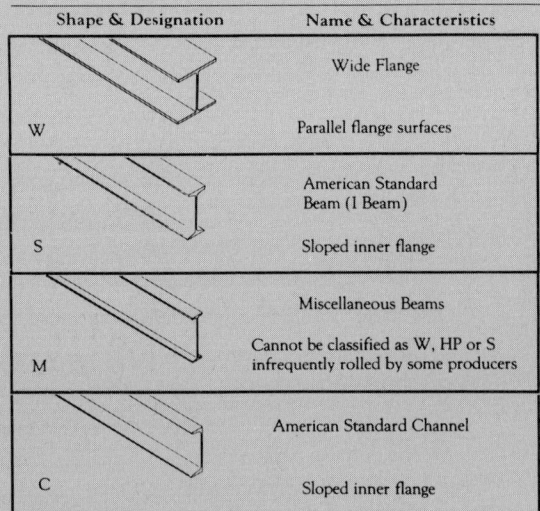
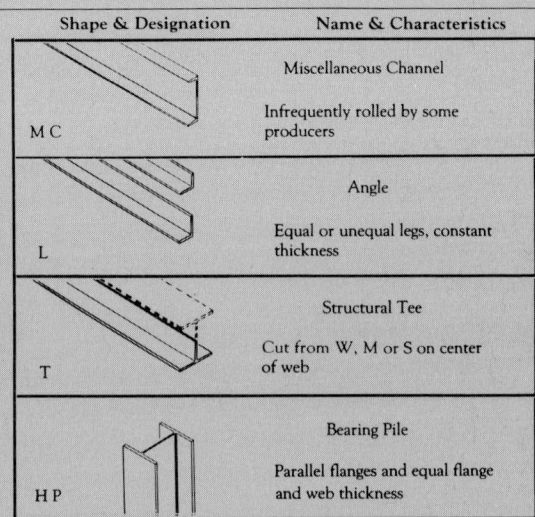

Common drawing designations follow.

Wide Flange
W 18 x 35 ← Weight in Pounds Per Foot
  └ Nominal Depth in Inches (Actual 17-3/4")

American Standard Beam
S 12 x 31.8
  │ └ Weight in Pounds Per Foot
  └ Depth in Inches

Miscellaneous Beam
M 8 x 6.5
  │ └ Weight in Pounds Per Foot
  └ Depth in Inches

American Standard Channel
C 8 x 11.5
  │ └ Weight in Pounds Per Foot
  └ Depth in Inches

Miscellaneous Channel
MC 8 x 22.8
  │ └ Weight in Pounds Per Foot
  └ Depth in Inches

Angle
  │ ┌ Length of One Leg in Inches
  L 6 x 3-1/2 x 3/8 ← Thickness of Each Leg in Inches
      └ Length of Other Leg in Inches

Tee Cut From W16 x 100
WT 8 x 50
  │ └ Weight in Pounds Per Foot
  └ Nominal Depth in Inches (Actual 8-1/2")

Tee Cut From S 12 x 35
ST 6 x 17.5
  │ └ Weight in Pounds Per Foot
  └ Depth in Inches

Tee Cut From M 10 x 9
MT 5 x 4.5
  │ └ Weight in Pounds Per Foot
  └ Depth in Inches

Bearing Pile
HP 12 x 84
  │ └ Weight in Pounds Per Foot
  └ Nominal Depth in Inches (Actual 12-1/4")

Hot rolled structural shapes are generally available in the following ASTM specifications.

Steel Type	ASTM Designation	Minimum Yield Stress KSI	Characteristics
Carbon	A36	36	
	A529	42	
High-Strength Low Alloy	A441	50	Structural Manganese Vanadium Steel
	A572	42	Columbium-Vanadium Steel
		50	
		60	
		65	
Corrosion Resistant High Strength Low Alloy	A242	50	Corrosion Resistant
	A588	50	Corrosion Resistant to 4" Thick

Figure 21.1

Angles
Equal Legs and Unequal Legs
Properties for Designing

Size and Thickness (in.)		Weight per Foot (lb.)	Size and Thickness (in.)		Weight per Foot (lb.)
L 9 x 4 x	5/8	26.3	L 4 x 4 x	3/4	18.5
	9/16	23.8		5/8	15.7
	1/2	21.3		1/2	12.8
L 8 x 8 x	1⅛	56.9		7/16	11.3
	1	51.0		3/8	9.8
	7/8	45.0		5/16	8.2
	3/4	38.9		1/4	6.6
	5/8	32.7	L 4 x 3½ x	5/8	14.7
	9/16	29.6		1/2	11.9
	1/2	26.4		7/16	10.6
L 8 x 6 x	1	44.2		3/8	9.1
	7/8	39.1		5/16	7.7
	3/4	33.8		1/4	6.2
	5/8	28.5	L 3 x 2½ x	1/2	8.5
	9/16	25.7		7/16	7.6
	1/2	23.0		3/8	6.6
	7/16	20.2		5/16	5.6
L 8 x 4 x	1	37.4		1/4	4.5
	3/4	28.7		3/16	3.39
	9/16	21.9	L 3 x 2 x	1/2	7.7
	1/2	19.6		7/16	6.8
L 7 x 4 x	3/4	26.2		3/8	5.9
	5/8	22.1		5/16	5.0
	1/2	17.9		1/4	4.1
	3/8	13.6		3/16	3.07
L 5 x 3½ x	3/4	19.8	L 2½ x 2½ x	1/2	7.7
	5/8	16.8		3/8	5.9
	1/2	13.6		5/16	5.0
	7/16	12.0		1/4	4.1
	3/8	10.4		3/16	3.07
	5/16	8.7	L 2½ x 2 x	3/8	5.3
	1/4	7.0		5/16	4.5
L 5 x 3 x	5/8	15.7		1/4	3.62
	1/2	12.8		3/16	2.75
	7/16	11.3	L 2 x 2 x	3/8	4.7
	3/8	9.8		5/16	3.92
	5/16	8.2		1/4	3.19
	1/4	6.6		3/16	2.44
				1/8	1.65

Figure 21.2

This estimate does not include overhead and profit. Priming and finish painting are to be accomplished by the painting contractor. Estimates for field fabricated items may be priced using the appropriate sections (1.5 and 5.4) of *Means Mechanical Cost Data*.

Units of Measure: Piping supports, anchors, and guides may be listed as each according to pipe size. Equipment supports may be listed as each. These items may all be totalled, however, and the material priced by the pound.

Material Units: Costs for these field fabricated items, in addition to the plate, rods, and shapes, include oxygen and acetylene for cutting and shaping, electrodes for welding, bolts, nuts, and other anchoring devices.

Labor Units: The following factors contribute to the labor units for supports, anchors, and guides:
- Physical unloading of the steel
- Cutting to desired lengths
- Relocation of the assembled units into place
- Responsibility for prime or finished coats of paint

Takeoff Procedure: From piping and equipment takeoffs, the estimator can determine where auxiliary steel is required. Mechanical sections and details, when shown, will also specify where supports, anchors, and guides are required. It is usually left to the estimator's judgement to size these fabrications, often estimated using rough sketches.

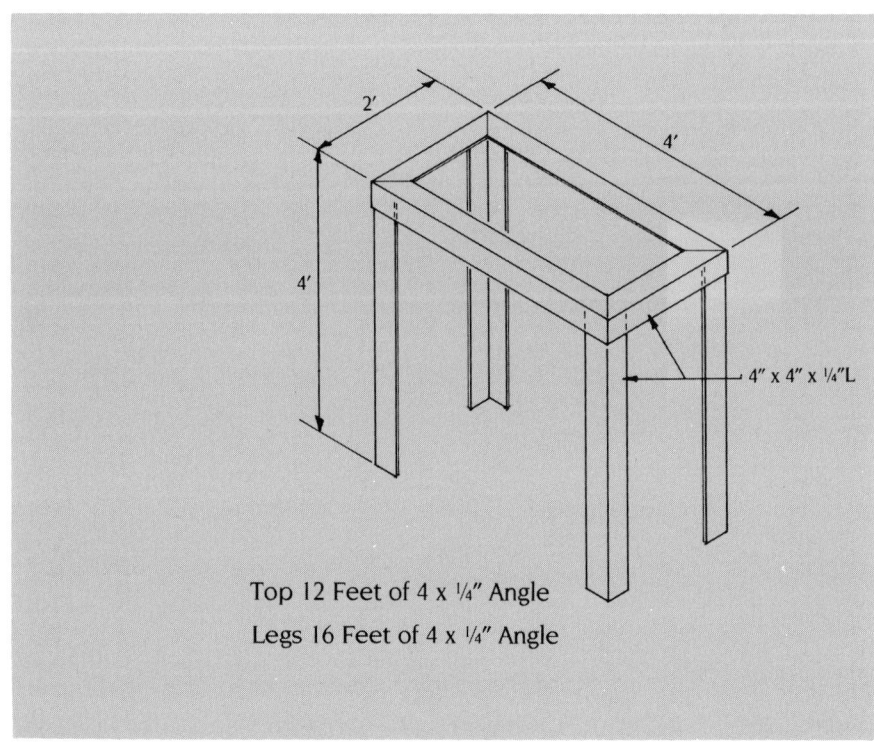

Figure 21.3

Part III
SAMPLE ESTIMATES

Chapter 22
USING MEANS MECHANICAL COST DATA

Users of *Means Mechanical Cost Data* are chiefly interested in obtaining quick, reasonable, average prices for mechanical construction items. This is the primary purpose of the annual book — to eliminate guesswork when pricing unknowns. Many persons use the cost data, however, for bids, verification of quotations, or budgets, without being fully aware of how the prices are obtained and derived. Without this knowledge, this resource is not being used to fullest advantage. In addition to the basic cost data, the book also contains a wealth of information to aid the estimator, the contractor, the designer, and the owner to better plan and manage mechanical construction projects. Productivity data is provided in order to assist with scheduling. National labor rates are analyzed. Tables and charts for location and time adjustments are included and help the estimator tailor the prices to a specific location. The costs in *Means Mechanical Cost Data* consist of over 18,000 unit price line items, as well as prices for 3000 mechanical systems. This information, organized according to the Construction Specification Institute Masterformat Divisions, also provides an invaluable checklist to the construction professional to assure that all required items are included in a project.

Format and Data

The major portion of *Means Mechanical Cost Data* is the Unit Price section, Section A. This is the primary source of unit cost data and is organized according to the CSI division index. This index was developed by representatives of all parties concerned with the building construction industry and has been accepted by the American Institute of Architects (AIA), the Associated General Contractors of America, Inc. (AGC), and the Construction Specifications Institute, Inc. (CSI). In *Means Mechanical Cost Data*, relevant parts of other divisions are included along with Division 15 — Mechanical. For example, items from Divisions 1, 2, 3, 4, 5, 6, 7, 8, 10, 11, 13, and 16 all appear in addition to the Division 15 entries.

CSI Masterformat Divisions:
Division 1 - General Requirements
Division 2 - Site Work
Division 3 - Concrete
Division 4 - Masonry
Division 5 - Metals
Division 6 - Wood & Plastics
Division 7 - Moisture-Thermal Control

Division 8 - Doors, Windows & Glass
Division 9 - Finishes
Division 10 - Specialties
Division 11 - Equipment
Division 12 - Furnishings
Division 13 - Special Construction
Division 14 - Conveying Systems
Division 15 - Mechanical
Division 16 - Electrical

In addition to the sixteen CSI divisions of the Unit Price section, Division 17, Square Foot and Cubic Foot Costs, presents consolidated data from over 10,500 actual reported construction projects and provides information based on total project costs as well as costs for major components. Division 18 contains data pertinent to repair and remodeling projects. Division 18 also contains percentage "factors" that may be applied to account for added costs where the normal progression of work may be altered due to the restrictive conditions of existing structures.

Section B, Assemblies Cost Tables, contains over 3,000 costs for mechanical and appropriate related assemblies, or systems. Components of the systems are fully detailed and accompanied by illustrations.

Section C contains tables and reference charts. It also provides estimating procedures and explanations of cost development which support and supplement the unit price and systems cost data. Also in Section C are the City Cost Indexes, representing the compilation of construction data for 162 major U. S. and Canadian cities. Cost factors are given for each city, by trade, relative to the national average.

The prices presented in *Means Mechanical Cost Data* are national averages. Material and equipment costs are developed through annual contact with manufacturers, dealers, distributors, and contractors throughout the United States and Canada. Means' staff of engineers is constantly updating prices and keeping abreast of changes and fluctuations within the industry. Labor rates are the national average of each trade as determined from union agreements from thirty major U. S. cities. Throughout the calendar year, as new wage agreements are negotiated, labor costs should be factored accordingly.

Following is a list of factors and assumptions on which the costs presented in *Means Mechanical Cost Data* have been based:

- **Quality:** The costs are based on methods, materials, and workmanship in accordance with U. S. Government standards and represent good, sound construction practice.
- **Overtime:** The costs, as presented, include *no* allowance for overtime. If overtime or premium time is anticipated, labor costs must be factored accordingly.
- **Productivity:** The daily output and man-hour figures are based on an eight hour workday, during daylight hours. The charts in Figures 11.7 and 11.8 shows that as the number of hours worked per day (over eight) increases, and as the days per week (over five) increase, production efficiency decreases. (See Chapter 11, "Cost Control and Analysis".)
- **Size of Project:** Costs in *Mechanical Cost Data* are based on commercial and industrial buildings for which total project costs are $400,000 and up. Large residential projects are also included.

- **Local Factors:** Weather conditions, season of the year, local union restrictions, and unusual building code requirements can all have a significant impact on construction costs. The availability of a skilled labor force, sufficient materials, and even adequate energy and utilities will also affect costs. These factors vary in impact and are not necessarily dependent upon location. They must be reviewed for each project in every area.

In presenting prices in *Means Mechanical Cost Data*, certain rounding rules are employed to make the numbers easy to use without significantly affecting accuracy. The rules are used consistently and are as follows:

Prices From	To	Rounded to nearest
$ 0.01	$ 5.00	0.01
5.01	20.00	0.05
20.01	100.00	1.00
100.01	1,000.00	5.00
1,000.01	10,000.00	25.00
10,000.01	50,000.00	100.00
50,000.01	up	500.00

Section A — Unit Price Costs

The Unit Price section of *Means Mechanical Cost Data* contains a great deal of information in addition to the unit cost for each construction component. Figure 22.1 is a typical page showing costs for fire valves. Note that prices are included for several types of valves, each in a wide range of size and capacity ratings. In addition, appropriate crews, workers, and productivity data are indicated. The information and cost data is broken down and itemized in this way to provide for the most detailed pricing possible. Both the unit price and the systems sections include detailed illustrations. The reference numbers enclosed in the squares refer the user to an appropriate assemblies section or reference table.

Within each individual line item, there is a description of the construction component, information regarding typical crews designated to perform the work, and productivity shown as daily output and as man-hours. Costs are presented in two ways: "bare", or unburdened costs, and costs with mark-ups for overhead and profit. Figure 22.2 is a graphic representation of how to use the Unit Price section as presented in *Means Mechanical Cost Data*.

Line Numbers

Every construction item in the Means unit price cost data books has a unique line number. This line number acts as an address so that each item can be quickly located and/or referenced. The numbering system is based on the CSI Masterformat classification by division. In Figure 22.2, note the bold number in reverse type, "15.1". This number represents the major subdivision, in this case "Pipe and Fittings", of the major CSI Division 15 — Mechanical. Within each subdivision, the data is broken down into major classifications. These major classifications are listed alphabetically and are designated by bold type for both numbers and descriptions. Each item, or line, is further defined by an individual number. As shown in Figure 22.2, the full line number for each item consists of: a major CSI subdivision number — a major classification number — an item line number. Each full line number describes a unique construction element. For example, in Figure 22.1, the line number for a 2", slow close, bronze butterfly valve is 15.4-500-1180.

15.4 Fire Extinguishing Systems

		CREW	DAILY OUTPUT	MAN-HOURS	UNIT	BARE COSTS MAT.	BARE COSTS LABOR	BARE COSTS EQUIP.	BARE COSTS TOTAL	TOTAL INCL O&P	
400	3900	2000 GPM, 100 psi, 200 HP, 1770 RPM, 6" pump	Q-13	.34	94.120	Ea.	16,950	2,150		19,100	21,800	400
	3950	2000 GPM, 150 psi, 300 HP, 1770 RPM, 6" pump		.28	114		24,550	2,600		27,150	30,800	
	4000	2500 GPM, 100 psi, 250 HP, 1770 RPM, 8" pump		.30	107		23,500	2,425		25,925	29,400	
	4040	2500 GPM, 135 psi, 350 HP, 1770 RPM, 8" pump		.26	123		26,800	2,800		29,600	33,600	
	4100	3000 GPM, 100 psi, 250 HP, 1770 RPM, 8" pump		.28	114		25,500	2,600		28,100	31,900	
	4150	3000 GPM, 140 psi, 450 HP, 1770 RPM, 10" pump		.24	133		34,950	3,050		38,000	42,900	
	4200	3500 GPM, 100 psi, 300 HP, 1770 RPM, 10" pump		.26	123		29,200	2,800		32,000	36,200	
	4250	3500 GPM, 140 psi, 450 HP, 1770 RPM, 10" pump		.24	133		33,200	3,050		36,250	41,000	
	5000	For jockey pump 1", 3 HP, add	Q-12	2	8		1,250	170		1,420	1,625	
500	0010	**FIRE VALVES**										500
	0020	Angle, combination pressure adjustable/restricting, rough brass										
	0030	1-½"	1 Spri	12	.667	Ea.	30.50	15.90		46.40	57	
	0040	2-½"	"	7	1.140		69	27		96	115	
	0050	For polished brass, add					30%					
	0060	For polished chrome, add					40%					
	0080	Wheel handle, 300 lb., 1-½"	1 Spri	12	.667		21	15.90		36.90	46	
	0090	2-½"	"	7	1.140		48	27		75	93	
	0100	For polished brass, add					35%					
	0110	For polished chrome, add					50%					
	1000	Ball drip, automatic, rough brass, ½"	1 Spri	20	.400		6	9.55		15.55	21	
	1010	¾"	"	20	.400		9	9.55		18.55	24	
	1100	Butterfly, 175 lb., sprinkler system, FM/UL, threaded, bronze										
	1120	Slow close										
	1150	1" size	1 Spri	19	.421	Ea.	42.80	10.05		52.85	62	
	1160	1-¼" size		15	.533		48.40	12.70		61.10	72	
	1170	1-½" size		13	.615		62.50	14.70		77.20	90	
	1180	2" size		11	.727		81.70	17.35		99.05	115	
	1190	2-½" size	Q-12	15	1.070		117	23		140	160	
	1230	For supervisory switch kit, all sizes										
	1240	One circuit, add	1 Spri	48	.167	Ea.	41.30	3.98		45.28	51	
	1250	Two circuits, add	"	40	.200	"	49	4.77		53.77	61	
	1280	Quarter turn for trim										
	1300	½" size	1 Spri	22	.364	Ea.	4.45	8.65		13.10	17.60	
	1310	¾" size		20	.400		6.80	9.55		16.35	21	
	1320	1" size		19	.421		8.85	10.05		18.90	24	
	1330	1-¼" size		15	.533		14.75	12.70		27.45	35	
	1340	1-½" size		13	.615		18.40	14.70		33.10	42	
	1350	2" size		11	.727		22.75	17.35		40.10	50	
	1400	Caps, polished brass with chain, ¾"					21			21	23	
	1420	1"					21			21	23	
	1440	1-½"					7			7	7.70	
	1460	2-½"					11			11	12.10	
	1480	3"					14.50			14.50	15.95	
	3000	Gate, hose, wheel handle, N.R.S., rough brass, 1-½"	1 Spri	12	.667		60	15.90		75.90	89	
	3040	2-½", 300 lb.	"	7	1.140		78	27		105	125	
	3080	For polished brass, add					40%					
	3090	For polished chrome, add					50%					
	3800	Hydrant, screw type, crank handle, brass										
	3840	2-½" size	Q-12	11	1.450	Ea.	185	31		216	250	
	3880	For chrome, same price										
	4200	Hydrolator, vent and draining, rough brass, 1-½"	1 Spri	12	.667		34.10	15.90		50	61	
	4220	2-½"	"	7	1.140		80	27		107	130	
	4280	For polished brass, add					25%					
	4290	For polished chrome, add					35%					
	5000	Pressure restricting, adjustable rough brass, 1-½"	1 Spri	12	.667		52.75	15.90		68.65	81	
	5020	2-½"	"	7	1.140		89.25	27		116.25	140	
	5080	For polished brass, add					30%					
	5090	For polished chrome, add					45%					
	6000	Roof manifold, horiz., brass, with valves & caps										

Figure 22.1

HOW TO USE UNIT PRICE PAGES

Important
Prices in this section are listed in two ways: as bare costs and as costs including overhead and profit of the installing contractor. In most cases, if the work is to be subcontracted, it is best for a general contractor to add an additional 10% to the figures found in the column titled **"TOTAL INCL. O&P"**.

Unit
The unit of measure listed here reflects the material being used in the line item. For example: values are expressed as each (Ea.).

Productivity
The daily output represents typical total daily amount of work that the designated crew will produce. Man-hours are a unit of measure for the labor involved in performing a task. To derive the total man-hours for a task, multiply the quantity of the item the involved times the man-hour figure shown.

Line Number Determination
Major subdivision is **15.1** (two digits plus decimal point plus last digit)

Major classification within subdivision is **820** (three digits)

Item line number is **2320** (four digits)

Complete line number is **15.1-820-2320**

Description
The meaning of this line item shows an iron body gate valve will be installed by a Q-2 crew at a rate of 2.5 per day. (9.6 man-hours per valve).

Crew Q-2

Crew No.	Bare Costs				Incl. Subs O & P		Cost Per Man-hour	
	Hr	Daily		Hr	Daily		Bare Costs	Incl O&P
Crew Q-2								
2 Plumbers	$23.00	$368.00		$33.55	$536.80		$21.46	$31.31
1 Plumber Apprentice	18.40	147.20		26.85	214.80			
24 M.H. Daily Totals		$515.20			$751.60		$21.46	$31.31

Bare Costs are developed as follows for line no. **15.1-820-2320**
Mat. is **Bare Material Cost ($515.00)**
Labor for Crew Q2 = Man-hour Cost (**$21.46**) × Man-hour Units (**9.600**) = **$205.00**
Equip. for Crew Q2 = Equip. Hour Cost (**$0**) × Man-hour Units (**9.600**) = **$0**
Total = **Mat. Cost ($515)** + **Labor Cost ($205.00)** + **Equip. Cost ($0)** = **$720.00**.
(**Note:** Equipment and Labor costs are derived from the Crew Tables. See example at the top of this page.)

Total Costs Including O&P are developed as follows:
Mat. is **Bare Material Cost** + 10% = **$515.00** + **$51.50** = **$566.50**
Labor for Crew Q2 = Man-hour Cost (**$31.31**) × Man-hour Units (**9.600**) = **$300.58**
Equip. for Crew Q2 = Equip. Hour Cost (**$0**) × Man-hour Units (**9.600**) = **$0**
Total = **Mat. Cost ($566.50)** + **Labor Cost ($300.58)** + **Equip. Cost ($0)** = **$867.08** (Rounded to **$865.00**)
(**Note:** Equipment and Labor costs are derived from the Crew Tables. See example at top of this page. Total line follows the rounding rules.)

15.1 Pipe & Fittings

			CREW	DAILY OUTPUT	MAN-HOURS	UNIT	MAT.	LABOR	EQUIP.	TOTAL	TOTAL INCL O&P	
820	0010	VALVES, IRON BODY										820
	0020	For grooved joint, see Division 15.1 610										
	1020	Butterfly, wafer type, lever actuator										
	1030	2" size	1 Plum	14	.571	Ea.	49	13.15		62.15	73	
	1040	2-½" size	Q-1	9	1.780		57	37		94	115	
	1050	3" size		8	2		59	41		100	125	
	1060	4" size		5	3.200		75	66		141	180	
	1070	5" size	Q-2	5	4.800		95	105		200	255	
	1080	6" size		5	4.800		120	105		225	280	
	1090	8" size		4.50	5.330		160	115		275	345	
	1100	10" size		4	6		220	130		350	430	
	1110	12" size		3	8		300	170		470	580	
	1180	For gear actuator, add					65%					
	1200	Lug type, lever actuator										
	1220	2" size	1 Plum	14	.571	Ea.	59	13.15		72.15	84	
	1230	2-½" size	Q-1	9	1.780		66	37		103	125	
	1240	3" size		8	2		72	41		113	140	
	1250	4" size		5	3.200		86	66		152	190	
	1260	5" size	Q-2	5	4.800		115	105		220	275	
	1270	6" size		5	4.800		135	105		240	300	
	1280	8" size		4.50	5.330		190	115		305	375	
	1290	10" size		4	6		245	130		375	455	
	1300	12" size		3	8		370	170		540	660	
	1320	For gear actuator, add					60%					
	1400	Diverter, 150 lb. flanged, bronze or iron plugs										
	1440	2" pipe size	Q-1	2	8	Ea.	730	165		895	1,050	
	1450	3" pipe size	"	1.50	10.670	"	1,025	220		1,245	1,450	
15	1650	Gate, 125 lb., N.R.S., threaded										
	1700	2" size	1 Plum	11	.727	Ea.	120	16.75		136.75	155	
	1740	2-½" size	Q-1	15	1.070		130	22		152	175	
	1760	3" size		13	1.230		155	25		180	210	
	1780	4" size		10	1.600		200	33		233	270	
	2150	Flanged										
	2200	2" size	1 Plum	5	1.600	Ea.	97	37		134	160	
	2240	2-½" size	Q-1	5	3.200		110	66		176	220	
	2260	3" size		4.50	3.560		125	74		199	245	
	2280	4" size		3	5.330		180	110		290	360	
	2290	5" size	Q-2	3.40	7.050		285	150		435	535	
	2300	6" size		3	8		285	170		455	565	
	2320	8" size		2.50	9.600		515	205		720	865	

Figure 22.2

Line Description

Each line has a text description of the item for which costs are listed. The description may be self-contained and all inclusive or, if indented, the complete description for a line is dependent upon the information provided above. All indented items are delineations (by size, color, material, etc.) or breakdowns of previously described items. An index is provided in the back of *Means Mechanical Cost Data* to aid in locating particular items.

Crew

For each construction element, (each line item), a minimum typical crew is designated as appropriate to perform the work. The crew may include one or more trades, foremen, craftsmen and helpers, and any equipment required for proper installation of the described item. If an individual trade installs the item using only hand tools, the smallest efficient number of tradesmen will be indicated (1 Plum, 2 Spri, etc.). Abbreviations for trades are shown in Figure 22.3. If more than one trade is required to install the item and/or if powered equipment is needed, a crew number will be designated (Q-19, Q-8, etc.). A complete listing of crews is presented in the Foreword pages of *Means Mechanical Cost Data* (see Figure 22.4). On these pages, each crew is broken down into the following components:

1. Number and type of workers designated.
2. Number, size, and type of any equipment required.
3. Hourly labor costs listed two ways: "bare" — base rate including fringe benefits; and including installing contractor's overhead and profit — billing rate. (See Figure 22.3 from the inside back cover of *Means Mechanical Cost Data* for labor rate information).
4. Daily equipment costs, based on the weekly equipment rental cost divided by 5, plus the hourly operating cost, times 8 hours. This cost is listed two ways: as a bare cost and with a 10% percent markup to cover handling and management costs.
5. Labor and equipment are broken down further into: cost per man-hour for labor, and cost per man-hour for the equipment.
6. The total daily man-hours for the crew.
7. The total bare costs per day for the crew, including equipment.
8. The total daily cost of the crew including the installing contractor's overhead and profit.

The total daily cost of the required crew is used to calculate the unit installation cost for each item (for both bare costs and cost including overhead and profit).

The crew designation does not mean that this is the only crew that can perform the work. Crew size and content have been developed and chosen based on practical experience and feedback from contractors. These designations represent a labor and equipment make-up commonly found in the industry. The most appropriate crew for a given task is best determined based on particular project requirements. Unit costs may vary if crew sizes or content are significantly changed.

Figure 22.5 is a page from Division 1.5 of *Means Mechanical Cost Data*. This page lists the equipment costs used in the presentation and calculation of the crew costs and unit price data. Rental costs are shown as daily, weekly, and monthly rates. The Hourly Operating Cost represents the cost

Installing Contractor's Overhead & Profit

Below are the **average** installing contractor's percentage mark-ups applied to base labor rates to arrive at typical billing rates.

Column A: Labor rates are based on union wages averaged for 30 major U.S. cities. Base rates including fringe benefits are listed hourly and daily. These figures are the sum of the wage rate, employer-paid fringe benefits such as vacation pay, employer-paid health and welfare costs, pension costs, plus appropriate training and industry advancement funds costs.

Column B: Workers' Compensation rates are the national average of state rates established for each trade.

Column C: Column C lists average fixed overhead figures for all trades. Included are Federal and State Unemployment costs set at 5.5%; Social Security Taxes (FICA) set at 7.15%; Builder's Risk Insurance costs set at 1.14%; and Public Liability costs set at 0.82%. All the percentages except those for Social Security Taxes vary from state to state as well as from company to company.

Column D and E: Percentages in Columns D and E are based on the presumption that the installing contractor has annual billing of $500,000 and up. Overhead percentages may increase with smaller annual billing. The overhead percentages for any given contractor may vary greatly and depend on a number of factors, such as the contractor's annual volume, engineering and logistical support costs, and staff requirements. The figures for overhead and profit will also vary depending on the type of job, the job location, and the prevailing economic conditions. All factors should be examined very carefully for each job.

Column F: Column F lists the total of columns B, C, D, and E.

Column G: Column G is Column A (hourly base labor rate) multiplied by the percentage in Column F (O&P percentage).

Column H: Column H is the total of Column A (hourly base labor rate) plus Column G (Total O&P).

Column I: Column I is Column H multiplied by eight hours.

		A		B	C	D	E	F		G	H	I
		Base Rate Incl. Fringes		Workers' Comp. Ins.	Average Fixed Overhead	Overhead	Profit	Total Overhead & Profit			Rate with O & P	
Abbr.	Trade	Hourly	Daily					%	Amount		Hourly	Daily
Skwk	Skilled Workers Average (35 trades)	$20.80	$166.40	10.4%	14.6%	12.6%	10%	47.6%	$ 9.90		$30.70	$245.60
	Helpers Average (5 trades)	15.90	127.20	11.1		12.7		48.4	7.70		23.60	188.80
	Foremen Average, Inside (50¢ over trade)	21.30	170.40	10.4		12.6		47.6	10.15		31.45	251.60
	Foremen Average, Outside ($2.00 over trade)	22.80	182.40	10.4		12.6		47.6	10.85		33.65	269.20
Clab	Common Building Laborers	16.10	128.80	11.4		10.8		46.8	7.55		23.65	189.20
Asbe	Asbestos Workers	23.00	184.00	8.8		15.7		49.1	11.30		34.30	274.40
Boil	Boilermakers	23.00	184.00	7.2		16.0		47.8	11.00		34.00	272.00
Bric	Bricklayers	20.55	164.40	8.5		10.7		43.8	9.00		29.55	236.40
Brhe	Bricklayer Helpers	16.15	128.40	8.5		10.7		43.8	7.05		23.20	185.60
Carp	Carpenters	20.55	164.40	11.4		10.8		46.8	9.60		30.15	241.20
Cefi	Cement Finishers	19.70	157.60	6.6		10.8		42.0	8.30		27.95	223.60
Elec	Electricians	22.65	181.20	4.5		15.9		45.0	10.20		32.85	262.80
Elev	Elevator Constructors	23.05	184.40	6.0		15.9		46.5	10.70		33.75	270.00
Eqhv	Equipment Operators, Crane or Shovel	21.20	169.60	7.8		13.8		46.2	9.80		31.00	248.00
Eqmd	Equipment Operators, Medium Equipment	20.75	166.00	7.8		13.8		46.2	9.60		30.35	242.80
Eqlt	Equipment Operators, Light Equipment	19.60	156.80	7.8		13.8		46.2	9.05		28.65	229.20
Eqol	Equipment Operators, Oilers	17.55	140.40	7.8		13.8		46.2	8.10		25.65	205.20
Eqmm	Equipment Operators, Master Mechanics	22.00	176.00	7.8		13.8		46.2	10.15		32.15	257.20
Glaz	Glaziers	20.75	166.00	8.7		10.8		44.1	9.15		29.90	239.20
Lath	Lathers	20.50	164.00	7.2		10.8		42.6	8.75		29.25	234.00
Marb	Marble Setters	20.25	162.00	8.5		10.7		43.8	8.85		29.10	232.80
Mill	Millwrights	21.25	170.00	7.2		10.8		42.6	9.05		30.30	242.40
Mstz	Mosaic and Terrazzo Workers	20.05	160.40	6.0		10.7		41.3	8.30		28.35	226.80
Pord	Painters, Ordinary	19.55	156.40	8.9		10.8		44.3	8.65		28.20	225.60
Psst	Painters, Structural Steel	20.20	161.60	29.4		10.1		64.1	12.95		33.15	265.20
Pape	Paper Hangers	19.75	158.00	8.9		10.8		44.3	8.75		28.50	228.00
Pile	Pile Drivers	20.35	162.80	18.3		15.7		58.6	11.95		32.30	258.40
Plas	Plasterers	20.25	162.00	8.7		10.9		44.2	8.95		29.20	233.60
Plah	Plasterer Helpers	16.65	133.20	8.7		10.9		44.2	7.35		24.00	192.00
Plum	Plumbers	23.00	184.00	5.3		15.9		45.8	10.55		33.55	268.40
Rodm	Rodmen (Reinforcing)	22.10	176.80	18.6		13.3		56.5	12.50		34.60	276.80
Rofc	Roofers, Composition	19.15	153.20	20.8		10.4		55.8	10.70		29.85	238.80
Rots	Roofers, Tile & Slate	19.25	154.00	20.8		10.4		55.8	10.75		30.00	240.00
Rohe	Roofer Helpers (Composition)	14.20	113.60	20.8		10.4		55.8	7.90		22.10	176.80
Shee	Sheet Metal Workers	23.10	184.80	7.0		15.8		47.4	10.95		34.05	272.40
Spri	Sprinkler Installers	23.85	190.80	6.1		15.9		46.6	11.10		34.95	279.60
Stpi	Steamfitters or Pipefitters	23.30	186.40	5.3		15.9		45.8	10.70		33.95	271.60
Ston	Stone Masons	20.60	164.80	8.5		10.7		43.8	9.00		29.60	236.80
Sswk	Structural Steel Workers	22.10	176.80	21.5		13.7		59.8	13.20		35.30	282.40
Tilf	Tile Layers (Floor)	20.00	160.00	6.0		10.7		41.3	8.25		28.25	226.00
Tilh	Tile Layer Helpers	16.10	128.80	6.0		10.7		41.3	6.65		22.75	182.00
Trlt	Truck Drivers, Light	16.70	133.60	9.8		10.7		45.1	7.55		24.25	194.00
Trhv	Truck Drivers, Heavy	16.95	135.60	9.8		10.7		45.1	7.65		24.60	196.80
Sswl	Welders, Structural Steel	22.10	176.80	21.5		13.7		59.8	13.20		35.30	282.40
Wrck	*Wrecking	16.10	128.80	23.1		10.4		58.1	9.35		25.45	203.60

*Not included in Averages.

Figure 22.3

CREWS

Crew No.	Bare Costs		Incl. Subs O & P		Cost Per Man-hour	
Crew Q-6	Hr.	Daily	Hr.	Daily	Bare Costs	Incl. O&P
2 Steamfitters	$23.30	$372.80	$33.95	$543.20	$21.74	$31.70
1 Steamfitter Apprentice	18.64	149.12	27.20	217.60		
24 M.H., Daily Totals		$521.92		$760.80	$21.74	$31.70
Crew Q-7	Hr.	Daily	Hr.	Daily	Bare Costs	Incl. O&P
1 Steamfitter Foreman (ins)	$23.80	$190.40	$34.70	$277.60	$22.26	$32.45
2 Steamfitters	23.30	372.80	33.95	543.20		
1 Steamfitter Apprentice	18.64	149.12	27.20	217.60		
32 M.H., Daily Totals		$712.32		$1038.40	$22.26	$32.45
Crew Q-8	Hr.	Daily	Hr.	Daily	Bare Costs	Incl. O&P
1 Steamfitter Foreman (ins)	$23.80	$190.40	$34.70	$277.60	$22.26	$32.45
1 Steamfitter	23.30	186.40	33.95	271.60		
1 Welder (steamfitter)	23.30	186.40	33.95	271.60		
1 Steamfitter Apprentice	18.64	149.12	27.20	217.60		
1 Electric Welding Mach.		19.00		20.90	.59	.65
32 M.H., Daily Totals		$731.32		$1059.30	$22.85	$33.10
Crew Q-9	Hr.	Daily	Hr.	Daily	Bare Costs	Incl. O&P
1 Sheet Metal Worker	$23.10	$184.80	$34.05	$272.40	$20.79	$30.65
1 Sheet Metal Apprentice	18.48	147.84	27.25	218.00		
16 M.H., Daily Totals		$332.64		$490.40	$20.79	$30.65
Crew Q-10	Hr.	Daily	Hr.	Daily	Bare Costs	Incl. O&P
2 Sheet Metal Workers	$23.10	$369.60	$34.05	$544.80	$21.56	$31.78
1 Sheet Metal Apprentice	18.48	147.84	27.25	218.00		
24 M.H., Daily Totals		$517.44		$762.80	$21.56	$31.78
Crew Q-11	Hr.	Daily	Hr.	Daily	Bare Costs	Incl. O&P
1 Sheet Metal Foreman (ins)	$23.60	$188.80	$34.80	$278.40	$22.07	$32.53
2 Sheet Metal Workers	23.10	369.60	34.05	544.80		
1 Sheet Metal Apprentice	18.48	147.84	27.25	218.00		
32 M.H., Daily Totals		$706.24		$1041.20	$22.07	$32.53
Crew Q-12	Hr.	Daily	Hr.	Daily	Bare Costs	Incl. O&P
1 Sprinkler Installer	$23.85	$190.80	$34.95	$279.60	$21.46	$31.45
1 Sprinkler Apprentice	19.08	152.64	27.95	223.60		
16 M.H., Daily Totals		$343.44		$503.20	$21.46	$31.45
Crew Q-13	Hr.	Daily	Hr.	Daily	Bare Costs	Incl. O&P
1 Sprinkler Foreman (ins)	$24.35	$194.80	$35.70	$285.60	$22.78	$33.38
2 Sprinkler Installers	23.85	381.60	34.95	559.20		
1 Sprinkler Apprentice	19.08	152.64	27.95	223.60		
32 M.H., Daily Totals		$729.04		$1068.40	$22.78	$33.38
Crew Q-14	Hr.	Daily	Hr.	Daily	Bare Costs	Incl. O&P
1 Asbestos Worker	$23.00	$184.00	$34.30	$274.40	$20.70	$30.87
1 Asbestos Apprentice	18.40	147.20	27.45	219.60		
16 M.H., Daily Totals		$331.20		$494.00	$20.70	$30.87
Crew Q-15	Hr.	Daily	Hr.	Daily	Bare Costs	Incl. O&P
1 Plumber	$23.00	$184.00	$33.55	$268.40	$20.70	$30.20
1 Plumber Apprentice	18.40	147.20	26.85	214.80		
1 Electric Welding Mach.		19.00		20.90	1.18	1.30
16 M.H., Daily Totals		$350.20		$504.10	$21.88	$31.50
Crew Q-16	Hr.	Daily	Hr.	Daily	Bare Costs	Incl. O&P
2 Plumbers	$23.00	$368.00	$33.55	$536.80	$21.46	$31.31
1 Plumber Apprentice	18.40	147.20	26.85	214.80		
1 Electric Welding Mach.		19.00		20.90	.79	.87
24 M.H., Daily Totals		$534.20		$772.50	$22.25	$32.18

Crew No.	Bare Costs		Incl. Subs O & P		Cost Per Man-hour	
Crew Q-17	Hr.	Daily	Hr.	Daily	Bare Costs	Incl. O&P
1 Steamfitter	$23.30	$186.40	$33.95	$271.60	$20.97	$30.57
1 Steamfitter Apprentice	18.64	149.12	27.20	217.60		
1 Electric Welding Mach		19.00		20.90	1.18	1.30
16 M.H., Daily Totals		$354.52		$510.10	$22.15	$31.87
Crew Q-18	Hr.	Daily	Hr.	Daily	Bare Costs	Incl. O&P
2 Steamfitters	$23.30	$372.80	$33.95	$543.20	$21.74	$31.70
1 Steamfitter Apprentice	18.64	149.12	27.20	217.60		
1 Electric Welding Mach.		19.00		20.90	.79	.87
24 M.H., Daily Totals		$540.92		$781.70	$22.53	$32.57
Crew Q-19	Hr.	Daily	Hr.	Daily	Bare Costs	Incl. O&P
1 Steamfitter	$23.30	$186.40	$33.95	$271.60	$21.53	$31.33
1 Steamfitter Apprentice	18.64	149.12	27.20	217.60		
1 Electrician	22.65	181.20	32.85	262.80		
24 M.H., Daily Totals		$516.72		$752.00	$21.53	$31.33
Crew Q-20	Hr.	Daily	Hr.	Daily	Bare Costs	Incl. O&P
1 Sheet Metal Worker	$23.10	$184.80	$34.05	$272.40	$21.16	$31.09
1 Sheet Metal Apprentice	18.48	147.84	27.25	218.00		
.5 Electrician	22.65	90.60	32.85	131.40		
20 M.H., Daily Totals		$423.24		$621.80	$21.16	$31.09
Crew Q-21	Hr.	Daily	Hr.	Daily	Bare Costs	Incl. O&P
2 Steamfitters	$23.30	$372.80	$33.95	$543.20	$21.97	$31.98
1 Steamfitter Apprentice	18.64	149.12	27.20	217.60		
1 Electrician	22.65	181.20	32.85	262.80		
32 M.H., Daily Totals		$703.12		$1023.60	$21.97	$31.98
Crew R-1	Hr.	Daily	Hr.	Daily	Bare Costs	Incl. O&P
1 Electrician Foreman	$23.15	$185.20	$33.55	$268.40	$20.48	$29.88
3 Electricians	22.65	543.60	32.85	788.40		
2 Helpers	15.90	254.40	23.60	377.60		
48 M.H., Daily Totals		$983.20		$1434.40	$20.48	$29.88
Crew R-2	Hr.	Daily	Hr.	Daily	Bare Costs	Incl. O&P
1 Electrician Foreman	$23.15	$185.20	$33.55	$268.40	$20.58	$30.04
3 Electricians	22.65	543.60	32.85	788.40		
2 Helpers	15.90	254.40	23.60	377.60		
1 Equip. Oper. (crane)	21.20	169.60	31.00	248.00		
1 S.P. Crane, 5 Ton		175.00		192.50	3.12	3.43
56 M.H., Daily Totals		$1327.80		$1874.90	$23.70	$33.47
Crew R-3	Hr.	Daily	Hr.	Daily	Bare Costs	Incl. O&P
1 Electrician Foreman	$23.15	$185.20	$33.55	$268.40	$22.56	$32.76
1 Electrician	22.65	181.20	32.85	262.80		
.5 Equip. Oper. (crane)	21.20	84.80	31.00	124.00		
.5 S.P. Crane, 5 Ton		87.50		96.25	4.37	4.81
20 M.H., Daily Totals		$538.70		$751.45	$26.93	$37.57
Crew R-4	Hr.	Daily	Hr.	Daily	Bare Costs	Incl. O&P
1 Struc. Steel Foreman	$24.10	$192.80	$38.50	$308.00	$22.61	$35.45
3 Struc. Steel Workers	22.10	530.40	35.30	847.20		
1 Electrician	22.65	181.20	32.85	262.80		
1 Gas Welding Machine		54.75		60.25	1.36	1.50
40 M.H., Daily Totals		$959.15		$1478.25	$23.97	$36.95

xxv

Figure 22.4

of fuel, lubrication, and routine maintenance. Equipment costs used in the crews are calculated as follows:

Line Number: 01.5-150-7800
Equipment: Electric Welder, 300 Amp.
Rent per week: $67.00
Hourly Operating Cost: $0.70

$$\frac{\text{Weekly rental}}{5 \text{ days}} + (\text{Hourly Oper. Cost} \times 8 \text{ hrs/day}) = \text{Daily Equipment Cost}$$

$$\frac{\$67}{5} + (\$0.70 \times 8) = \$19.00 \text{ per day}$$

Note: Crew operating labor is not included.

Units

The unit column (see Figures 22.1 and 22.2) defines the component for which the costs have been calculated. It is this "unit" on which Unit Price Estimating is based. The units as used represent standard estimating and quantity takeoff procedures. However, the estimator should always check to be sure that the units taken off are the same as those priced. A list of standard abbreviations is included at the back of *Means Mechanical Cost Data*.

Bare Costs

The four columns listed under "Bare Costs" — "Material", "Labor", "Equipment", and "Total" represent the actual cost of construction items to the contractor. In other words, bare costs are those which do not include the overhead and profit of the installing contractor, whether it is a subcontractor or a general contracting company using its own crews.

Material: Material costs are based on the national average contractor purchase price delivered to the job site. Delivered costs are assumed to be within a 20 mile radius of metropolitan areas. No sales tax is included in the material prices because of variations from state to state.

The prices are based on quantities that would normally be purchased for complete buildings or projects costing $400,000 and up. Prices for small quantities must be adjusted accordingly. If more current costs for materials are available for the appropriate location, it is recommended that adjustments be made to the unit costs to reflect any cost difference.

Labor: Labor costs are calculated by multiplying the "Bare Labor Cost" per man-hour times the number of man-hours, from the "Man-Hours" column. The "Bare" labor rate is determined by adding the base rate plus fringe benefits. The base rate is the actual hourly wage of a worker used in figuring payroll. It is from this figure that employee deductions are taken (Federal withholding, FICA, State withholding). Fringe benefits include all employer-paid benefits, above and beyond the payroll amount (employer-paid health, vacation pay, pension, profit sharing). The "Bare Labor Cost" is, therefore, the actual amount that the contractor must pay directly for construction workers. Figure 22.3 shows labor rates for the 35 construction trades plus skilled worker, helper, and foreman averages. These rates are the averages of union wage agreements effective January 1 of the current year from 30 major cities in the United States. The "Bare Labor Cost" for each trade, as used in *Means Mechanical Cost Data*, is shown in column "A"

1.5 Contractor Equipment

		UNIT	HOURLY OPER. COST.	RENT PER DAY	RENT PER WEEK	TOTAL PER MONTH	CREW EQUIPMENT COST	
150	6100	Circular, hand held, electric, 7" diameter	Ea.	.14	16	47	140	10.50	150
	6200	12" diameter		.15	25	75	220	16.20	
	6410	Toilet, portable chemical				23	73	4.60	
	6420	Recycle flush type				29	88	5.80	
	6430	Toilet, fresh water flush, garden hose,					135	9	
	6440	Hoisted, non-flush, for high rise					72	4.80	
	6450	Toilet, trailers, minimum					175	11.65	
	6460	Maximum					225	15	
	6500	Trailers, platform, flush deck, 2 axle, 25 ton capacity		1.05	120	365	1,100	81.40	
	6600	40 ton capacity		1.25	175	530	1,590	116	
	6700	3 axle, 50 ton capacity		2.35	210	635	1,910	145.80	
	6800	75 ton capacity		2.98	310	935	2,800	210.85	
	7100	Truck, pickup, ¾ ton, 2 wheel drive		4.80	42	125	380	63.40	
	7200	4 wheel drive		4.95	58	175	530	74.60	
	7300	Tractor, 4 x 2, 30 ton capacity, 195 H.P.		8.85	285	860	2,575	242.80	
	7410	250 H.P.		13.15	310	935	2,950	292.20	
	7700	Welder, electric, 200 amp		.54	20	60	180	16.30	
	7800	300 amp		.70	22	67	200	19	
	7900	Gas engine, 200 amp		3.12	38	115	340	47.95	
	8000	300 amp		4.20	45	130	385	59.60	
	8100	Wheelbarrow, any size	↓		6	19	56	3.80	
200	0010	**LIFTING & HOISTING EQUIPMENT RENTAL**							200
	0100	without operators							
	0200	Crane, climbing, 106' jib, 6000 lb. capacity, 410 FPM	Ea.	12.25			8,125	639.65	
	0300	101' jib, 10,250 lb. capacity, 270 FPM	"	16.45			10,500	831.60	
	0400	Tower, static, 130' high, 106' jib,							
	0500	6200 lb. capacity at 400 FPM	Ea.	12.30			8,700	678.40	
	0650	Crawler, cable, ½ C.Y., 15 tons at 12' radius		11.95	315	940	2,825	283.60	
	0700	¾ C.Y., 20 tons at 12' radius		12.30	445	1,335	4,000	365.40	
	0800	1 C.Y., 25 tons at 12' radius		13.40	510	1,535	4,600	414.20	
	0900	1-½ C.Y., 40 tons at 12' radius		15.55	655	1,965	5,900	517.40	
	1000	2 C.Y., 50 tons at 12' radius		19	730	2,190	6,575	590	
	1100	3 C.Y., 75 tons at 12' radius		24	830	2,485	7,450	689	
	1200	100 ton capacity, standard boom		32	925	2,780	8,340	812	
	1300	165 ton capacity, standard boom		46	1,420	4,265	12,800	1,221	
	1400	200 ton capacity, 150' boom		52	1,890	5,665	17,000	1,549	
	1500	450' boom		55	2,145	6,440	19,320	1,728	
	1600	Truck mounted, cable operated, 6 x 4, 20 tons at 10' radius		9.55	395	1,185	3,550	313.40	
	1700	25 tons at 10' radius		11.65	510	1,530	4,590	399.20	
	1800	8 x 4, 30 tons at 10' radius		14.25	565	1,695	5,090	453	
	1900	40 tons at 12' radius		14.40	640	1,925	5,775	500.20	
	2000	8 x 4, 60 tons at 15' radius		18.90	710	2,125	6,375	576.20	
	2050	82 tons at 15' radius		20	790	2,370	7,115	634	
	2100	90 tons at 15' radius		24	1,035	3,105	9,315	813	
	2200	115 tons at 15' radius		28	1,180	3,535	10,600	931	
	2300	150 tons at 18' radius		34	1,305	3,915	11,750	1,055	
	2350	165 tons at 18' radius		35	1,365	4,090	12,275	1,098	
	2400	Truck mounted, hydraulic, 12 ton capacity		8.50	290	875	2,625	243	
	2500	25 ton capacity		12.70	510	1,530	4,595	407.60	
	2550	33 ton capacity		13.20	540	1,615	4,850	428.60	
	2600	55 ton capacity		17.90	690	2,065	6,195	556.20	
	2700	80 ton capacity		22	1,250	3,745	11,235	925	
	2800	Self-propelled, 4 x 4, with telescoping boom, 5 ton		5.25	220	665	2,000	175	
	2900	12-½ ton capacity		9.15	245	730	2,190	219.20	
	3000	15 ton capacity		9.40	310	935	2,800	262.20	
	3100	25 ton capacity		11.55	490	1,465	4,400	385.40	
	3200	Derricks, guy, 20 ton capacity, 60' boom, 75' mast		3.75	145	435	1,300	117	
	3300	100' boom, 115' mast		4.25	180	535	1,600	141	
	3400	Stiffleg, 20 ton capacity, 70' boom, 37' mast	↓	3.10	195	585	1,750	141.80	

For expanded coverage of these items see *Means Building Construction Cost Data 1987*

Figure 22.5

as the base rate including fringes. Refer to the "Crew" column to determine what rate is used to calculate the "Bare Labor Cost" for a particular line item.

Equipment: Equipment costs are calculated by multiplying the "Bare Equipment Cost" per man-hour, from the appropriate "Crew" listing, times the man-hours in the "Man-Hours" column. The calculation of the equipment portion of installation costs is outlined earlier in this chapter.

Total Bare Costs

This column simply represents the arithmetic sum of the bare material, labor, and equipment costs. This total is the average cost to the contractor for the particular item of construction, supplied and installed, or "in place". No overhead and/or profit is included.

Total Including Overhead and Profit

This column represents the total cost of an item including the installation contractor's overhead and profit. The installing contractor could be either the prime mechanical contractor or a subcontractor. If these costs are used for an item to be installed by a subcontractor, the prime mechanical contractor should include an additional percentage (usually 10% to 20%) to cover the expenses of supervision and management. Consideration must be given also to sub-subcontractors who often appear in mechanical contracting. An example might be an electrical sub to the Temperature Control Sub-contractor who, of course, is a sub to the HVAC contractor. Each must have their own overhead and profit markup.

The costs in the "Total Including Overhead and Profit" are the arithmetical sum of the following three calculations:
- Bare Material Cost plus 10%
- Labor Cost, including overhead and profit, per man-hour times the number of man-hours
- Equipment Costs, including overhead and profit, per man-hour times the number of man-hours

The Labor and Equipment Costs, including overhead and profit are found in the appropriate crew listings. The overhead and profit percentage factor for Labor is obtained from Column F in Figure 22.3. The overhead and profit for Equipment is 10% of "Bare" cost.

Labor costs are increased by percentages for overhead and profit, depending on trade as shown in Figure 22.3. The resulting rates are listed in the right hand columns of the same figure. Note that the percentage increase for overhead and profit for the average skilled worker is 47.6 % of the base rate. The following items are included in the increase for overhead and profit, as shown in Figure 22.3:

Workers' Compensation and Employer's Liability: Workers' Compensation and Employer's Liability Insurance rates vary from state to state and are tied into the construction trade safety records in that particular state. Rates also vary by trade according to the hazard involved. (See Figure 22.6, average insurance rates as of January, 1987.) The proper authorities will most likely keep the contractor well informed of the rates and obligations.

State and Federal Unemployment Insurance: The employer's tax rate is adjusted by a merit-rating system according to the number of former employees applying for benefits. Contractors who find it possible to offer a maximum of steady employment can enjoy a reduction in the unemployment tax rate.

Employer-Paid Social Security (FICA): The tax rate is adjusted annually by the federal government. It is a percentage of an employee's salary up to a maximum annual contribution.

Builder's Risk and Public Liability: These insurance rates vary according to the trades involved and the state in which the work is done.

Overhead: The column listed as "Overhead" provides percentages to be added for office or operating overhead. This is the cost of doing business. The percentages are presented as national averages by trade as shown in Figure 22.3. Note that the operating overhead costs are applied to *labor only* in Means Mechanical Cost Data.

Profit: This percentage is the fee added by the contractor to offer both a return on investment and an allowance to cover the risk involved in the type of construction being bid. The profit percentage may vary from 4% on large, straightforward projects to as much as 25% on smaller, high-risk jobs. Profit percentages are directly affected by economic conditions, the expected number of bidders, and the estimated risk involved in the project. For estimating purposes, *Means Mechanical Cost Data* assumes 10% (applied to labor) as a reasonable average profit factor.

Square Foot and Cubic Foot Costs

Division 17 in *Mechanical Cost Data* has been developed to facilitate the preparation of rapid preliminary budget estimates. The cost figures in this division are derived from more than 10,500 actual building projects contained in the R. S. Means data bank of construction costs and include the contractor's overhead and profit. The prices shown *do not* include architectural fees or land costs. The files are updated each year with costs for new projects. In no case are all subdivisions of a project listed.

These projects were located throughout the United States and reflect differences in square foot and cubic foot costs due to both the variations in labor and material costs, and the differences in the owners' requirements. For instance, a bank in a large city would have different features and costs than one in a rural area. This is true of all the different types of buildings analyzed. All individual cost items were computed and tabulated separately. Thus, the sum of the median figures for Plumbing, H.V.A.C. and Electrical will not normally add up to the total Mechanical and Electrical costs arrived at by separate analysis and tabulation of the projects.

The data and prices presented on a Division 17 page (as shown in Figure 22.7) are listed both as square foot or cubic foot costs and as a percentage of total costs. Each category tabulates the data in a similar manner. The median, or middle figure, is listed. This means that 50% of all projects had lower costs, and 50% had higher costs than the median figure. Figures in the "1/4" column indicate that 25% of the projects had lower costs and 75% had higher costs. Similarly, figures in the "3/4" column indicate that 75% had lower costs and 25% of the projects had higher costs.

The costs and figures represent all projects and do not take into account project size. As a rule, larger buildings (of the same type and relative location) will cost less to build per square foot than similar buildings of a smaller size. This cost difference is due to economies of scale as well as a lower exterior envelope-to-floor area ratio. A conversion is necessary to adjust project costs based on size relative to the norm. See Figure 23.6.

GENERAL REQUIREMENTS C10.2-200 Workers' Compensation

The table below tabulates the national averages for Workers' Compensation insurance rates by trade and type of building. The average "Insurance Rate" is multiplied by the "% of Building Cost" for each trade. This produces the "Workers' Compensation Cost" by % of total labor cost, to be added for each trade by building type to determine the weighted average Workers' Compensation rate for the building types analyzed.

Table 10.2-201 Insurance Rates by Trade

Trade	Insurance Rate (% of Labor Cost)		% of Building Cost			Workers' Compensation Cost		
	Range	Average	Office Bldgs.	Schools & Apts.	Mfg.	Office Bldgs.	Schools & Apts.	Mfg.
Excavation, Grading, etc.	2.4% to 28.9%	7.8%	4.8%	4.9%	4.5%	.37%	.38%	.35%
Piles & Foundations	4.9 to 47.6	18.3	7.1	5.2	8.7	1.30	.95	1.59
Concrete	2.7 to 28.2	10.3	5.0	14.8	3.7	.51	1.52	.38
Masonry	2.2 to 22.0	8.5	6.9	7.5	1.9	.59	.64	.16
Structural Steel	3.0 to 52.4	21.5	10.7	3.9	17.6	2.30	.84	3.78
Miscellaneous & Ornamental Metals	1.6 to 19.6	7.5	2.8	4.0	3.6	.21	.30	.27
Carpentry & Millwork	3.6 to 56.5	11.4	3.7	4.0	0.5	.42	.46	.06
Metal or Composition Siding	3.3 to 23.3	8.9	2.3	0.3	4.3	.20	.03	.38
Roofing	4.3 to 55.4	20.8	2.3	2.6	3.1	.48	.54	.64
Doors & Hardware	2.0 to 16.6	6.6	0.9	1.4	0.4	.06	.09	.03
Sash & Glazing	3.1 to 13.9	8.7	3.5	4.0	1.0	.30	.35	.09
Lath & Plaster	3.1 to 23.9	8.7	3.3	6.9	0.8	.29	.60	.07
Tile, Marble & Floors	1.3 to 23.3	6.0	2.6	3.0	0.5	.16	.18	.03
Acoustical Ceilings	1.8 to 16.0	7.2	2.4	0.2	0.3	.17	.01	.02
Painting	3.0 to 20.0	8.9	1.5	1.6	1.6	.13	.14	.14
Interior Partitions	3.6 to 56.5	11.4	3.9	4.3	4.4	.44	.49	.50
Miscellaneous Items	1.6 to 96.4	10.6	5.2	3.7	9.7	.55	.39	1.03
Elevators	2.1 to 16.2	6.0	2.1	1.1	2.2	.13	.07	.13
Sprinklers	2.0 to 17.5	6.1	0.5	—	2.0	.03	—	.12
Plumbing	1.5 to 13.9	5.3	4.9	7.2	5.2	.26	.38	.28
Heat., Vent., Air Conditioning	2.4 to 14.5	7.0	13.5	11.0	12.9	.95	.77	.90
Electrical	1.1 to 13.7	4.5	10.1	8.4	11.1	.45	.38	.50
Total	1.1% to 96.4%	—	100.0%	100.0%	100.0%	10.3%	9.51%	11.45%
Overall Weighted Average								10.42%

The table below lists the weighted average Workers' Compensation base rate for each state with a factor comparing this with the national average of 10.4%.

Table 10.2-202 Insurance Rates by States

State	Weighted Average	Factor	State	Weighted Average	Factor	State	Weighted Average	Factor
Alabama	7.1%	68	Kentucky	6.4%	62	North Dakota	6.8%	65
Alaska	13.5	130	Louisiana	8.4	81	Ohio	7.6	73
Arizona	11.4	110	Maine	13.1	126	Oklahoma	11.4	110
Arkansas	6.9	66	Maryland	14.6	140	Oregon	21.5	207
California	11.7	113	Massachusetts	13.1	126	Pennsylvania	11.4	110
Colorado	13.3	128	Michigan	11.3	109	Rhode Island	15.8	152
Connecticut	16.4	158	Minnesota	18.5	178	South Carolina	7.4	71
Delaware	10.6	102	Mississippi	6.1	59	South Dakota	7.2	69
District of Columbia	18.7	180	Missouri	6.3	61	Tennessee	5.8	56
Florida	13.9	134	Montana	16.7	161	Texas	7.3	70
Georgia	7.2	69	Nebraska	6.1	59	Utah	6.7	64
Hawaii	26.1	251	Nevada	10.8	104	Vermont	7.7	74
Idaho	8.5	82	New Hampshire	13.2	127	Virginia	8.3	80
Illinois	14.4	138	New Jersey	6.6	63	Washington	7.9	76
Indiana	2.9	28	New Mexico	14.4	138	West Virginia	7.7	74
Iowa	6.6	63	New York	9.4	90	Wisconsin	8.5	82
Kansas	6.8	65	North Carolina	5.4	52	Wyoming	5.4	52
Weighted Average for U.S. is 10.4% of payroll = 100								

Rates in the following table are the base or manual costs per $100 of payroll for Workers' Compensation in each state. Rates are usually applied to straight time wages only and not to premium time wages and bonuses.

The weighted average skilled worker rate for 35 trades is 10.4%. For bidding purposes, apply the full value of Workers' Compensation directly to total labor costs, or if labor is 32%, materials 48% and overhead and profit 20% of total cost, carry 32/80 x 10.4% = 4.2% of cost (before overhead and profit) into overhead. Rates vary not only from state to state but also with the experience rating of the contractor.

Rates are the most current available at the time of publication.

Figure 22.6

17.1 S.F., C.F. and % of Total Costs

				UNIT COSTS			% OF TOTAL			
			UNIT	¼	MEDIAN	¾	¼	MEDIAN	¾	
760	0010	SCHOOLS Junior High & Middle	S.F.	50.30	59.10	71.10				760
	0020	Total project costs	C.F.	3.13	3.89	4.57				
	2720	Plumbing	S.F.	2.96	3.57	4.64	5.40%	6.90%	8.10%	
	2770	Heating, ventilating, air conditioning		3.63	6.95	9.80	8.70%	11.50%	17.40%	
	2900	Electrical		4.86	5.85	6.90	8%	9.50%	10.60%	
	3100	Total: Mechanical & Electrical		10.65	14.65	21.55	19.30%	27.10%	32.50%	
	9000	Per pupil, total cost	Ea.	4,450	6,750	8,850				
780	0010	SCHOOLS Senior High	S.F.	49.85	58.70	79.43				780
	0020	Total project costs	C.F.	3.09	3.80	4.71				
	2720	Plumbing	S.F.	2.64	4.44	7.10	5%	6.50%	8%	
	2770	Heating, ventilating, air conditioning		6.05	6.80	9.80	8.90%	11.50%	14.20%	
	2900	Electrical		5.05	6.40	10.05	8.30%	10.10%	12.30%	
	3100	Total: Mechanical & Electrical		10.20	16.15	21.90	16.80%	23.80%	30%	
	9000	Per pupil, total cost	Ea.	5,150	8,300	10,800				
800	0010	SCHOOLS Vocational	S.F.	41.65	56.30	75.30				800
	0020	Total project costs	C.F.	2.62	3.62	4.91				
	2720	Plumbing	S.F.	2.79	4.11	5.90	5.40%	7%	8.50%	
	2770	Heating, ventilating, air conditioning		5.29	7.30	12.15	9.30%	12.50%	17%	
	2900	Electrical		4.52	6.10	9.10	9.50%	11.80%	13.90%	
	3100	Total: Mechanical & Electrical		10.20	14.80	22.45	21.10%	29.70%	36%	
	9000	Per pupil, total cost	Ea.	2,550	15,400	22,800				
	9500	Total: Mechanical & Electrical	"	845	2,050	5,250				
830	0010	SPORTS ARENAS	S.F.	36.55	46.05	58.75				830
	0020	Total project costs	C.F.	1.99	3.65	4.59				
	2720	Plumbing	S.F.	1.87	3.21	5.75	4.30%	6.30%	8.50%	
	2770	Heating, ventilating, air conditioning		3.88	5.40	7.05	5.80%	10.20%	13.50%	
	2900	Electrical		3.02	5.05	6.20	7.10%	9.70%	12.20%	
	3100	Total: Mechanical & Electrical		6.70	11.85	15.25	13.40%	22.50%	30.80%	
850	0010	SUPERMARKETS		33.20	38.45	44.30				850
	0020	Total project costs	C.F.	1.85	2.19	2.81				
	2720	Plumbing	S.F.	1.81	2.38	2.79	5%	6%	6.90%	
	2770	Heating, ventilating, air conditioning		2.78	3.31	4.04	8.50%	8.60%	9.50%	
	2900	Electrical		3.93	4.80	5.70	10.40%	12.50%	13.40%	
	3100	Total: Mechanical & Electrical		6.25	8.70	11.20	17.80%	21.70%	27.60%	
860	0010	SWIMMING POOLS		51.35	66.45	93.70				860
	0020	Total project costs	C.F.	4.48	5.20	6.10				
	2720	Plumbing	S.F.	3.60	5.90	8.25	4.60%	9.60%	12.40%	
	2900	Electrical		3.90	5.50	8.35	6.50%	7.60%	7.90%	
	3100	Total: Mechanical & Electrical		9.05	16.30	31.95	17.50%	24.90%	31%	
870	0010	TELEPHONE EXCHANGES	S.F.	74.25	102	135				870
	0020	Total project costs	C.F.	4.49	6.65	9.60				
	2720	Plumbing	S.F.	2.57	4.42	6.50	3.50%	5.70%	6.60%	
	2770	Heating, ventilating, air conditioning		6.25	14.35	17.80	11.70%	16%	18.40%	
	2900	Electrical		6.50	11.65	21.15	10.30%	14%	17.80%	
	3100	Total: Mechanical & Electrical		15.10	20.70	40.85	19.80%	27.40%	35%	
	0010	TERMINALS Bus		33.80	51.30	61.40				890
	0020	Total project costs	C.F.	1.67	2.89	3.70				
	2720	Plumbing	S.F.	1.12	2.53	3.89	2.30%	7.20%	8.80%	
	2900	Electrical		1.74	2.33	6.65	3.90%	7.50%	11.80%	
	3100	Total: Mechanical & Electrical		1.77	5.44	8.65	8.30%	16.90%	19.50%	
910	0010	THEATERS	S.F.	42.75	54.30	84.70				910
	0020	Total project costs	C.F.	2.23	3.17	4.60				
	2720	Plumbing	S.F.	1.42	1.65	4.95	2.90%	4.60%	6.10%	
	2770	Heating, ventilating, air conditioning		3.95	5.30	6.10	7.30%	11.60%	13.30%	
	2900	Electrical		4	5.40	8.35	8%	9.90%	11.80%	
	3100	Total: Mechanical & Electrical		8.75	11.20	20.55	16%	24.90%	27.30%	

For expanded coverage of these items see *Means Square Foot Cost Data 1987*

Figure 22.7

There are two stages of project development when square foot cost estimates are most useful. The first is during the conceptual stage when few, if any details are available. At this time, square foot costs are appropriate for ballpark budget purposes. As soon as details become available in the project design, the square foot approach should be discontinued and the project priced more accurately. After the estimate is completed, square foot costs can be used again — this time for verification and as a check against gross errors.

When using the figures in Division 17, it is recommended that the median cost column be consulted for preliminary figures if no additional information is available. When costs have been converted for location (see City Cost Indexes) the median numbers (as shown in Figure 22.7) should provide a fairly accurate base figure. This figure should then be adjusted according to the estimator's experience, local economic conditions, code requirements, and the owner's particular project requirements. There is no need to factor the percentage figures, as these should remain relatively constant from city to city.

Repair and Remodeling
Cost figures in *Means Mechanical Cost Data* are based on new construction utilizing the most cost-effective combination of labor, equipment, and material. The work is scheduled in the proper sequence to allow the various trades to accomplish their tasks in an efficient manner. Figure 22.8 (from Division 18 of *Means Mechanical Cost Data*) shows factors that can be used to adjust figures in other sections of the book for repair and remodeling projects. For expanded coverage, see Means' *Repair and Remodeling Cost Data*.

Section B — Assemblies Cost Tables

Means' systems data are divided into twelve "uniformat" divisions, which organize the components of construction into logical groupings. The Systems or Assemblies approach was devised to provide quick and easy methods for estimating even when only preliminary design data are available. The groupings, or systems, are presented in such a way so that the estimator can substitute one system for another. This is extremely useful when adapting to budget, design, or other considerations. Figure 22.9 shows how the data are presented in the Systems section.

Each system is illustrated and accompanied by a detailed description. The book lists the components and sizes of each system. Each individual component is found in the Unit Price Section.

Quantity
A unit of measure is established for each system. For example, sprinkler systems are measured by the square foot of floor area; plumbing fixture systems are measured by "each"; HVAC systems are measured by the square foot of floor area. Within each system, the components are measured by industry standard, using the same units as in the Unit Price section.

Material
The cost of each component in the Material column is the "Bare Material Cost", plus 10% handling, for the unit and quantity as defined in the "Quantity" column.

Installation
Installation costs as listed in the Systems pages contain both labor and equipment costs. The labor rate includes the "Bare Labor Cost" plus the

17.1 S.F., C.F. and % of Total Costs

				UNIT COSTS			% OF TOTAL			
			UNIT	¼	MEDIAN	¾	¼	MEDIAN	¾	
940	0010	TOWN HALLS City Halls & Municipal Buildings	S.F.	51.75	64.95	85.05				940
	0020	Total project costs	C.F.	3.49	5.15	6.55				
	2720	Plumbing	S.F.	1.81	3.55	6.20	4.20%	5.90%	7.90%	
	2770	Heating, ventilating, air conditioning		3.85	7.65	8.75	7%	9%	13.20%	
	2900	Electrical		4.10	6.10	8.65	7.90%	9.40%	11.60%	
	3100	Total: Mechanical & Electrical		8.60	13.75	21.15	15.60%	21.10%	30%	
970	0010	WAREHOUSES And Storage Buildings		18.60	25.80	39.55				970
	0020	Total project costs	C.F.	.97	1.54	2.55				
990	0010	WAREHOUSE & OFFICES Combination	S.F.	22.35	29.05	40.80				990
	0020	Total project costs	C.F.	1.17	1.74	2.55				
	2720	Plumbing	S.F.	.89	1.52	2.35	3.60%	4.60%	6.20%	
	2770	Heating, ventilating, air conditioning		1.43	2.22	3.10	5%	5.60%	9.50%	
	2900	Electrical		1.55	2.29	3.56	5.90%	7.90%	9.90%	
	3100	Total: Mechanical & Electrical		3.01	5	7.70	11.50%	16%	21.30%	

For expanded coverage of these items see *Means Square Foot Cost Data 1987*

18.1 Repair & Remodeling

			CREW	DAILY OUTPUT	MAN-HOURS	UNIT	BARE COSTS MAT.	LABOR	EQUIP.	TOTAL	TOTAL INCL O&P	
120	0010	CUTOUTS Openings to 5 S.F., interior walls, not incl. re-framing										120
	6010											
	6100	Drywall to ⅝" thick C18.1-100	F-1	24	.333	Ea.		6.85	.44	7.29	10.55	
	6200	Paneling to ¾" thick		20	.400			8.20	.53	8.73	12.65	
	6300	Plaster on gypsum lath		20	.400			8.20	.53	8.73	12.65	
	6340	On wire lath		14	.571			11.75	.76	12.51	18.05	
	7100	Wood framing, opening to 5 S.F., not including re-framing										
	7200	Floors, incl. subfloor, underlayment, wood or resilient flooring										
	7210	up to 2" thick	F-1	5	1.600	Ea.		33	2.12	35.12	51	
	7300	Roofs, sheathing to 1" thick		6	1.330			27	1.77	28.77	42	
	7400	Walls, sheathing to 1" thick		7	1.140			23	1.51	24.51	36	
200	0010	FACTORS To adjust figures in other sections of this										200
	0020	book for repair and remodeling projects:										
	0504	Cut & patch to match existing construction, add, minimum C18.1-100				Costs	2%	3%				
	0550	Maximum					5%	9%				
	0800	Dust protection, add, minimum					1%	2%				
	0850	Maximum					4%	11%				
	1100	Equipment usage curtailment, add, minimum					1%	1%				
	1150	Maximum					3%	10%				
	1400	Material handling & storage limitation, add, minimum					1%	1%				
	1450	Maximum					6%	7%				
	1700	Protection of existing work, add, minimum					2%	2%				
	1750	Maximum					5%	7%				
	2000	Shift work requirements, add, minimum						5%				
	2050	Maximum						30%				
	2300	Temporary shoring and bracing, add, minimum					2%	5%				
	2350	Maximum					5%	12%				
250	0010	CLEANING MASONRY No staging included										
	0200	Chemical cleaning, brush and wash, average	D-1	1,000	.016	S.F.	.04	.29		.33	.46	
	0400	High pressure, water only, average	B-9	3,000	.013			.22	.05	.27	.38	
	0800	Water and chemical, average	"	1,250	.032		.03	.53	.12	.68	.94	
300	0010	CAULKING MASONRY No staging included										300
	0020	Re-caulk only, ½" x ½" joint										
	0050	Oil base	1 Bric	225	.036	L.F.	.14	.73		.87	1.20	
	0100	Butyl		205	.039		.27	.80		1.07	1.45	
	0200	Polysulfide		200	.040		.40	.82		1.22	1.62	
	0300	Silicone rubber		195	.041		.56	.84		1.40	1.83	

For expanded coverage of these items see *Means Repair & Remodeling Cost Data 1987*

Figure 22.8

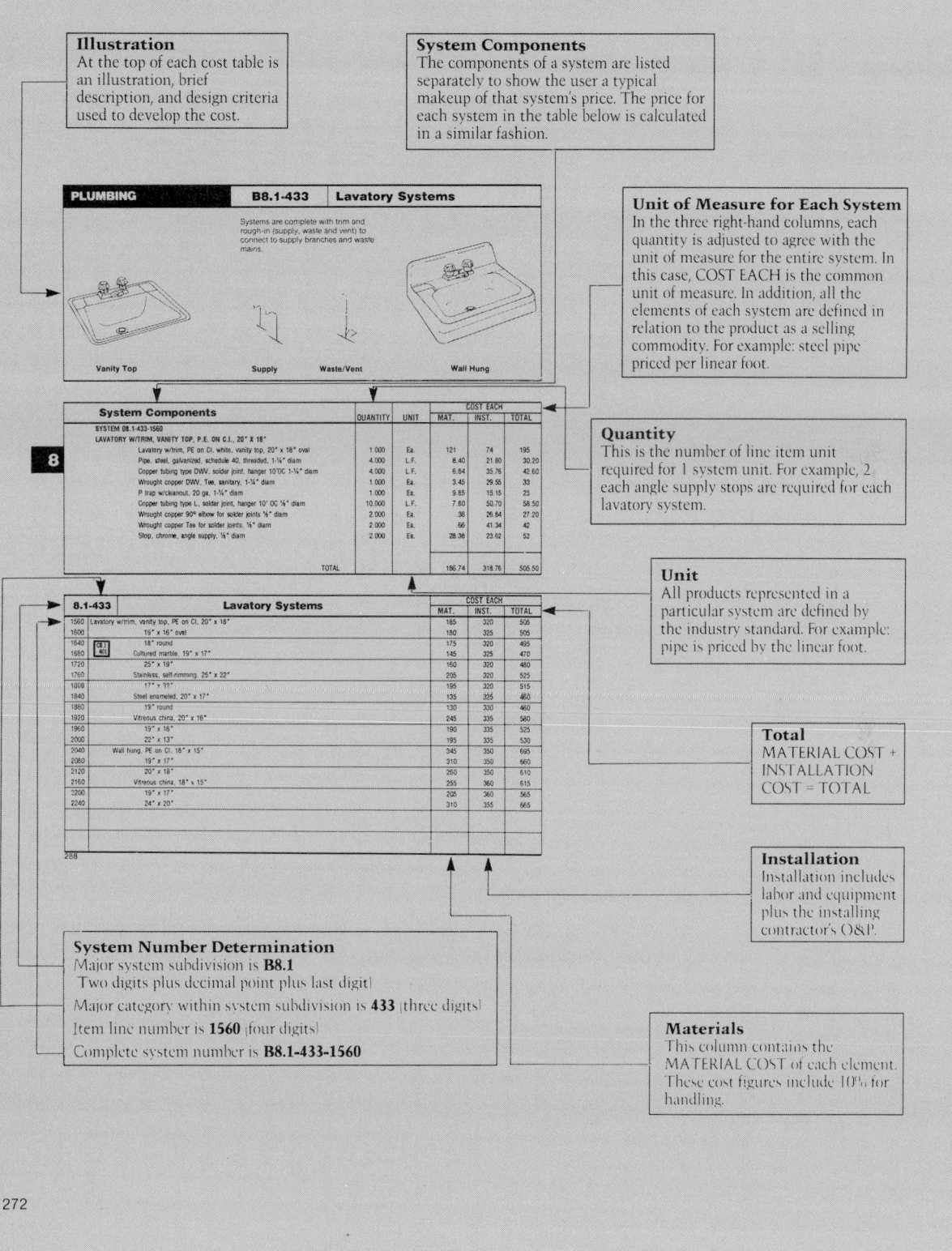

Figure 22.9

Section C — Estimating References

installing contractor's overhead and profit. These rates are shown in Figure 22.3. The equipment rate is the "Bare Equipment Cost", plus 10%.

Throughout the Unit Price and Systems sections are reference numbers highlighted with bold squares. These numbers serve as footnotes, referring the reader to illustrations, charts, and estimating reference tables in Section C, as well as to related information in Sections A and B. Figure 22.10 shows an example reference number (for plumbing fixtures) as it appears on a Unit Price page. Figure 22.11 shows the corresponding reference page from Section B. The development of unit costs for many items is explained in these reference tables. Design criteria for many types of mechanical systems are also included to aid the designer/estimator in making appropriate choices.

15.2 Plumbing Fixtures

			CREW	DAILY OUTPUT	MAN-HOURS	UNIT	BARE COSTS MAT.	LABOR	EQUIP.	TOTAL	TOTAL INCL O&P	
660	0930	Elongated bowl	1 Plum	24	.333	Ea.	16.35	7.65		24	29	660
	1000	Solid plastic, white										
	1030	Industrial, w/o cover, open front, regular bowl	1 Plum	24	.333	Ea.	12.70	7.65		20.35	25	
	1080	Extra heavy, concealed check hinge		24	.333		18.10	7.65		25.75	31	
	1100	Self-sustaining hinge		24	.333		13.80	7.65		21.45	26	
	1150	Elongated bowl		24	.333		12.80	7.65		20.45	25	
	1170	Concealed check		24	.333		13.20	7.65		20.85	26	
	1190	Self-sustaining hinge, concealed check		24	.333		14.10	7.65		21.75	27	
	1220	Residential, with cover, closed front, regular bowl		24	.333		17.70	7.65		25.35	31	
	1240	Elongated bowl		24	.333		22.25	7.65		29.90	36	
	1260	Open front, regular bowl		24	.333		17.70	7.65		25.35	31	
	1280	Elongated bowl		24	.333		22.25	7.65		29.90	36	
680	0010	**URINALS**										680
	3000	Wall hung, vitreous china, with hanger & self-closing valve	Q-1	3	5.330	Ea.	300	110		410	490	
	3300	Rough-in, supply, waste & vent		1.99	8.040		53.36	165		218.36	300	
	5000	Stall type, vitreous china, includes valve		2.50	6.400		370	130		500	600	
	5100	3" seam cover, add		12	1.330		93	28		121	145	
	5200	6" seam cover, add		12	1.330		130	28		158	185	
	6980	Rough-in, supply, waste and vent		1.99	8.040		68.61	165		233.61	320	
720	0010	**WASH CENTER** Prefabricated, stainless steel, semirecessed										720
	0050	Lavatory, storage cabinet, mirror, light & switch, electric										
	0060	outlet, towel dispenser, waste receptacle & trim										
	0100	Foot water valve, cup & soap dispenser, 16" W x 54-¾" H	Q-1	8	2	Ea.	960	41		1,001	1,125	
	0200	Handicap, wrist blade handles, 17" W x 66-½" H		8	2		710	41		751	840	
	0220	20" W x 67-⅜" H		8	2		945	41		986	1,100	
	0300	Push button metering & thermostatic mixing valves										
	0320	Handicap 17" W x 27-½" H	Q-1	8	2	Ea.	570	41		611	685	
	0400	Rough-in, supply, waste and vent	"	2.10	7.620	"	34.65	160		194.65	270	
760	0010	**WASH FOUNTAINS** Rigging not included										760
	1900	Group, foot control										
	2000	Precast terrazzo, circular, 36" diam., 5 or 6 persons	Q-2	3	8	Ea.	915	170		1,085	1,250	
	2100	54" diameter for 8 or 10 persons		2.50	9.600		1,065	205		1,270	1,475	
	2400	Semi-circular, 36" diam. for 3 persons		3	8		835	170		1,005	1,175	
	2420	36" diam. for 3 persons in wheelchairs		3	8		1,370	170		1,540	1,750	
	2500	54" diam. for 4 or 5 persons		2.50	9.600		1,025	205		1,230	1,425	
	2520	54" diam. for 4 persons in wheelchairs		2.50	9.600		1,475	205		1,680	1,925	
	2700	Quarter circle (corner), 54" for 3 persons		3.50	6.860		1,085	145		1,230	1,400	
	2720	54" diam. for 3 persons in wheelchairs		3.50	6.860		1,610	145		1,755	1,975	
	3000	Stainless steel, circular, 36" diameter		3.50	6.860		1,090	145		1,235	1,425	
	3100	54" diameter		2.80	8.570		1,415	185		1,600	1,825	
	3400	Semi-circular, 36" diameter		3.50	6.860		940	145		1,085	1,250	
	3500	54" diameter		2.80	8.570		1,220	185		1,405	1,600	
	5000	Thermoplastic, circular, 36" diameter		3.50	6.860		1,050	145		1,195	1,375	
	5100	54" diameter		2.80	8.570		1,370	185		1,555	1,775	
	5400	Semi-circular, 36" diameter		3.50	6.860		905	145		1,050	1,200	
	5600	54" diameter		2.80	8.570		1,175	185		1,360	1,550	
	5700	Rough-in, supply, waste and vent for above wash fountains	Q-1	1.38	11.590			240		240	350	
	6200	Duo for small washrooms, stainless steel		2	8		505	165		670	795	
	6400	Bowl with backsplash		2	8		590	165		755	890	
	6500	Rough-in, supply, waste & vent for duo fountains		2.02	7.920		30.18	165		195.18	270	
800	0010	**WATER CLOSETS**										800
	0020	For seats, see 15.2-660										
	0150	Tank type, vitreous china, incl. seat, supply pipe w/stop										
	0200	Wall hung, one piece	Q-1	5.30	3.020	Ea.	470	62		532	610	
	0400	Two piece, close coupled		5.30	3.020		305	62		367	425	
	0960	For rough-in, supply, waste, vent and carrier		2.24	7.140		121.74	150		271.74	350	
	1000	Floor mounted, one piece		5.30	3.020		370	62		432	500	
	1020	One piece, low profile		5.30	3.020		625	62		687	780	

Figure 22.10

PLUMBING — B8.1-560 Group Wash Fountains

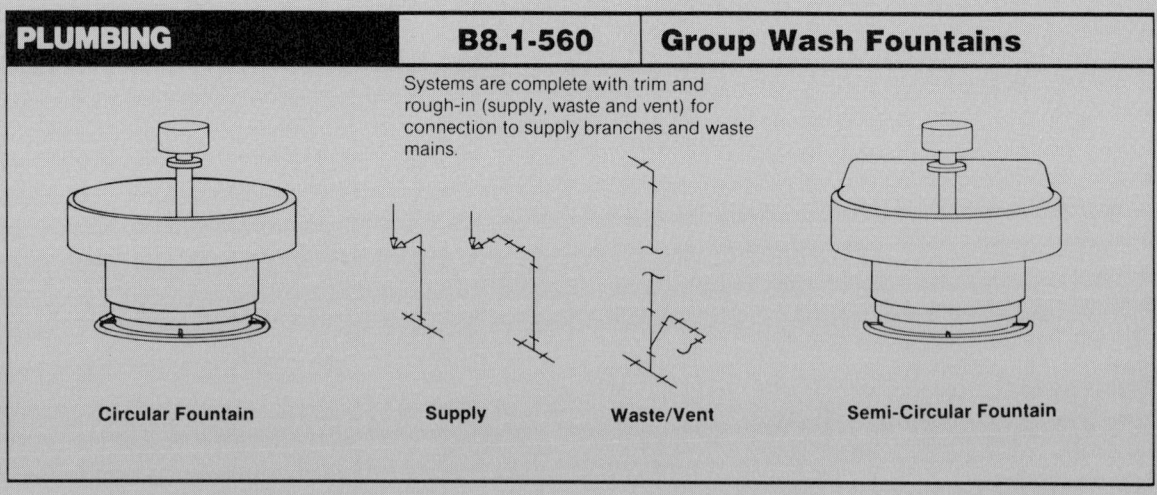

Systems are complete with trim and rough-in (supply, waste and vent) for connection to supply branches and waste mains.

Circular Fountain — Supply — Waste/Vent — Semi-Circular Fountain

System Components	QUANTITY	UNIT	COST EACH MAT.	INST.	TOTAL
SYSTEM 08.1-560-1760					
GROUP WASH FOUNTAIN, PRECAST TERRAZZO					
CIRCULAR, 36" DIAMETER					
Wash fountain, group, precast terrazzo, foot control 36" diam	1.000	Ea.	1,006.50	243.50	1,250
Copper tubing type DWV, solder joint, hanger 10'OC, 2" diam	10.000	L.F.	27.40	111.60	139
P trap, standard, copper, 2" diam	1.000	Ea.	16.78	18.22	35
Wrought copper, Tee, sanitary, 2" diam	1.000	Ea.	5.89	38.11	44
Copper tubing type L, solder joint, hanger 10' OC ½" diam	20.000	L.F.	15.60	101.40	117
Wrought copper 90° elbow for solder joints ½" diam	3.000	Ea.	.54	40.26	40.80
Wrought copper Tee for solder joints, ½" diam	2.000	Ea.	.66	41.34	42
TOTAL			1,073.37	594.43	1,667.80

8.1-560	Group Wash Fountain Systems	MAT.	INST.	TOTAL
1740	Group wash fountain, precast terrazzo			
1760	Circular, 36" diameter	1,075	595	1,670
1800	54" diameter	1,250	655	1,905
1840	Semi-circular, 36" diameter	985	605	1,590
1880	54" diameter	1,200	650	1,850
1960	Stainless steel, circular, 36" diameter	1,275	575	1,850
2000	54" diameter	1,625	620	2,245
2040	Semi-circular, 36" diameter	1,100	565	1,665
2080	54" diameter	1,400	610	2,010
2160	Thermoplastic, circular, 36" diameter	1,225	570	1,795
2200	54" diameter	1,575	620	2,195
2240	Semi-circular, 36" diameter	1,050	555	1,605
2280	54" diameter	1,350	610	1,960

Figure 22.11

City Cost Indexes

The unit prices in *Means Mechanical Cost Data* are national averages. When they are to be applied to a particular location, these prices must be adjusted to local conditions. R.S. Means has developed the City Cost Indexes for just that purpose. Section C of *Means Mechanical Cost Data* contains tables of indexes for 162 U.S. and Canadian cities based on a 30 major city average of 100. The figures are broken down into material and installation for all trades, as shown in Figure 22.12. Please note that for each city there is a weighted average based on total project costs. This average is based on the relative contribution of each division to the construction process as a whole.

In addition to adjusting the figures in *Means Mechanical Cost Data* for particular locations, the City Cost Index can also be used to adjust costs from one city to another. For example, the price of the mechanical work for a particular building type is known for City A. In order to budget the costs of the same building type in City B, the following calculation can be made:

$$\frac{\text{City B Index}}{\text{City A Index}} \times \text{City A Cost} = \text{City B Cost}$$

While City Cost Indexes provide a means to adjust prices for location, the Historical Cost Index, (also included in *Means Mechanical Cost Data* and shown in Figure 22.13) provides a means to adjust for time. Using the same principle as above, a time-adjustment factor can be calculated:

$$\frac{\text{Index for Year X}}{\text{Index for Year Y}} \times \text{Time-adjustment Factor}$$

This time-adjustment factor can be used to determine the budget costs for a particular building type in Year X, based on costs for a similar building type known from Year Y. Used together, the two indexes allow for cost adjustments from one city during a given year to another city in another year (the present or otherwise). For example, an office building built in San Francisco in 1974 originally cost $1,000,000. How much will a similar building cost in Phoenix in 1987? Adjustment factors are developed as shown above using data from Figures 22.12 and 22.13:

$$\frac{\text{Phoenix index}}{\text{San Francisco index}} = \frac{92.8}{123.4} = 0.75$$

$$\frac{1986 \text{ index}}{1974 \text{ index}} = \frac{195.8}{94.7} = 2.07$$

Original cost x location adjustment x time adjustment =

Proposed new cost $1,000,000 x 0.75 x 2.07 = $1,552,500

CITY COST INDEXES — C13.1-100 City Modifiers

DIVISION		ALABAMA										ALASKA			ARIZONA		
		BIRMINGHAM			HUNTSVILLE			MOBILE			MONTGOMERY			ANCHORAGE			PHOENIX		
		MAT.	INST.	TOTAL	MAT.	INST.	TOTAL	MAT.	INST.	TOTAL	MAT.	INST.	TOTAL	MAT.	INST.	TOTAL	MAT.	INST.	TOTAL
2	SITE WORK	96.7	91.2	94.3	115.6	92.1	105.1	118.5	88.9	105.3	88.1	87.7	87.9	154.7	131.7	144.4	89.9	95.0	92.2
3.1	FORMWORK	90.5	72.6	76.5	92.7	73.4	77.6	97.0	78.7	82.7	102.1	70.2	77.2	114.1	146.3	139.3	108.4	91.8	95.4
3.2	REINFORCING	94.6	78.8	88.0	95.8	73.3	86.4	83.0	77.9	80.9	83.0	78.8	81.2	117.8	140.5	127.3	111.7	105.7	109.2
3.3	CAST IN PLACE CONC.	89.4	92.0	91.0	102.0	93.6	96.8	100.1	94.6	96.7	101.4	90.8	94.8	226.0	115.0	157.7	107.7	93.4	98.9
3	CONCRETE	90.7	83.2	85.9	98.8	83.9	89.2	95.7	86.9	90.1	97.5	81.7	87.3	180.4	129.5	147.7	108.7	93.9	99.2
4	MASONRY	79.9	76.7	77.5	86.1	69.8	73.7	92.2	84.5	86.3	85.2	56.9	63.6	137.0	149.3	146.4	93.3	69.5	75.1
5	METALS	95.8	84.0	91.6	100.4	80.1	93.1	93.7	84.2	90.3	96.1	83.9	91.7	116.7	130.9	121.7	98.7	101.1	99.6
6	WOOD & PLASTICS	91.7	73.7	81.5	104.1	73.6	86.8	92.0	80.4	85.5	101.4	73.9	85.9	117.9	141.2	131.1	99.6	90.7	94.6
7	MOISTURE PROTECTION	84.6	68.4	79.5	92.0	69.7	85.0	87.1	71.7	82.3	88.3	68.2	82.0	102.5	143.5	115.3	92.7	89.6	91.7
8	DOORS, WINDOWS, GLASS	91.0	74.3	82.3	101.5	67.4	83.8	90.9	78.8	88.5	98.3	72.1	84.7	129.4	136.8	133.3	103.8	89.3	96.3
9.1	LATH & PLASTER	96.1	71.2	77.2	87.0	73.3	76.6	91.9	86.0	87.5	108.4	73.7	82.1	120.4	148.3	141.6	93.1	93.5	93.4
9.2	DRYWALL	100.5	73.4	87.7	108.5	71.9	91.2	92.5	81.5	87.3	100.7	74.5	88.3	121.9	143.8	132.3	90.8	90.1	90.5
9.5	ACOUSTICAL WORK	98.7	73.3	84.9	101.1	73.2	86.0	94.2	80.0	86.5	94.2	73.0	82.7	125.6	142.7	134.9	103.1	89.8	95.9
9.6	FLOORING	112.0	79.6	103.5	97.4	70.7	90.4	114.0	86.4	106.8	100.4	50.7	87.4	117.3	149.3	125.6	93.0	74.5	88.2
9.8	PAINTING	104.5	71.9	78.4	110.7	70.5	78.5	121.5	81.6	89.5	119.7	80.2	88.0	123.2	146.3	141.7	96.5	91.2	92.3
9	FINISHES	103.3	73.2	87.1	105.1	71.5	87.0	100.4	82.0	90.5	102.2	74.7	87.4	121.3	145.3	134.2	92.9	89.6	91.1
10-14	TOTAL DIV. 10-14	100.0	75.4	92.6	100.0	73.2	91.9	100.0	82.7	94.8	100.0	71.4	91.4	100.0	143.2	112.9	100.0	89.8	96.9
15	MECHANICAL	96.5	77.7	86.9	99.6	77.4	88.3	97.5	79.8	88.4	99.2	74.6	86.6	107.5	136.6	122.4	98.5	85.4	91.8
16	ELECTRICAL	94.1	78.1	82.9	92.1	77.7	82.0	90.0	82.1	84.5	90.9	66.4	73.7	108.4	147.0	135.5	105.0	88.9	93.7
1-16	WEIGHTED AVERAGE	94.4	78.6	85.8	99.6	76.9	87.3	97.3	82.9	89.5	96.3	73.1	83.7	124.8	139.2	132.6	99.1	87.5	92.8

DIVISION		ARIZONA			ARKANSAS						CALIFORNIA								
		TUCSON			FORT SMITH			LITTLE ROCK			ANAHEIM			BAKERSFIELD			FRESNO		
		MAT.	INST.	TOTAL	MAT.	INST.	TOTAL	MAT.	INST.	TOTAL	MAT.	INST.	TOTAL	MAT.	INST.	TOTAL	MAT.	INST.	TOTAL
2	SITE WORK	106.6	97.0	102.3	96.6	91.8	94.4	103.3	94.1	99.2	101.0	113.1	106.4	93.1	111.2	101.2	91.7	120.5	104.5
3.1	FORMWORK	100.7	89.7	92.1	102.5	65.7	73.7	95.7	70.9	76.3	94.9	129.7	122.1	112.8	129.7	126.0	99.6	123.5	118.3
3.2	REINFORCING	95.1	105.7	99.5	124.6	71.3	102.3	117.8	69.1	97.5	99.3	130.1	112.2	96.1	130.1	110.3	106.5	130.1	116.3
3.3	CAST IN PLACE CONC.	105.7	97.7	100.7	90.6	91.9	91.4	98.5	92.3	94.7	109.5	109.2	109.3	103.4	109.4	107.1	93.1	107.3	101.9
3	CONCRETE	102.4	95.2	97.8	100.4	79.8	87.2	102.2	81.9	89.2	104.4	119.1	113.8	103.6	119.2	113.6	97.3	115.7	109.1
4	MASONRY	90.2	69.4	74.3	93.5	78.4	81.9	89.8	78.8	81.4	106.6	127.3	122.4	98.8	115.8	111.8	114.7	111.9	112.6
5	METALS	91.1	102.7	95.3	96.9	79.0	90.5	106.6	77.7	96.3	99.5	122.1	107.8	99.7	122.4	107.8	95.3	123.5	105.3
6	WOOD & PLASTICS	107.5	87.9	96.4	106.9	65.8	83.6	94.4	72.2	81.9	95.9	126.5	113.2	95.2	126.5	112.9	96.9	120.1	110.0
7	MOISTURE PROTECTION	105.6	77.5	96.8	84.7	67.3	79.2	84.2	68.3	79.2	108.2	131.6	115.5	84.7	117.6	95.0	107.7	108.2	107.8
8	DOORS, WINDOWS, GLASS	88.4	88.3	88.4	93.2	61.6	76.8	95.7	65.3	79.9	93.8	123.1	107.1	100.4	123.4	112.4	101.4	120.6	111.4
9.1	LATH & PLASTER	109.1	93.8	97.5	93.0	76.0	80.1	98.5	78.4	83.3	97.1	132.4	123.8	95.4	110.3	106.7	101.9	126.3	120.4
9.2	DRYWALL	82.1	88.7	85.2	95.1	64.6	80.7	114.8	71.3	94.2	97.4	127.9	111.9	98.0	121.7	109.2	98.7	124.0	110.7
9.5	ACOUSTICAL WORK	114.9	87.4	99.9	84.5	64.5	73.6	84.5	71.2	77.3	82.2	127.5	106.8	94.1	127.5	112.3	97.4	121.1	110.3
9.6	FLOORING	109.9	74.0	100.5	89.3	78.8	86.6	88.5	80.1	86.3	117.3	125.5	119.7	111.9	112.5	112.1	88.5	106.3	93.2
9.8	PAINTING	98.6	84.1	87.0	111.1	52.9	64.4	104.7	66.8	74.3	108.3	123.9	120.8	120.2	108.1	110.5	108.0	121.8	119.0
9	FINISHES	93.3	86.3	89.5	94.5	62.2	77.1	105.0	70.8	86.5	101.8	126.6	115.2	103.0	116.1	110.1	97.3	121.9	110.6
10-14	TOTAL DIV. 10-14	100.0	87.4	96.2	100.0	74.6	92.4	100.0	75.4	92.6	100.0	128.4	108.5	100.0	125.2	107.5	100.0	146.2	113.8
15	MECHANICAL	98.9	85.4	92.0	97.5	69.3	83.1	97.1	73.6	85.1	96.9	126.2	111.9	95.1	107.5	101.4	92.7	120.9	107.1
16	ELECTRICAL	102.3	87.9	92.2	99.3	76.6	83.3	93.6	80.2	84.2	98.7	139.7	127.5	106.0	103.4	104.2	109.5	101.0	103.5
1-16	WEIGHTED AVERAGE	98.5	86.9	92.2	96.7	74.2	84.5	98.5	77.2	86.9	100.6	126.1	114.5	98.4	114.7	107.3	98.9	117.0	108.7

DIVISION		CALIFORNIA																	
		LOS ANGELES			OXNARD			RIVERSIDE			SACRAMENTO			SAN DIEGO			SAN FRANCISCO		
		MAT.	INST.	TOTAL	MAT.	INST.	TOTAL	MAT.	INST.	TOTAL	MAT.	INST.	TOTAL	MAT.	INST.	TOTAL	MAT.	INST.	TOTAL
2	SITE WORK	96.1	115.7	104.9	97.9	106.2	101.6	95.1	111.7	102.5	82.9	105.4	93.0	94.9	108.3	100.9	102.6	114.6	108.0
3.1	FORMWORK	112.9	130.1	126.4	90.1	130.0	121.3	102.5	129.7	123.8	101.0	123.7	118.7	105.2	127.5	122.6	104.9	133.7	127.4
3.2	REINFORCING	64.4	130.1	91.8	99.3	130.1	112.2	124.6	130.1	126.9	99.3	130.1	112.2	119.3	130.1	123.8	124.1	130.1	126.6
3.3	CAST IN PLACE CONC.	97.9	112.3	106.8	102.5	109.9	107.1	102.5	109.6	106.9	116.1	106.5	110.2	100.4	105.3	103.4	105.6	116.3	112.1
3	CONCRETE	93.5	120.8	111.1	99.4	119.6	112.4	107.3	119.3	115.0	109.5	115.3	113.2	105.5	116.1	112.3	109.5	124.3	119.0
4	MASONRY	108.8	127.3	122.9	98.4	120.9	115.7	102.8	121.1	116.8	101.4	105.6	104.6	111.3	118.4	116.7	128.6	143.4	139.9
5	METALS	101.8	123.1	109.1	105.6	122.4	111.6	99.6	122.2	107.7	111.4	123.4	115.7	99.0	121.1	106.9	104.2	126.5	112.2
6	WOOD & PLASTICS	100.6	127.6	115.9	92.7	127.1	112.2	94.8	126.5	112.8	78.4	120.4	102.2	97.1	123.8	112.2	94.0	132.6	115.9
7	MOISTURE PROTECTION	103.9	133.5	113.2	89.8	132.4	103.1	90.3	130.7	102.9	85.1	122.6	96.8	94.4	120.1	102.4	100.5	128.7	109.3
8	DOORS, WINDOWS, GLASS	102.2	127.3	115.2	103.0	127.3	115.6	103.0	127.3	115.9	92.2	120.1	106.7	108.1	123.6	116.1	113.4	132.3	123.2
9.1	LATH & PLASTER	95.8	132.5	123.6	97.6	134.0	125.1	97.6	128.3	120.9	98.9	120.1	115.0	102.2	119.7	115.5	101.5	147.1	136.0
9.2	DRYWALL	89.1	127.9	107.5	98.7	130.0	113.5	94.7	127.9	110.5	97.3	120.5	108.3	101.4	124.4	112.3	81.1	136.6	107.4
9.5	ACOUSTICAL WORK	98.7	127.5	114.4	88.4	127.5	109.7	88.4	127.5	109.7	86.6	121.1	105.4	100.7	124.8	113.8	100.7	134.2	118.9
9.6	FLOORING	96.3	126.5	104.2	95.6	126.5	103.7	85.8	126.5	103.7	85.8	133.3	98.3	98.5	121.7	104.6	107.4	140.0	115.9
9.8	PAINTING	84.3	126.3	118.0	92.2	117.8	112.8	100.8	123.9	119.3	112.3	132.4	128.4	92.1	127.9	120.8	102.8	143.5	135.5
9	FINISHES	91.2	127.5	110.7	96.5	125.5	112.1	95.1	126.4	112.0	95.4	125.6	111.6	99.8	125.2	113.5	91.3	139.7	117.4
10-14	TOTAL DIV. 10-14	100.0	128.9	108.6	100.0	128.3	108.5	100.0	128.3	108.5	100.0	147.9	114.3	100.0	126.8	108.0	100.0	153.0	115.9
15	MECHANICAL	97.7	128.8	113.6	98.8	127.4	113.4	96.6	130.9	114.2	98.2	124.1	111.5	103.1	127.6	115.6	100.5	158.4	130.2
16	ELECTRICAL	101.8	132.5	123.4	98.7	157.9	140.3	98.2	134.9	124.0	109.5	99.1	102.2	105.8	108.9	108.0	107.9	151.9	138.8
1-16	WEIGHTED AVERAGE	98.7	126.4	113.7	98.8	127.4	114.3	99.0	125.4	113.3	99.2	116.5	108.6	101.9	119.6	111.5	103.9	139.9	123.4

Figure 22.12

CITY COST INDEXES — C13.1-100 General

Historical Cost Indexes (Div. 1.1-160)

The table below lists both the Means City Cost Index based on Jan. 1, 1975 = 100 as well as the computed value of an index based on January 1, 1987 costs. Since the Jan. 1, 1987 figure is estimated, space is left to write in the actual index figures as they become available thru either the quarterly "Means Construction Cost Indexes" or as printed in the "Engineering News-Record". To compute the actual index based on Jan. 1, 1987 = 100, divide the Quarterly City Cost Index for a particular year by the actual Jan. 1, 1987 Quarterly City Cost Index. Space has been left to advance the index figures as the year progresses.

Year	"Quarterly City Cost Index" Jan. 1, 1975 = 100		Current Index Based on Jan. 1, 1987 = 100		Year	"Quarterly City Cost Index" Jan. 1, 1975 = 100	Current Index Based on Jan. 1, 1987 = 100		Year	"Quarterly City Cost Index" Jan. 1, 1975 = 100	Current Index Based on Jan. 1, 1987 = 100	
	Est.	Actual	Est.	Actual		Actual	Est.	Actual		Actual	Est.	Actual
Oct. 1987					July 1974	94.7	49.3		July 1958	43.0	22.4	
July 1987					1973	86.3	44.9		1957	42.2	22.0	
April 1987					1972	79.7	41.5		1956	40.4	21.0	
Jan. 1987	195.8		100.0	100.0	1971	73.5	38.3		1955	38.1	19.8	
July 1986		192.8	100.6		1970	65.8	34.3		1954	36.7	19.1	
1985		189.1	98.5		1969	61.6	32.1		1953	36.2	18.9	
1984		187.6	97.7		1968	56.9	29.6		1952	35.3	18.4	
1983		183.5	95.6		1967	53.9	28.1		1951	34.4	17.9	
1982		174.3	90.8		1966	51.9	27.0		1950	31.4	16.4	
1981		160.2	83.4		1965	49.7	25.9		1949	30.4	15.8	
1980		144.0	75.0		1964	48.6	25.3		1948	30.4	15.8	
1979		132.3	68.9		1963	47.3	24.6		1947	27.6	14.4	
1978		122.4	63.8		1962	46.2	24.1		1946	23.2	12.1	
1977		113.3	59.0		1961	45.4	23.6		1945	20.2	10.5	
1976		107.3	55.9		1960	45.0	23.4		1944	19.3	10.1	
1975		102.6	53.4		1959	44.2	23.0		1943	18.6	9.7	

City Cost Indexes (Div. 1.1-060)

Tabulated on the following pages are average construction cost indexes for 162 major U.S. and Canadian cities. Index figures for both material and installation are based on the 30 major city average of 100 and represent the cost relationship as of April 1, 1986. The index for each division is computed from representative material and labor quantities for that division. The weighted average for each city is a weighted total of the components listed above it, but does not include relative productivity between trades or cities.

The material index for the weighted average includes about 100 basic construction materials with appropriate quantities of each material to represent typical "average" building construction projects.

The installation index for the weighted average includes the contribution of about 30 construction trades with their representative man-days in proportion to the material items installed. Also included in the installation costs are the representative equipment costs for those items requiring equipment.

Since each division of the book contains many different items, any particular item multiplied by the particular city index may give incorrect results. However, when all the book costs for a particular division are summarized and then factored, the result should be very close to the actual costs for that particular division for that city.

If a project has a preponderance of materials from any particular division (say structural steel), then the weighted average index should be adjusted in proportion to the value of the factor for that division.

414

Figure 22.13

Chapter 23
SQUARE FOOT AND SYSTEMS ESTIMATING EXAMPLES

Often the contractor is faced with the need to develop a preliminary estimate for a project. The estimate may be for the entire project or for only a portion of the work, which is often the case for the mechanical contractor. This chapter contains two complete sample estimates for the mechanical portion of an office building: a square foot estimate and a systems estimate. In these step-by-step examples, forms are filled out and calculations made according to the techniques described in Part I, "The Estimating Process". All cost and reference tables are from the annual *Means Mechanical Cost Data*.

Project Description

A mechanical contractor has been invited by a familiar general contractor to submit a budget estimate on an office building. The project is a three story office building with a penthouse and garage. Area calculations for this building are shown in Figure 23.1.

For purposes of illustration, a budget estimate will first be developed as a Square Foot Estimate based on the bare minimum of information supplied. A sketch of the building concept with overall dimensions is shown in Figure 23.1.

Square Foot Estimating Example

There are two occasions when square foot cost estimates are useful. The first is during the conceptual stage when few, if any, details are available. At this time, square foot costs make a useful starting point for ballpark budget purposes. The second instance where square foot costs are used is after bids are received. S.F. costs may be used at this time to check on the accuracy or competitiveness of the bids. As soon as details become available in the project design, the square foot approach should be discontinued and the project priced by its particular assemblies or unit costs.

The dimensions of the project (based on Figure 23.1) and the mathematical procedures to obtain the areas of the office building and the parking garage are shown in Figure 23.2.

Once the areas of the proposed building are known, size modification calculations can be made as shown in Figure 23.3 taken from *Means Mechanical Cost Data*, 1987. The appropriate building type is located and its typical size provides the denominator for the modifier equation. Dividing through produces the factors of 6.8 for the office and .12 for the garage. It will be seen that 6.8 is off the horizontal scale to the right and .12 is off the scale to the left. In both cases use the maximum value indicated on the vertical scale at its intersection. In other words, the modifier for any size factor less than .5 will be 1.1 and for any size factor greater than 3.5 will be .90 as the example problem illustrates.

Square foot costs are then found in Division 17 of *Means Mechanical Cost Data*, a page of which appears in Figure 23.4, or taken from your own cost records. These square foot costs are multiplied by the appropriate modified areas, and costs for plumbing, heating, ventilating, and air

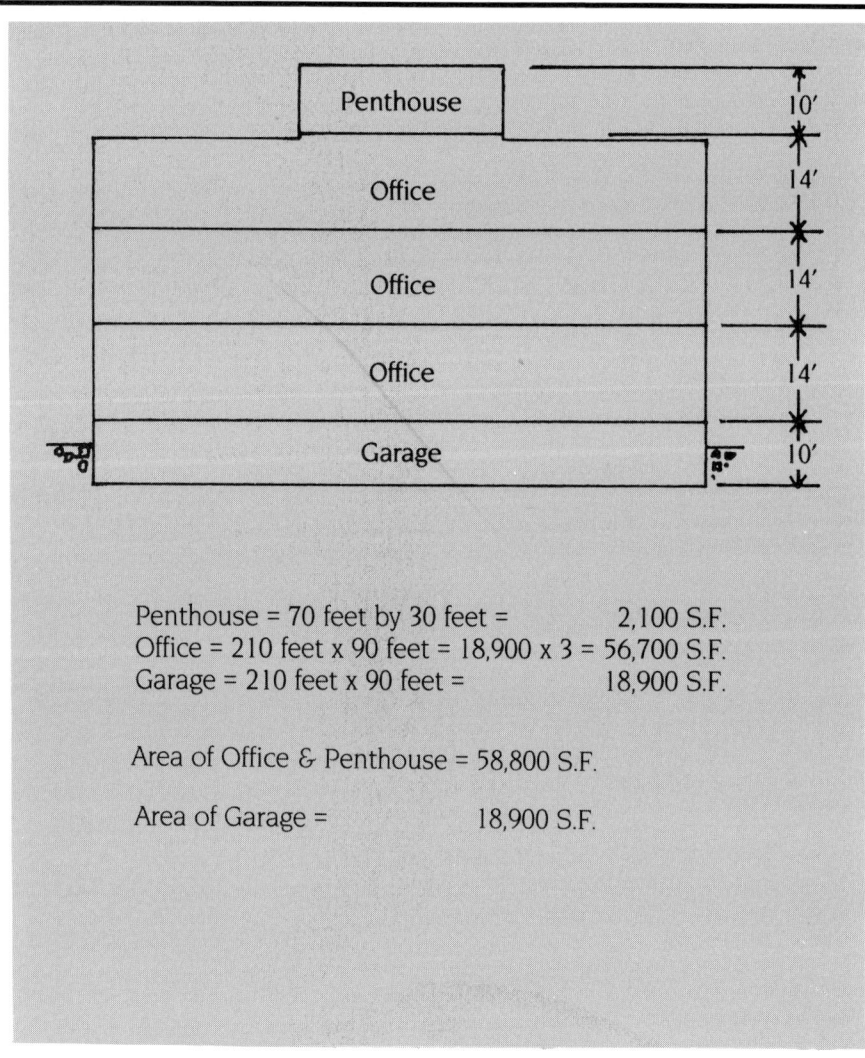

Figure 23.1

conditioning are determined. The total mechanical cost is then modified for geographic area (see Chapter 22, "How to Use *Means Mechanical Cost Data*", Figure 22.12).

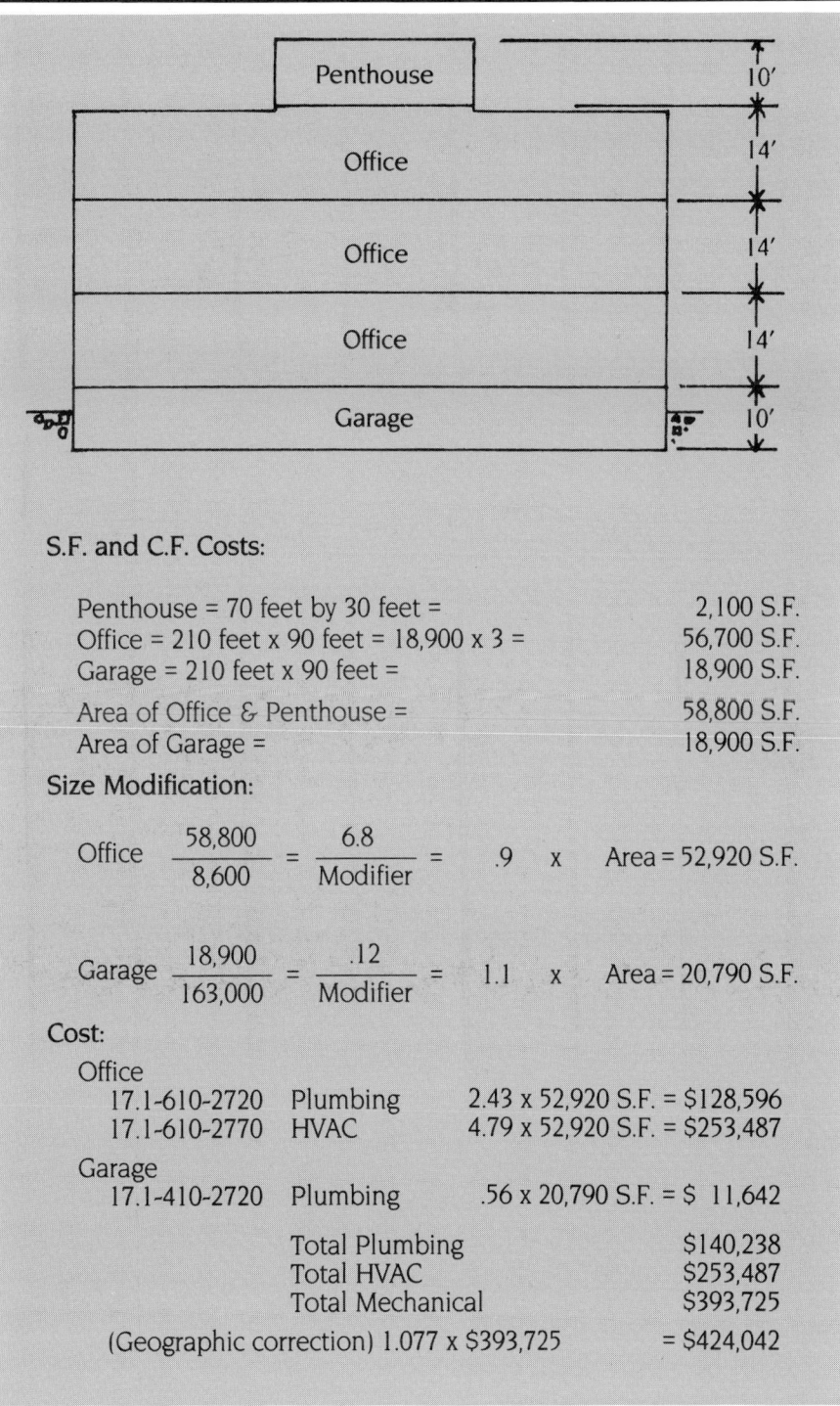

Figure 23.2

SQUARE FOOT — C14.1-100 — S.F., C.F., & % of Total Costs

Square Foot Project Size Modifier (Div. 17.1)

One factor that affects the S.F. cost of a particular building is the size. In general, for buildings built to the same specifications in the same locality, the larger building will have the lower S.F. Cost. This is due mainly to the decreasing contribution of the exterior walls plus the economy of scale usually achievable in larger buildings. The Area Conversion Scale shown below will give a factor to convert costs for the typical size building to an adjusted cost for the particular project.

The Square Foot Base Size lists the median costs, most typical project size in our accumulated data and the range in size of the projects.

The Size Factor for your project is determined by dividing your project area in S.F. by the typical project size for the particular Building Type. With this factor, enter the Area Conversion Scale at the appropriate Size Factor and determine the appropriate cost multiplier for your building size.

Example: Determine the cost per S.F. for a 100,000 S.F. Mid-rise apartment building.

$$\frac{\text{Proposed building area} = 100,000 \text{ S.F.}}{\text{Typical size from below} = 50,000 \text{ S.F.}} = 2.00$$

Enter Area Conversion scale at 2.0, intersect curve, read horizontally the appropriate cost multiplier of 0.94. Size adjusted cost becomes 0.94 × $50.00 = $47.00 based on national average costs.

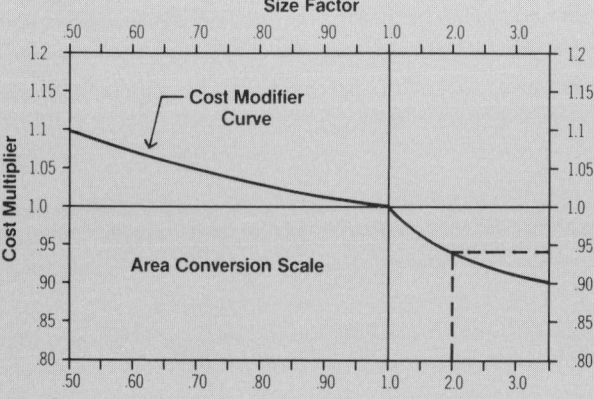

Square Foot Base Size						
Building Type	Median Cost per S.F.	Typical Size Gross S.F.	Typical Range Gross S.F.	Building Type	Median Cost per S.F.	Typical Size Gross S.F.	Typical Range Gross S.F.
Apartments, Low Rise	$ 39.60	21,000	9,700 - 37,200	Jails	$117.00	13,700	7,500 - 28,000
Apartments, Mid Rise	50.00	50,000	32,000 - 100,000	Libraries	71.00	12,000	7,000 - 31,000
Apartments, High Rise	54.85	310,000	100,000 - 650,000	Medical Clinics	68.30	7,200	4,200 - 15,700
Auditoriums	66.65	25,000	7,600 - 39,000	Medical Offices	64.55	6,000	4,000 - 15,000
Auto Sales	42.15	20,000	10,800 - 28,600	Motels	49.05	27,000	15,800 - 51,000
Banks	92.00	4,200	2,500 - 7,500	Nursing Homes	68.15	23,000	15,000 - 37,000
Churches	61.00	9,000	5,300 - 13,200	Offices, Low Rise	54.15	8,600	4,700 - 19,000
Clubs, Country	59.15	6,500	4,500 - 15,000	Offices, Mid Rise	58.25	52,000	31,300 - 83,100
Clubs, Social	58.20	10,000	6,000 - 13,500	Offices, High Rise	72.30	260,000	151,000 - 468,000
Clubs, YMCA	62.55	28,300	12,800 - 39,400	Police Stations	90.20	10,500	4,000 - 19,000
Colleges (Class)	80.80	50,000	23,500 - 98,500	Post Offices	68.15	12,400	6,800 - 30,000
Colleges (Science Lab)	94.30	45,600	16,600 - 80,000	Power Plants	450.00	7,500	1,000 - 20,000
College (Student Union)	86.90	33,400	16,000 - 85,000	Religious Education	50.60	9,000	6,000 - 12,000
Community Center	63.15	9,400	5,300 - 16,700	Research	88.65	19,000	6,300 - 45,000
Court Houses	84.25	32,400	17,800 - 106,000	Restaurants	79.95	4,400	2,800 - 6,000
Dept. Stores	37.20	90,000	44,000 - 122,000	Retail Stores	39.05	7,200	4,000 - 17,600
Dormitories, Low Rise	60.00	24,500	13,400 - 40,000	Schools, Elementary	59.30	41,000	24,500 - 55,000
Dormitories, Mid Rise	75.40	55,600	36,100 - 90,000	Schools, Jr. High	59.10	92,000	52,000 - 119,000
Factories	31.80	26,400	12,900 - 50,000	Schools, Sr. High	58.70	101,000	50,500 - 175,000
Fire Stations	64.80	5,800	4,000 - 8,700	Schools, Vocational	56.30	37,000	20,500 - 82,000
Fraternity Houses	57.45	12,500	8,200 - 14,800	Sports Arenas	46.05	15,000	5,000 - 40,000
Funeral Homes	57.55	7,800	4,500 - 11,000	Supermarkets	38.45	20,000	12,000 - 30,000
Garages, Commercial	43.45	9,300	5,000 - 13,600	Swimming Pools	66.45	13,000	7,800 - 22,000
Garages, Municipal	46.15	8,300	4,500 - 12,600	Telephone Exchange	102.00	4,500	1,200 - 10,600
Garages, Parking	19.90	163,000	76,400 - 225,300	Terminals, Bus	51.30	11,400	6,300 - 16,500
Gymnasiums	55.05	19,200	11,600 - 41,000	Theaters	54.30	10,500	8,800 - 17,500
Hospitals	113.00	55,000	27,200 - 125,000	Town Halls	64.95	10,800	4,800 - 23,400
House (Elderly)	56.00	37,000	21,000 - 66,000	Warehouses	25.80	25,000	8,000 - 72,000
Housing (Public)	47.05	36,000	14,400 - 74,400	Warehouse & Office	29.05	25,000	8,000 - 72,000
Ice Rinks	44.75	29,000	27,200 - 33,600				

425

Figure 23.3

17.1 S.F., C.F. and % of Total Costs

			UNIT	UNIT COSTS ¼	UNIT COSTS MEDIAN	UNIT COSTS ¾	% OF TOTAL ¼	% OF TOTAL MEDIAN	% OF TOTAL ¾	
360	2720	Plumbing	S.F.	3.12	4.87	6.60	5.90%	7.30%	9.50%	360
	2770	Heating, ventilating, air conditioning		2.68	4.31	6.85	4.80%	7.30%	9.20%	
	2900	Electrical		3.63	6.10	8.75	7.20%	9.70%	12.10%	
	3100	Total: Mechanical & Electrical		9.15	14.35	19.60	17.50%	22.60%	27.50%	
370	0010	FRATERNITY HOUSES And Sorority Houses		48.15	57.45	64.10				370
	0020	Total project costs	C.F.	4.60	5.60	6.20				
	2720	Plumbing	S.F.	3.63	4.24	5.65	5.90%	8%	10.80%	
	2900	Electrical		3.17	4.18	7.60	6.50%	8.80%	10.40%	
	3100	Total: Mechanical & Electrical		9	12.75	15.40	14.60%	20.70%	24.20%	
380	0010	FUNERAL HOMES	S.F.	45.80	57.55	84.35				380
	0020	Total project costs	C.F.	3.23	4.65	5.50				
	2720	Plumbing	S.F.	1.81	2.52	2.75	4.10%	4.40%	4.70%	
	2770	Heating, ventilating, air conditioning		4.03	4.10	4.91	7%	9.20%	10.40%	
	2900	Electrical		3.02	3.76	5.90	6.20%	7.70%	11%	
	3100	Total: Mechanical & Electrical		8.25	10.90	12.65	18.80%	20.80%	27.20%	
390	0010	GARAGES, COMMERCIAL		27.45	43.45	57.65				390
	0020	Total project costs	C.F.	1.75	2.55	3.64				
	2720	Plumbing	S.F.	1.77	2.79	5.60	4.90%	7.30%	11%	
	2730	Heating & ventilating		2.61	3.58	4.33	5.20%	7.20%	9.50%	
	2900	Electrical		2.46	4.05	5.65	7.10%	9%	11.40%	
	3100	Total: Mechanical & Electrical		5.70	10.10	14.60	15.70%	21.90%	27.80%	
400	0010	GARAGES, MUNICIPAL		29.50	46.15	64.45				400
	0020	Total project costs	C.F.	2.11	2.86	3.83				
	2720	Plumbing	S.F.	1.84	3.47	5.70	4.10%	6.90%	8.60%	
	2730	Heating & ventilating		2.18	3.69	6.20	4.90%	7.60%	11.90%	
	2900	Electrical		2.58	4.03	5.50	6.30%	8%	10.10%	
	3100	Total: Mechanical & Electrical		5.90	11.75	17.40	15.50%	24.10%	31.50%	
410	0010	GARAGES, PARKING		15.30	19.90	33.50				410
	0020	Total project costs	C.F.	1.37	1.82	2.99				
	2720	Plumbing	S.F.	.28	.56	.80	2.10%	2.80%	3.80%	
	2900	Electrical		.67	.97	1.58	4.20%	5.20%	6.50%	
	3100	Total: Mechanical & Electrical		.99	1.56	2.23	6.80%	8.30%	9.50%	
	9000	Per car, total cost	Car	5,075	6,925	9,750				
	9500	Total: Mechanical & Electrical	"	365	545	670				

	2720	Plumbing	S.F.	4.57	5.60	8.30	8.30%	10.30%	14.10%	
	2770	Heating, ventilating, air conditioning		4.89	7.05	8.30	10.60%	11.70%	11.80%	
	2900	Electrical		5.25	6.65	8.20	9.70%	11%	12.50%	
	3100	Total: Mechanical & Electrical		12	16.35	23.60	22%	28.10%	33.20%	
	9000	Per bed or person, total cost	Bed	20,500	26,500	32,900				
610	0010	OFFICES Low-Rise (1 to 4 story)	S.F.	42.15	54.15	71.30				610
	0020	Total project costs	C.F.	3.10	4.34	5.80				
	2720	Plumbing	S.F.	1.61	2.43	3.48	3.60%	4.50%	6.10%	
	2770	Heating, ventilating, air conditioning		3.47	4.79	7	7.20%	10.40%	11.90%	
	2900	Electrical		3.53	4.90	6.76	7.40%	9.60%	11%	
	3100	Total: Mechanical & Electrical		7.25	10.80	15.85	14.50%	20.50%	26.90%	

For expanded coverage of these items see *Means Square Foot Cost Data 1987*

Figure 23.4

Systems Estimating Example

The next phase of the estimating process, after the preliminary Square Foot Estimate, may be to complete a Systems Estimate. However, the Systems Estimate, like the Square Foot Estimate, is not a substitute for a Unit Price Estimate. A Systems Estimate is most often prepared during the conceptual stage of a project when certain parameters are known, but the building plans are not yet completed. This preliminary estimate may help the designer to keep the project within the owner's budget. The sketch and dimensions of the proposed building would be identical to that used for the square foot estimate, however, in addition to the areas, the occupied volume is calculated. (see Figure 23.5).

Once the physical area and occupied volume have been defined, the number of occupants needs to be determined. Occupancy is based on the net floor area. Figures 23.6 and 23.7 can be used to determine the occupancy requirements. The floor area ratio table in Figure 23.6 indicates a net to gross ratio for offices as 75% on a per occupied floor basis. Therefore, 18,900 S.F. (area of each office floor as shown in Figure 23.5) is

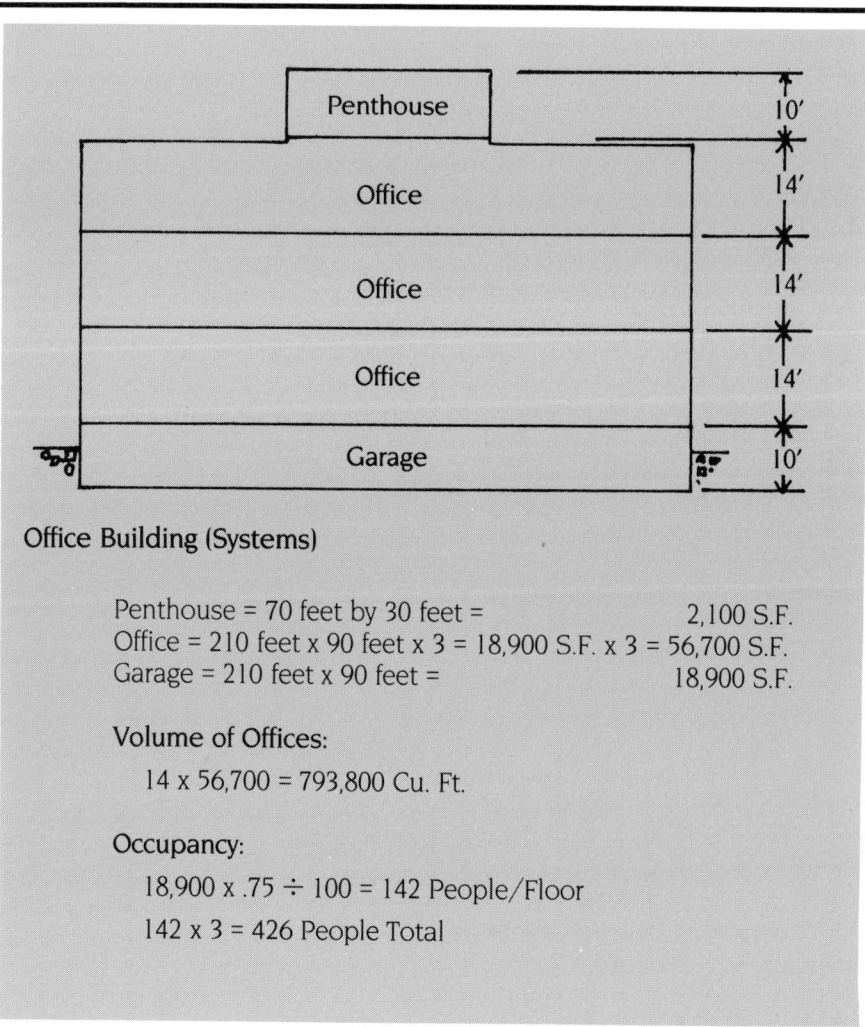

Office Building (Systems)

Penthouse = 70 feet by 30 feet = 2,100 S.F.
Office = 210 feet x 90 feet x 3 = 18,900 S.F. x 3 = 56,700 S.F.
Garage = 210 feet x 90 feet = 18,900 S.F.

Volume of Offices:

14 x 56,700 = 793,800 Cu. Ft.

Occupancy:

18,900 x .75 ÷ 100 = 142 People/Floor
142 x 3 = 426 People Total

Figure 23.5

Floor Area Ratios
Commonly Used Gross-to-Net Area and Net-to-Gross Area Ratios
Expressed in % for Various Building Types

Building Type	Gross to Net Ratio	Net to Gross Ratio	Building Type	Gross to Net Ratio	Net to Gross Ratio
Apartment	156	64	School Buildings (campus type)		
Bank	140	72	Administrative	150	67
Church	142	70	Auditorium	142	70
Courthouse	162	61	Biology	161	62
Department Store	123	81	Chemistry	170	59
Garage	118	85	Classroom	152	66
Hospital	183	55	Dining Hall	138	72
Hotel	158	63	Dormitory	154	65
Laboratory	171	58	Engineering	164	61
Library	132	76	Fraternity	160	63
Office	135	75	Gymnasium	142	70
Restaurant	141	70	Science	167	60
Warehouse	108	93	Service	120	83
			Student Union	172	59

The gross area of a building is the total floor area based on outside dimensions. The net area of a building is the usable floor area for the function intended and excludes such items as stairways, corridors and mechanical rooms. In the case of commerical building, it might be considered as the "leaseable area."

Figure 23.6

Occupancy Determinations

Description		S.F. Required Per Person*			
		BBC	BOCA	SBC	UBC
Assembly Areas	Fixed Seats	6	**	6	7
	Movable Seats	15		15	15
	Concentrated		7		
	Unconcentrated		15		
	Standing Space		3		
Educational	Unclassified	40			
	Classrooms		20	40	20
	Shop Areas		50	100	50
Institutional	Unclassified	150		125	
	In-Patient Areas		240		
	Sleeping Areas		120		
Mercantile	Basement	30	30	30	20
	Ground Floor	30	30	30	30
	Upper Floors	60	60	60	50
Office		100	100	100	100

*BBC = Basic Building Code
BOCA = Building Officials & Code Administrators
SBC = Southern Building Code
UBC = Uniform Building Code

**The occupancy load for assembly area with fixed seats shall be determined by the number of fixed seats installed.

Figure 23.7

reduced to 14,175 S.F. net area. This net or reduced area has excluded the space occupied by stairwells, corridors, mechanical rooms, etc. In a commercial building, the net area might be referred to as the "leasable area". The occupancy determinations table in Figure 23.7 indicates the occupancy recommendations of four major building code authorities for several building classifications. All agree that for office buildings, the maximum density is one person per 100 S.F. Based on this calculation, the net floor area of 14,175 S.F. allows a maximum of 142 persons per floor. Without any data to the contrary, a male female ratio of 50:50 should be assumed. With the basic requirements of the building established, the next step is to develop costs using the systems estimating method.

In a systems estimate, individual components are grouped together into systems that reflect the way buildings are constructed. The grouping of many components into a single system allows the estimator to compare various systems and select the one best suited to accommodate costs, usage, compatibility, and any special requirements for the particular project.

Prior to beginning the estimate, all pertinent data must be assembled and analyzed. A typical preprinted form to record the available data is shown in Figure 23.8.

In a systems estimate the uniformat grouping of twelve construction divisions is used, instead of the sixteen Construction Specifications Institute Masterformat divisions used for Unit Price Estimates. The mechanical data is found in Division 8 of the Uniformat data. (See Chapter 2 for more information on the Uniformat Divisions.)

It is desirable that the building dimensions, the owner's/occupants' requirements, local building codes, existing utilities, local labor sources, and anticipated economic conditions be known. With this information and the inclusion of any contingencies, a reasonable estimate may be arrived at, which can be updated as more design parameters are developed. For this example, the proposed building is a regional office building of good quality for an insurance company located in the Boston area.

A systems estimate for the mechanical trades is accomplished by obtaining all the available data, including physical area, function, occupancy, etc., and selecting the appropriate mechanical systems to cover all the requirements for such a building. The figures and other pertinent information to compile systems and their costs should come from the estimator's own experience and records, or from a reliable cost data source such as *Means Mechanical Cost Data*, which contains 24 plumbing, 8 fire protection, and 25 heating and cooling systems. Some of these systems are priced by the square foot area of a building (e.g., sprinkler systems). Others are priced by the required number of units (e.g., bathrooms, water heaters, etc.). All systems costs include an allowance for overhead and profit.

Plumbing

The table shown in Figure 23.9, "Minimum Plumbing Fixture Requirements," from *Means Mechanical Cost Data* is the basis for determining the number of plumbing fixtures required depending on building use and the number of occupants. It has been determined (see Figure 23.3) that this office building will house 142 persons per floor, with a 50:50 male female ratio.

Means Forms

ASSEMBLY NUMBER	DESCRIPTION	QTY	UNIT	TOTAL COST UNIT	TOTAL COST TOTAL	COST PER S.F.
8.0	**Mechanical System**					
9.0	**Electrical**					

Page 4 of 6

Figure 23.8

Based on this determined occupancy of 142 persons per floor, the minimum plumbing fixture requirements mandate the following fixture count for business offices:

Fixture	Men	Women	Handicapped	Misc.	Totals Per Floor
water closets	2	3	1	—	6
lavatories	3	3	1	—	7
urinal	2	—	—	—	2
water coolers			1	2	3
service sink				1	1

As shown in Figure 23.9, 1/3 of the water closets can be replaced with urinals. This would change the required three men's water closets to two with one urinal. A second urinal has been arbitrarily added based on an estimate of what an occupant would require. Three handicapped fixtures per floor have also been included in the estimate to conform with local and national codes.

In addition to the fixtures and appliances (water coolers), the plumbing estimate would not be complete without water heaters and roof drains. The roof area is 18,900 S.F. and the building height is 52 feet (see Figure 23.10). An additional four feet is allowed to bury the storm drain. With these measurements, the roof drain system can be sized using the chart shown in Figure 23.10 from *Means Mechanical Cost Data*. Six 4" drains are capable of handling up to 20,760 S.F.

Hot water consumption rates are determined by using the table shown in Figure 23.11 from *Means Mechanical Cost Data*. This table shows that an office building has a maximum hourly demand of 0.4 gallons per person. The office building in this estimating example, having an occupancy of 142 persons per floor or 426 total, requires a 170 gallon minimum hourly demand (gallons per hour - GPH) water heater or series of heaters.

All the plumbing fixtures and appliances should be listed by categories for pricing on a plumbing estimate worksheet, as shown in Figure 23.12. Using the assemblies cost tables shown in Figure 23.13 through 23.19 from *Means Mechanical Cost Data*, 1987, a systems table number is assigned for each fixture. The line number for pricing is also recorded. The lavatories, for example, can be priced using Figure 23.13. Vanity top oval lavatories, 19" x 16", are selected and priced according to line 1600. The same procedure is followed for the remaining fixtures and appliances. The complete plumbing systems estimate is shown in Figure 23.20.

To compensate for materials not included in the systems groupings certain percentages must be added to the plumbing system. These percentages can be found in table C8.1-031 from *Means Mechanical Cost Data* (reproduced in Figure 23.21). The first of these is for "water control" and covers water meter, backflow preventer, shock absorbers, and vacuum breakers. An additional 10 to 15 percent should be added to the subtotal for these items.

PLUMBING — C8.1-400 — Individual Fixture Systems

Table 8.1-401 Minimum Plumbing Fixture Requirements

TYPE OF BUILDING/USE	WATER CLOSETS		URINALS		LAVATORIES		BATHTUBS OR SHOWERS		DRINKING FOUNTAIN	OTHER
	Persons	Fixtures	Persons	Fixtures	Persons	Fixtures	Persons	Fixtures	Fixtures	Fixtures
Assembly Halls Auditoriums, Theater, Public assembly	1-100 / 101-200 / 201-400	1 / 2 / 3	1-200 / 201-400 / 401-600	1 / 2 / 3	1-200 / 201-400 / 401-750	1 / 2 / 3			1 for each 1000 persons	1 service sink
	Over 400 add 1 fixt. for ea. 500 men; 1 fixt. for ea. 300 women		Over 600 add 1 fixture for each 300 men		Over 750 add 1 fixture for each 500 persons					
Assembly Public Worship	300 men / 150 women	1 / 1	300 men	1	men / women	1 / 1			1	
Dormitories	Men: 1 for each 10 persons; Women: 1 for each 8 persons		1 for each 25 men, over 150 add 1 fixture for each 50 men		1 for ea. 12 persons; 1 separate dental lav. for each 50 persons recom.		1 for ea. 8 persons; For women add 1 additional for each 30. Over 150 persons add 1 for each 20.		1 for each 75 persons	Laundry trays 1 for each 50 serv. sink 1 for ea. 100
Dwellings Apartments and homes	1 fixture for each unit				1 fixture for each unit		1 fixture for each unit			
Hospitals Indiv. Room, Ward, Waiting room	8 persons	1 / 1 / 1			10 persons	1 / 1 / 1	20 persons	1	1 for 100 patients	1 service sink per floor
Industrial Mfg. plants, Warehouses	1-10 / 11-25 / 26-50 / 51-75 / 76-100	1 / 2 / 3 / 4 / 5	0-30 / 31-80 / 81-160 / 161-240	1 / 2 / 3 / 4	1-100 / over 100	1 for ea. 10 / 1 for ea. 15	1 Shower for each 15 persons subject to excessive heat or occupational hazard		1 for each 75 persons	
	1 fixture for each additional 30 persons									
Public Buildings Businesses, Offices	1-15 / 16-35 / 36-55 / 56-80 / 81-110 / 111-150	1 / 2 / 3 / 4 / 5 / 6	Urinals may be provided in place of water closets but may not replace more than 1/3 required number of men's water closets		1-15 / 16-35 / 36-60 / 61-90 / 91-125	1 / 2 / 3 / 4 / 5			1 for each 75 persons	1 service sink per floor
	1 fixture for ea. additional 40 persons				1 fixture for ea. additional 45 persons					
Schools Elementary	1 for ea. 30 boys; 1 for ea. 25 girls		1 for ea. 25 boys		1 for ea. 35 boys; 1 for ea. 35 girls		For gym or pool shower room 1/5 of a class		1 for each 40 pupils	
Schools Secondary	1 for ea. 40 boys; 1 for ea. 30 girls		1 for ea. 25 boys		1 for ea. 40 boys; 1 for ea. 40 girls		For gym or pool shower room 1/5 of a class		1 for each 50 pupils	

Figure 23.9

Pipe and fittings is the next category and includes the interconnecting piping (mains) between the several systems. For pipe and fittings, 30 to 60 percent should be added to the subtotal. This percentage depends on the building design. For example a building similar in size to the office building in this example, but having toilet rooms at each of the far ends of the building, would have a much higher percentage than the 30 percent indicated for this plumbing estimate. In extreme cases, this percentage might reach 100%.

Quality complexity is the last additive. It should be applied to fire protection, HVAC, and plumbing estimates to compensate for quality/complexity over and above the basic level assumed in the Means Assemblies prices. Percentages for quality/complexity are 0 to 5% for economy installation, 5 to 15% for good quality medium complexity installation, and 15 to 25% for above average quality and complexity.

Construction and estimating experience is necessary to determine the proper percentages to be added. Keeping cost records of completed projects will contribute to this experience factor.

Fire Protection

Fire protection systems are usually composed of sprinkler systems, hose standpipes, or combinations thereof. System classifications according to hazard and design have been developed by the National Fire Protection Association (NFPA) and are discussed in Chapter 16, "Fire Protection". Based on these criteria, a firehose standpipe and sprinkler system for the example office building can be estimated.

Roof Drain Systems		
Pipe Diameter	Max. S.F. Roof Area	Gallons per Min.
2"	544	23
3"	1,610	67
4"	3,460	144
5"	6,280	261
6"	10,200	424
8"	22,000	913

Design Assumptions: Vertical conductor size is based on a maximum rate of rainfall of 4" per hour. To convert roof area to other rates multiply "Max S.F. Roof Area" shown by four and divide the result by desired rate. The answer is the local roof area that may be handled by the indicated pipe diameter.

Basic cost is for roof drain, 10' of vertical leader and 10' of horizontal, plus connection to the main.

Figure 23.10

PLUMBING — C8.1-100 — Hot Water

Table 8.1-101 Hot Water Consumption Rates

Type of Building	Size Factor	Maximum Hourly Demand	Average Day Demand
Apartment Dwellings	No. of Apartments:		
	Up to 20	12.0 Gal. per apt.	42.0 Gal. per apt.
	21 to 50	10.0 Gal. per apt.	40.0 Gal. per apt.
	51 to 75	8.5 Gal. per apt.	38.0 Gal. per apt.
	76 to 100	7.0 Gal. per apt.	37.0 Gal. per apt.
	101 to 200	6.0 Gal. per apt.	36.0 Gal. per apt.
	201 up	5.0 Gal. per apt.	35.0 Gal. per apt.
Dormitories	Men	3.8 Gal. per man	13.1 Gal. per man
	Women	5.0 Gal. per woman	12.3 Gal. per woman
Hospitals	Per bed	23 Gal. per patient	90 Gal. per patient
Hotels	Single room with bath	17 Gal. per unit	50 Gal. per unit
	Double room with bath	27 Gal. per unit	80 Gal. per unit
Motels	No. of units:		
	Up to 20	6.0 Gal. per unit	20.0 Gal. per unit
	21 to 100	5.0 Gal. per unit	14.0 Gal. per unit
	101 Up	4.0 Gal. per unit	10.0 Gal. per unit
Nursing Homes		4.5 Gal. per bed	18.4 Gal. per bed
Office buildings		0.4 Gal. per person	1.0 Gal. per person
Restaurants	Full meal type	1.5 Gal./max. meals/hr.	2.4 Gal. per meal
	Drive-in snack type	0.7 Gal./max. meals/hr.	0.7 Gal. per meal
Schools	Elementary	0.6 Gal. per student	0.6 Gal. per student
	Secondary & High	1.0 Gal. per student	1.8 Gal. per student

For evaluation purposes, recovery rate and storage capacity are inversely proportional. Water heaters should be sized so that the maximum hourly demand anticipated can be met in addition to allowance for the heat loss from the pipes and storage tank.

Table 8.1-102 Fixture Demands in Gallons Per Fixture Per Hour

Table below is based on 140° F final temperature except for dishwashers in public places (*) where 180° F water is mandatory.

Fixture	Apartment House	Club	Gym	Hospital	Hotel	Indust. Plant	Office	Private Home	School
Bathtubs	20	20	30	20	20			20	
Dishwashers, automatic*	15	50-150		50-150	50-200	20-100		15	20-100
Kitchen sink	10	20		20	30	20	20	10	20
Laundry, stationary tubs	20	28		28	28			20	
Laundry, automatic wash	75	75		100	150			75	
Private lavatory	2	2	2	2	2	2	2	2	2
Public lavatory	4	6	8	6	8	12	6		15
Showers	30	150	225	75	75	225	30	30	225
Service sink	20	20		20	30	20	20	15	20
Demand factor	0.30	0.30	0.40	0.25	0.25	0.40	0.30	0.30	0.40
Storage capacity factor	1.25	0.90	1.00	0.60	0.80	1.00	2.00	0.70	1.00

To obtain the probable maximum demand multiply the total demands for the fixtures (gal./fixture/hour) by the demand factor. The heater should have a heating capacity in gallons per hour equal to this maximum. The storage tank should have a capacity in gallons equal to the probable maximum demand multiplied by the storage capacity factor.

Figure 23.11

Means Forms

COST ANALYSIS

PROJECT: Plumbing Systems Worksheet	SHEET NO.
ARCHITECT	ESTIMATE NO.
	DATE

TAKE OFF BY: QUANTITIES BY: PRICES BY: EXTENSIONS BY: CHECKED BY:

Fixture	Table number	Qty.	Unit Cost	Total Costs	Calculations
Bathtubs					
Drinking Fountain					
Kitchen Sink					
Laundry Sink					
Lavatory		21			
Service Sink		3			
Urinal		6			
Water Cooler		9			
Water Closet		18			
Wash Fount. Group					
Water Heater		1			
Roof Drains		6			
Subtotal					
Water Control					
Pipe & Fittings					
Quality Complexity					
Total					

Figure 23.12

PLUMBING — B8.1-433 — Lavatory Systems

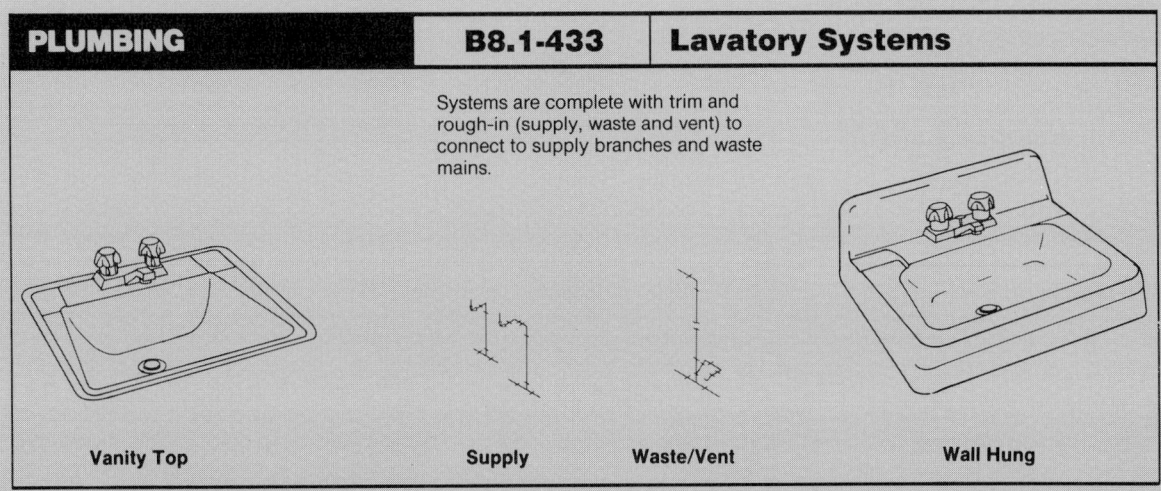

Systems are complete with trim and rough-in (supply, waste and vent) to connect to supply branches and waste mains.

Vanity Top — Supply — Waste/Vent — Wall Hung

System Components	QUANTITY	UNIT	COST EACH MAT.	COST EACH INST.	COST EACH TOTAL
SYSTEM 08.1-433-1560					
LAVATORY W/TRIM, VANITY TOP, P.E. ON C.I., 20" X 18"					
Lavatory w/trim, PE on CI, white, vanity top, 20" x 18" oval	1.000	Ea.	121	74	195
Pipe, steel, galvanized, schedule 40, threaded, 1-¼" diam	4.000	L.F.	8.40	21.80	30.20
Copper tubing type DWV, solder joint, hanger 10'OC 1-¼" diam	4.000	L.F.	6.84	35.76	42.60
Wrought copper DWV, Tee, sanitary, 1-¼" diam	1.000	Ea.	3.45	29.55	33
P trap w/cleanout, 20 ga, 1-¼" diam	1.000	Ea.	9.85	15.15	25
Copper tubing type L, solder joint, hanger 10' OC ½" diam	10.000	L.F.	7.80	50.70	58.50
Wrought copper 90° elbow for solder joints ½" diam	2.000	Ea.	.36	26.84	27.20
Wrought copper Tee for solder joints, ½" diam	2.000	Ea.	.66	41.34	42
Stop, chrome, angle supply, ½" diam	2.000	Ea.	28.38	23.62	52
TOTAL			186.74	318.76	505.50

8.1-433	Lavatory Systems	MAT.	INST.	TOTAL
1560	Lavatory w/trim, vanity top, PE on CI, 20" x 18"	185	320	505
1600	19" x 16" oval	180	325	505
1640	18" round	175	320	495
1680	Cultured marble, 19" x 17"	145	325	470
1720	25" x 19"	160	320	480
1760	Stainless, self-rimming, 25" x 22"	205	320	525
1800	17" x 22"	195	320	515
1840	Steel enameled, 20" x 17"	135	325	460
1880	19" round	130	330	460
1920	Vitreous china, 20" x 16"	245	335	580
1960	19" x 16"	190	335	525
2000	22" x 13"	195	335	530
2040	Wall hung, PE on CI, 18" x 15"	345	350	695
2080	19" x 17"	310	350	660
2120	20" x 18"	260	350	610
2160	Vitreous china, 18" x 15"	255	360	615
2200	19" x 17"	205	360	565
2240	24" x 20"	310	355	665

288

Figure 23.13

PLUMBING — B8.1-434 Service Sink Systems

Service sink systems are complete with trim and rough-in (supply, waste and vent) to connect to supply branches and waste mains.

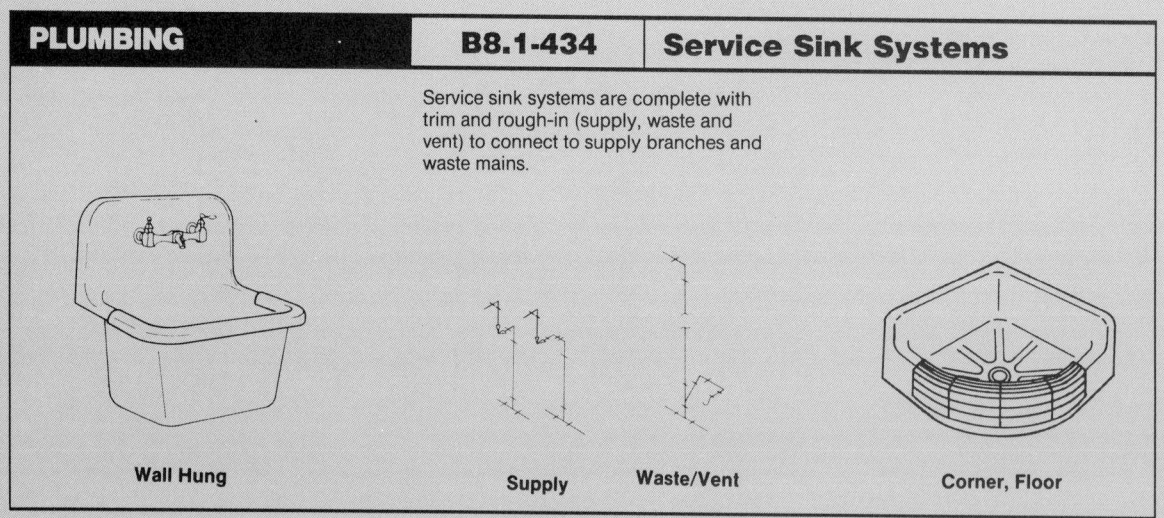

Wall Hung · Supply · Waste/Vent · Corner, Floor

System Components	QUANTITY	UNIT	COST EACH MAT.	COST EACH INST.	COST EACH TOTAL
SYSTEM 08.1-434-4260					
SERVICE SINK, PE ON CI, CORNER FLOOR, 28"X28", W/RIM GUARD & TRIM					
Service sink, corner floor, PE on CI, 28" x 28", w/rim guard & trim	1.000	Ea.	357.50	122.50	480
Copper tubing type DWV, solder joint, hanger 10'OC 3" diam	6.000	L.F.	28.44	90.66	119.10
Copper tubing type DWV, solder joint, hanger 10'OC 2" diam	4.000	L.F.	10.96	44.64	55.60
Wrought copper DWV, Tee, sanitary, 3" diam	1.000	Ea.	12.10	68.90	81
P trap with cleanout & slip joint, copper 3" diam	1.000	Ea.	31.90	24.10	56
Copper tubing, type L, solder joints, hangers 10' OC, ½" diam	10.000	L.F.	7.80	50.70	58.50
Wrought copper 90° elbow for solder joints ½" diam	2.000	Ea.	.36	26.84	27.20
Wrought copper Tee for solder joints, ½" diam	2.000	Ea.	.66	41.34	42
Stop, angle supply, chrome, ½" diam	2.000	Ea.	28.38	23.62	52
TOTAL			478.10	493.30	971.40

8.1-434	Service Sink Systems	MAT.	INST.	TOTAL
4260	Service sink w/trim, PE on CI, corner floor, 28" x 28", w/rim guard	480	495	975
4300	Wall hung w/rim guard, 22" x 18"	450	585	1,035
4340	24" x 20"	475	585	1,060
4380	Vitreous china, wall hung 22" x 20"	490	580	1,070

Figure 23.14

PLUMBING — B8.1-450 Urinal Systems

Systems are complete with trim, flush valve and rough-in (supply, waste and vent) for connection to supply branches and waste mains.

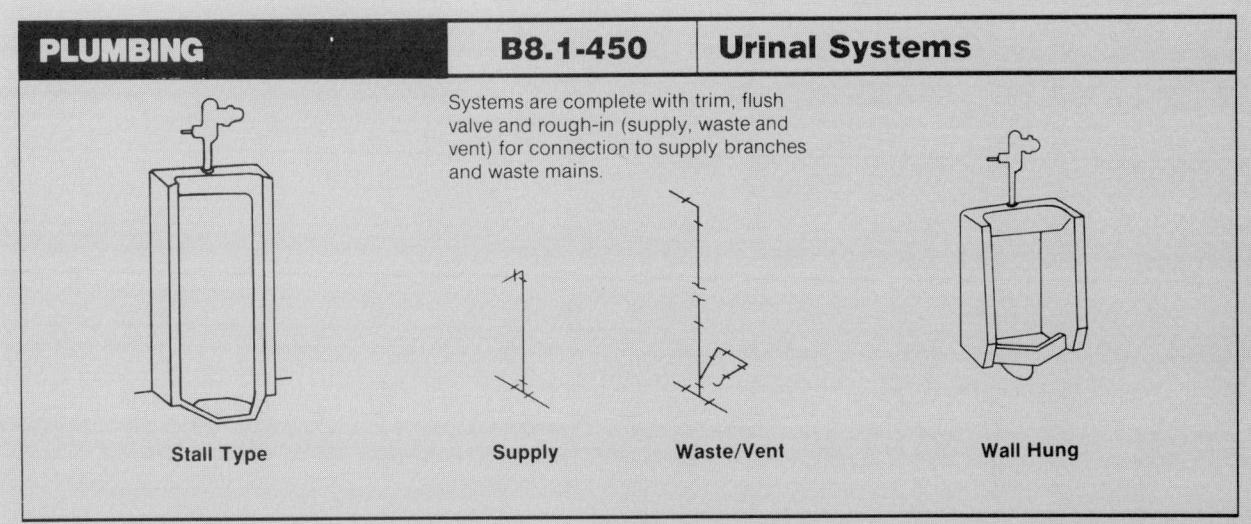

Stall Type • Supply • Waste/Vent • Wall Hung

System Components	QUANTITY	UNIT	COST EACH MAT.	INST.	TOTAL
SYSTEM 08.1-450-2000					
URINAL, VITREOUS CHINA, WALL HUNG					
Urinal, wall hung, vitreous china, incl. hanger	1.000	Ea.	330	160	490
Pipe, steel, galvanized, schedule 40, threaded, 1-½" diam	5.000	L.F.	12.80	30.20	43
Copper tubing type DWV, solder joint, hangers 10'OC, 2" diam	3.000	L.F.	8.22	33.48	41.70
Combination Y & ⅛ bend for CI soil pipe, no hub, 3" diam	1.000	Ea.	6.15		6.15
Pipe, CI, no hub, cplg 10' OC, hanger 5' OC, 3" diam	4.000	L.F.	16.12	33.28	49.40
Pipe coupling standard, CI soil, no hub, 3" diam	3.000	Ea.	8.58	60.42	69
Copper tubing type L, solder joint, hanger 10' OC ¾" diam	5.000	L.F.	5.65	30.60	36.25
Wrought copper 90° elbow for solder joints ¾" diam	1.000	Ea.	.43	14.12	14.55
Wrought copper Tee for solder joints, ¾" diam	1.000	Ea.	.73	22.27	23
TOTAL			388.68	384.37	773.05

8.1-450	Urinal Systems	MAT.	INST.	TOTAL
2000	Urinal, vitreous china, wall hung	390	385	775
2040	Stall type	480	435	915
C8.1-401				

Figure 23.15

PLUMBING — B8.1-460 — Water Cooler Systems

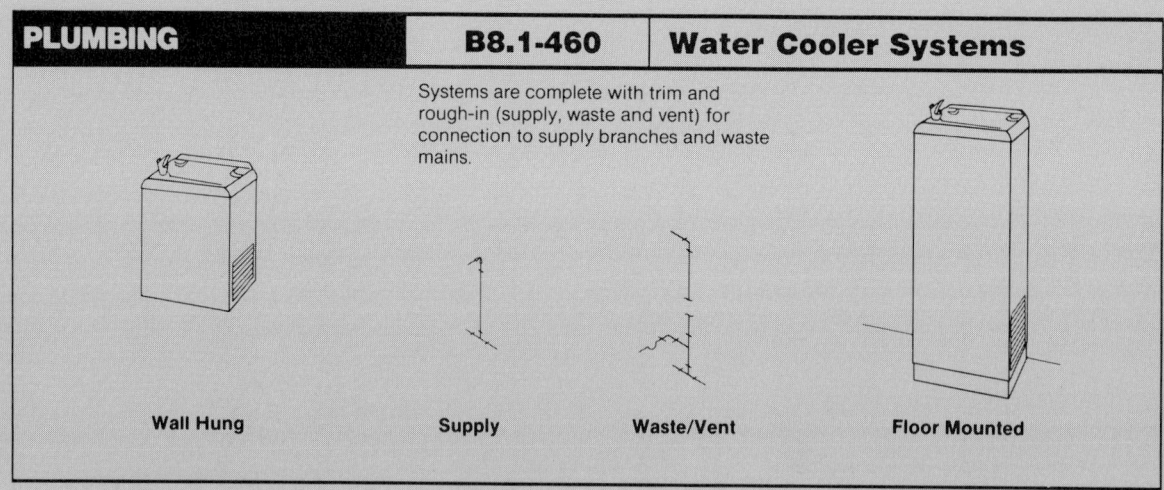

Systems are complete with trim and rough-in (supply, waste and vent) for connection to supply branches and waste mains.

Wall Hung | Supply | Waste/Vent | Floor Mounted

System Components	QUANTITY	UNIT	MAT.	INST.	TOTAL
SYSTEM 08.1-460-1840					
WATER COOLER, ELECTRIC, SELF CONTAINED, WALL HUNG, 8.2 GPH					
Water cooler, wall mounted, 8.2 GPH	1.000	Ea.	368.50	121.50	490
Copper tubing type DWV, solder joint, hanger 10'OC 1-¼" diam	4.000	L.F.	6.84	35.76	42.60
Wrought copper DWV, Tee, sanitary 1-¼" diam	1.000	Ea.	3.45	29.55	33
P trap, copper drainage, 1-¼" diam	1.000	Ea.	9.85	15.15	25
Copper tubing type L, solder joint, hanger 10' OC ⅜" diam	5.000	L.F.	3.30	23.95	27.25
Wrought copper 90° elbow for solder joints ⅜" diam	1.000	Ea.	.48	12.22	12.70
Wrought copper Tee for solder joints, ⅜" diam	1.000	Ea.	1.18	18.82	20
Stop and waste, straightway, bronze, solder, ⅜" diam	1.000	Ea.	3.96	11.19	15.15
TOTAL			397.56	268.14	665.70

8.1-460	Water Cooler Systems	MAT.	INST.	TOTAL
1840	Water cooler, electric, wall hung, 8.2 GPH	400	270	670
1880	Dual height, 14.3 GPH	555	275	830
1920	Wheelchair type, 7.5 G.P.H.	980	270	1,250
1960	Semi recessed, 8.1 G.P.H.	515	270	785
2000	Full recessed, 8 G.P.H.	840	285	1,125
2040	Floor mounted, 14.3 G.P.H.	420	235	655
2080	Dual height, 14.3 G.P.H.	590	280	870
2120	Refrigerated compartment type, 1.5 G.P.H.	760	235	995
2160	Cafeteria type, dual glass fillers, 27 G.P.H.	1,875	335	2,210

Figure 23.16

PLUMBING — B8.1-470 — Water Closet Systems

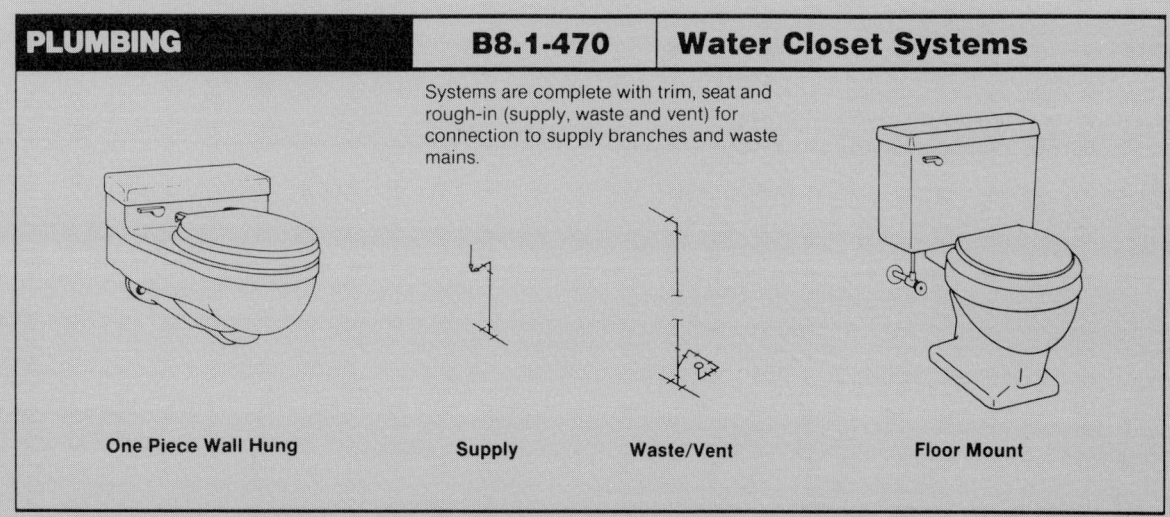

Systems are complete with trim, seat and rough-in (supply, waste and vent) for connection to supply branches and waste mains.

One Piece Wall Hung · Supply · Waste/Vent · Floor Mount

System Components	QUANTITY	UNIT	MAT.	INST.	TOTAL
SYSTEM 08.1-470-1840					
WATER CLOSET, VITREOUS CHINA, ELONGATED					
TANK TYPE, WALL HUNG, ONE PIECE					
Wtr closet tank type vit china wall hung 1 pc w/seat supply & stop	1.000	Ea.	517	93	610
Pipe steel galvanized, schedule 40, threaded, 2" diam	4.000	L.F.	13.44	30.16	43.60
Pipe, CI soil, no hub, cplg 10' OC, hanger 5' OC, 4" diam	2.000	L.F.	10.40	18.20	28.60
Pipe, coupling, standard coupling, CI soil, no hub, 4" diam	2.000	Ea.	6.82	43.18	50
Copper tubing type L, solder joint, hanger 10'OC, ½" diam	6.000	L.F.	4.68	30.42	35.10
Wrought copper 90° elbow for solder joints ½" diam	2.000	Ea.	.36	26.84	27.20
Wrought copper Tee for solder joints ½" diam	1.000	Ea.	.33	20.67	21
Support/carrier, for water closet, siphon jet, horiz, single, 4" waste	1.000	Ea.	97.90	47.10	145
TOTAL			650.93	309.57	960.50

8.1-470	Water Closet Systems	MAT.	INST.	TOTAL
1800	Water closet, vitreous china, elongated			
1840	Tank type, wall hung, one piece	650	310	960
1880	Close coupled two piece	470	305	775
1920	Floor mount, one piece	500	340	840
1960	One piece low profile	780	340	1,120
2000	Two piece close coupled	215	340	555
2040	Bowl only with flush valve			
2080	Wall hung	400	320	720
2120	Floor mount	335	350	685

Figure 23.17

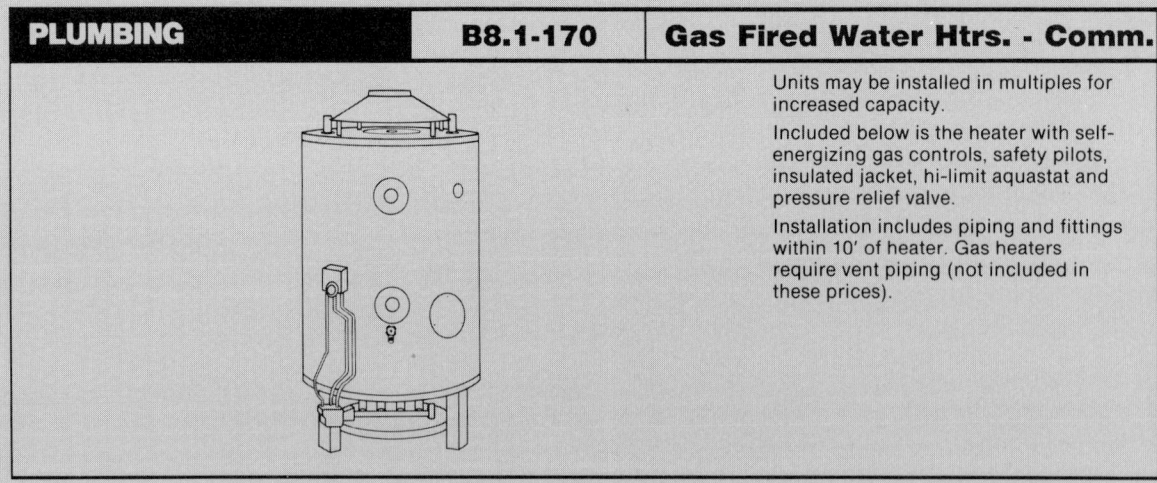

PLUMBING	B8.1-170	Gas Fired Water Htrs. - Comm.

Units may be installed in multiples for increased capacity.

Included below is the heater with self-energizing gas controls, safety pilots, insulated jacket, hi-limit aquastat and pressure relief valve.

Installation includes piping and fittings within 10' of heater. Gas heaters require vent piping (not included in these prices).

System Components	QUANTITY	UNIT	COST EACH		
			MAT.	INST.	TOTAL
SYSTEM 08.1-170-1780 GAS FIRED WATER HEATER, COMMERCIAL, 100° F RISE 75.5 MBH INPUT, 63 GPH					
Water heater, commercial, gas, 75.5 MBH, 63 GPH	1.000	Ea.	726	194	920
Copper tubing, type L, solder joint, hanger 10' OC, 1-¼" diam	30.000	L.F.	63.30	277.20	340.50
Wrought copper 90° elbow for solder joints 1-¼" diam	4.000	Ea.	7.80	71.60	79.40
Wrought copper Tee for solder joints, 1-¼" diam	2.000	Ea.	8.14	59.86	68
Wrought copper union for soldered joints, 1-¼" diam	2.000	Ea.	10.70	39.30	50
Valve, gate, bronze, 125 lb, NRS, soldered 1-¼" diam	2.000	Ea.	52.80	35.20	88
Relief valve, bronze, press & temp, self-close, ¾" IPS	1.000	Ea.	34.10	9.90	44
Copper tubing, type L, solder joints, ¾" diam	8.000	L.F.	9.04	48.96	58
Wrought copper 90° elbow for solder joints ¾" diam	1.000	Ea.	.43	14.12	14.55
Wrought copper, adapter, CTS to MPT, ¾" IPS	1.000	Ea.	.64	12.76	13.40
Pipe steel black, schedule 40, threaded, ¾" diam	10.000	L.F.	10.70	43.80	54.50
Pipe, 90° elbow, malleable iron black, 150 lb threaded, ¾" diam	2.000	Ea.	1.56	38.34	39.90
Pipe, union with brass seat, malleable iron black, ¾" diam	1.000	Ea.	2.83	20.17	23
Valve, gas stop w/o check, brass, ¾" IPS	1.000	Ea.	7.76	12.19	19.95
TOTAL			935.80	877.40	1,813.20

8.1-170	Gas Fired Water Heaters - Commercial Systems	COST EACH		
		MAT.	INST.	TOTAL
1760	Gas fired water heater, commercial, 100° F rise			
1780	75.5 MBH input, 63 GPH	935	875	1,810
1860	100 MBH input, 91 GPH	1,150	900	2,050
1980	155 MBH input, 150 GPH	1,650	1,050	2,700
2060	200 MBH input, 192 GPH	2,150	1,225	3,375
2140	300 MBH input, 278 GPH	3,350	1,500	4,850
2180	390 MBH input, 374 GPH	3,750	1,525	5,275
2220	500 MBH input, 480 GPH	4,975	1,625	6,600
2260	600 MBH input, 576 GPH	5,650	1,750	7,400
2300	800 MBH input, 768 GPH	6,650	1,900	8,550
2340	1000 MBH input, 960 GPH	7,825	1,950	9,775
2420	1500 MBH input, 1440 GPH	12,000	2,425	14,425
2460	1800 MBH input, 1730 GPH	13,400	2,700	16,100
2500	2450 MBH input, 2350 GPH	17,600	3,150	20,750
2540	3000 MBH input, 2880 GPH	21,300	3,750	25,050
2580	3750 MBH input, 3600 GPH	26,600	3,850	30,450

Figure 23.18

PLUMBING — B8.1-310 — Storm Drainage - Roof Drains

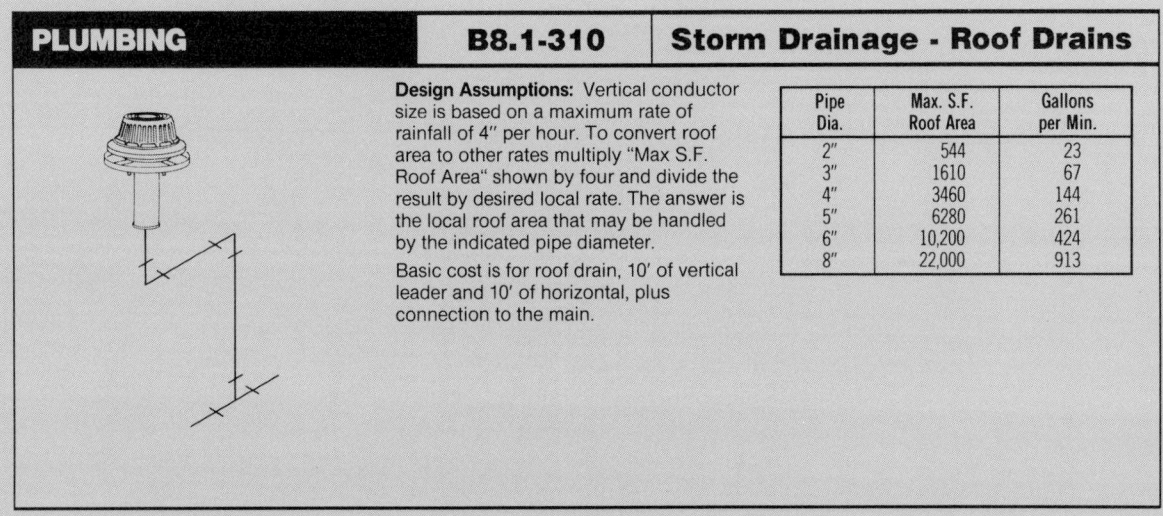

Design Assumptions: Vertical conductor size is based on a maximum rate of rainfall of 4" per hour. To convert roof area to other rates multiply "Max S.F. Roof Area" shown by four and divide the result by desired local rate. The answer is the local roof area that may be handled by the indicated pipe diameter.

Basic cost is for roof drain, 10' of vertical leader and 10' of horizontal, plus connection to the main.

Pipe Dia.	Max. S.F. Roof Area	Gallons per Min.
2"	544	23
3"	1610	67
4"	3460	144
5"	6280	261
6"	10,200	424
8"	22,000	913

System Components	QUANTITY	UNIT	MAT.	INST.	TOTAL
SYSTEM 081-310-1880					
ROOF DRAIN, DWV PVC PIPE, 2" DIAM., 10' HIGH					
Drain, roof, main, PVC, dome type 2" pipe size	1.000	Ea.	35.20	34.80	70
Clamp, roof drain, underdeck	1.000	Ea.	9.30	19.70	29
Pipe, Tee, PVC DWV, schedule 40, 2" pipe size	1.000	Ea.	1.06	27.94	29
Pipe, PVC, DWV, schedule 40, 2" diam.	20.000	L.F.	20.20	163.80	184
Pipe, elbow, PVC schedule 40, 2" diam.	2.000	Ea.	1.70	34.50	36.20
TOTAL			67.46	280.74	348.20

8.1-310	Roof Drain Systems	MAT.	INST.	TOTAL
1880	Roof drain, DWV PVC, 2" diam., piping, 10' high	67	280	347
1920	For each additional foot add	1.01	8.20	9.21
1960	3" diam., 10' high	87	335	422
2000	For each additional foot add	1.73	9.10	10.83
2040	4" diam., 10' high	115	380	495
2080	For each additional foot add	2.35	10.05	12.40
2120	5" diam., 10' high	340	405	745
2160	For each additional foot add	7.25	11.20	18.45
2200	6" diam., 10' high	345	485	830
2240	For each additional foot add	4.71	12.40	17.11
2280	8" diam., 10' high	660	760	1,420
2320	For each additional foot add	12.75	15.25	28
3940	C.I., soil, single hub, service wt., 2" diam. piping, 10' high	180	335	515
3980	For each additional foot add	3.28	9.65	12.93
4120	3" diam., 10' high	210	360	570
4160	For each additional foot add	4.10	10.05	14.15
4200	4" diam., 10' high	255	390	645
4240	For each additional foot add	5.50	10.95	16.45
4280	5" diam., 10' high	325	430	755
4320	For each additional foot add	7.15	12.15	19.30
4360	6" diam., 10' high	455	465	920
4400	For each additional foot add	10.15	12.85	23
4440	8" diam., 10' high	720	905	1,625
4480	For each additional foot add	15.30	21	36.30
6040	Steel galv. sch 40 threaded, 2" diam. piping, 10' high	190	285	475
6080	For each additional foot add	3.36	7.55	10.91

Figure 23.19

Means Forms

COST ANALYSIS

PROJECT: Plumbing Systems Worksheet

Fixture	Source/Dimensions Table number	Qty.	Unit Cost	Total Costs	Calculations
Bathtubs					
Drinking Fountain					
Kitchen Sink					
Laundry Sink					
Lavatory	B8.1-433-1600	21	505	10605	3 Flrs x 7 Lavs.
Service Sink	B8.1-434-4300	3	1035	3105	3 Flrs x 1 Sink
Urinal	B8.1-450-2000	6	775	4650	3 Flrs x 2 Urinals
Water Cooler	B8.1-460-1840 / 1920	6/3	670 / 1250	4020 / 3750	3 Flrs x 2 Coolers / 3 Flrs x 1 C Cool.
Water Closet	B8.1-470-2080	18	720	12960	3 Flrs x W=3 / M=3 / H=? Total
Wash Fount. Group					
Water Heater					.46 gal. x 426 People
Gas Fired	B8.1-170-2060	1	3375	3375	170 GPH
Roof Drains	B8.1-310-4200	6	645	3870	6-4" Cast Iron
Additional Piping	B8.1-310-4240	46'x6=276'	16.45	4540	C.I. Soil Pipe from roof to 4' below garage floor
Subtotal				50875	
Water Control	C8.1-031	10%		5088	meter, backflow etc.
Pipe & Fittings	C8.1-031	30%		15263	underground to 5'-0 outside building
Quality Complexity	C8.1-031	5%		2544	Economy
Total				73770	

Figure 23.20

PLUMBING — C8.1-030 — Piping

Table 8.1-031 Plumbing Approximations for Quick Estimating

Water Control
Water Meter; Backflow Preventer;
Shock Absorbers; Vacuum Breakers; } .... 10 to 15% of Fixtures
Mixer.
Pipe And Fittings: .... 30 to 60% of Fixtures

> **Note:** Lower percentage for compact buildings or larger buildings with plumbing in one area.
> Larger percentage for large buildings with plumbing spread out.
> In extreme cases pipe may be more than 100% of fixtures.
> Percentages **do not** include special purpose or process piping.

Plumbing Labor:
1 & 2 Story Residential .... Rough-in Labor = 80% of Materials
Apartment Buildings .... Rough-in Labor = 90 to 100% of Materials
Labor for handling and placing fixtures is approximately 25 to 30% of fixtures.

Quality/Complexity Multiplier (For all installations)
Economy installation, add .... 0 to 5%
Good quality, medium complexity, add .... 5 to 15%
Above average quality and complexity, add .... 15 to 25%

Table 8.1-032 Pipe Material Consideration

1. Malleable fittings should be used for gas service.
2. Malleable fittings are used where there are stresses/strains due to expansion and vibration.
3. Cast fittings may be broken as an aid to disassembling of heating lines frozen by long use, temperature and minerals.
4. Cast iron pipe is extensively used for underground and submerged service.
5. Type M (light wall) copper tubing is available in hard temper only and is used for nonpressure and less severe applications than K and L.
6. Type L (medium wall) copper tubing, available hard or soft for interior service.
7. Type K (heavy wall) copper tubing, available in hard or soft temper for use where conditions are severe. For underground and interior service.
8. Hard drawn tubing requires fewer hangers or supports but should not be bent. Silver brazed fittings are recommended, however 50/50 and 95/5 are occasionally used.

Table 8.1-033 Domestic/Imported Pipe and Fitting Cost

The prices shown in this publication for steel/cast iron pipe and steel, cast iron, malleable iron fittings are based on domestic production sold at the normal trade discounts. The above listed items of foreign manufacture may be available at prices of 1/3 to 1/2 those shown. Some imported items after minor machining or finishing operations are being sold as domestic to further complicate the system.

Caution: Most pipe prices in this book also include a coupling and pipe hangers which for the larger sizes can add considerable to the per foot cost and should be taken into account when comparing "book cost" with quoted supplier's cost.

Figure 23.21

Depending on the system's design, the firehose standpipe could be taken off and priced as part of the sprinkler system or the two could be taken off and priced separately. For this estimate, the firehose standpipe system and the sprinkler system are taken off and priced separately.

Sprinkler System: The system classification for an office building is "Light Hazard Occupancy". A conventional "wet type" distribution is selected for this heated building, and the designer has chosen Schedule #40 black steel pipe and threaded fittings as the piping material.

Using the previously determined area data (18,900 S.F. per floor and 14 L.F. floor heights), and the cost data from *Means Mechanical Cost Data*, 1987, shown in Figure 23.22a through 23.22c, the mathematical procedures shown in Figure 23.23 can be carried out. Line number 8.2-110-0620 (from Figure 23.22a) indicates a cost per square foot for a light hazard, steel pipe installation for the first floor. This price is for a 10,000 S.F. floor, which is closer to the 18,900 S.F. than the next line number which prices a 50,000 S.F. floor area. For the second and third floors, which do not require the alarm valve, water motor, and other sprinkler systems specialties, a square foot cost is selected. Line number 8.2-110-0740 indicates .96 cents per square foot. These prices for the first and upper floors are transferred to the fire protection worksheet shown in Figure 23.23.

Fire Standpipe System: To estimate the cost of a fire standpipe system for the proposed office building, the system classification and type should first be determined. Table 8.2-302 from *Means Mechanical Cost Data* (shown in Figure 23.24) indicates the several choices based on the NFPA 14 basic standpipe design. The building is Class III type usage, which allows a system to be operated by either occupants or by fire department personnel. A wet pipe system, a typical system for an office building, is selected. Two standpipes, at the building stairwells, will adequately cover the areas to be protected. The pipe size is 4", because the building height is less than 100 feet.

Figures 23.25a and 23.25b, from *Means Mechanical Cost Data*, 1987, illustrate a valve standpipe and lists its components, material, and labor costs for the three classifications. Line number 8.2-310-1540 (shown in Figure 23.25b) indicates a cost of $2,975 each for the first floor standpipes, including all the system components listed. Note that these components include 20 feet of pipe, a 10 foot riser and a 10 foot allowance for horizontal feed.

For the three additional floors at 14 feet each, multiplied by two, 84 additional feet of pipe is required. Line number 8.2-310-1560 (Figure 23.25b) indicates $715 for one 10 foot floor, or $71.50 per foot. The 84 additional feet totals $6,006. This additional footage includes a valved hose connection at each floor. The totals for two standpipes should be entered on the fire protection worksheet shown in Figure 23.23.

Six firehose and cabinet assemblies are priced from *Means Mechanical Cost Data*, 1987 (shown in Figure 23.26), line number 8.2-390-8400, and also entered on the worksheet.

The sprinkler and standpipe costs are totalled and a quality complexity percentage is added using the parameters from table number 8.2-303, shown in Figure 23.24.

FIRE PROTECTION — B8.2-110 — Wet Pipe Sprinkler Systems

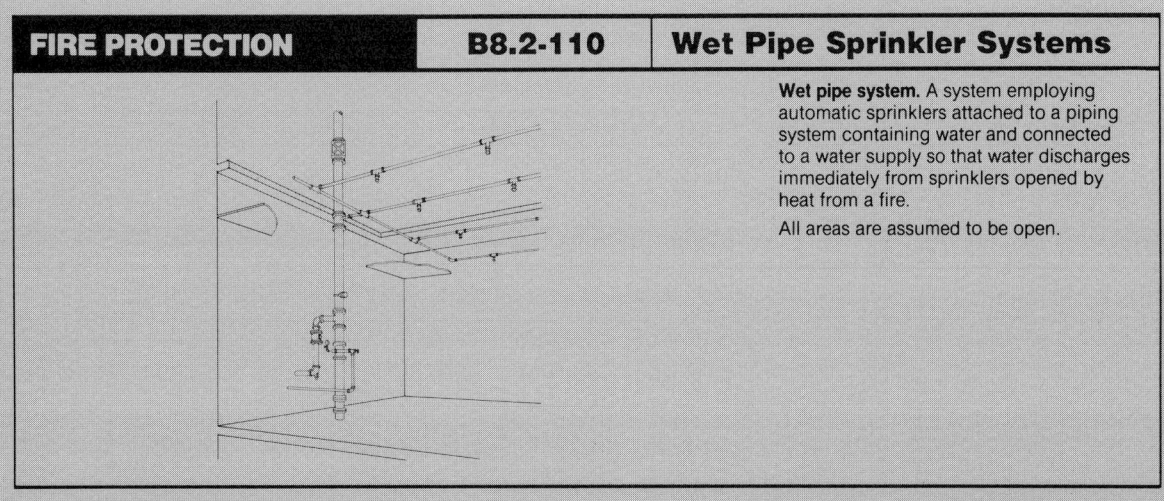

Wet pipe system. A system employing automatic sprinklers attached to a piping system containing water and connected to a water supply so that water discharges immediately from sprinklers opened by heat from a fire.

All areas are assumed to be open.

System Components	QUANTITY	UNIT	MAT.	INST.	TOTAL
SYSTEM 08.2-110-0580					
WET PIPE SPRINKLER, STEEL, BLACK, SCH. 40 PIPE					
LIGHT HAZARD, ONE FLOOR, 2000 S.F.					
Valve, gate, iron body, 125 lb, OS&Y, flanged, 4" diam	1.000	Ea.	148.50	121.50	270
Valve, swing check, bronze, 125 lb, regrinding disc, 2-½" pipe size	1.000	Ea.	64.35	25.65	90
Valve, angle, bronze, 150 lb, rising stem, threaded, 2" diam	1.000	Ea.	80.03	17.48	97.51
*Alarm valve, 2-½" pipe size	1.000	Ea.	400.13	124.88	525.01
Alarm, water motor, complete with gong	1.000	Ea.	79.61	51.64	131.25
Valve, swing check, w/balldrip Cl with brass trim 4" pipe size	1.000	Ea.	84.98	125.03	210.01
Pipe, steel, black, schedule 40, 4" diam	10.000	L.F.	56.78	100.73	157.51
*Flow control valve, trim & gauges, 4" pipe size	1.000	Set	787.88	280.88	1,068.76
Fire alarm horn, electric	1.000	Ea.	21.45	29.55	51
Pipe, steel, black, schedule 40, threaded, cplg & hngr 10'OC, 2-½" diam	20.000	L.F.	74.85	144.90	219.75
Pipe, steel, black, schedule 40, threaded, cplg & hngr 10'OC, 2" diam	12.500	L.F.	28.50	70.88	99.38
Pipe, steel, black, schedule 40, threaded, cplg & hngr 10'OC, 1-¼" diam	37.500	L.F.	53.72	153	206.72
Pipe steel, black, schedule 40, threaded cplg & hngr 10'OC, 1" diam	112.000	L.F.	128.52	425.88	554.40
Pipe Tee, malleable iron black, 150 lb threaded, 4" pipe size	2.000	Ea.	57.75	182.25	240
Pipe Tee, malleable iron black, 150 lb threaded, 2-½" pipe size	2.000	Ea.	22.28	81.23	103.51
Pipe Tee, malleable iron black, 150 lb threaded, 2" pipe size	1.000	Ea.	5	33.26	38.26
Pipe Tee, malleable iron black, 150 lb threaded, 1-¼" pipe size	5.000	Ea.	13.84	128.66	142.50
Pipe Tee, malleable iron black, 150 lb threaded, 1" pipe size	4.000	Ea.	6.81	101.19	108
Pipe 90° elbow, malleable iron black, 150 lb threaded, 1" pipe size	6.000	Ea.	6.57	92.43	99
Sprinkler head, standard spray, brass 135°-286°F ½" NPT, ⅜" orifice	12.000	Ea.	27.27	251.73	279
Valve, gate, bronze, NRS, class 150, threaded, 1" pipe size	1.000	Ea.	13.53	10.47	24
*Standpipe connection, wall, single, flush w/plug & chain 2-½"x2-½"	1.000	Ea.	70.95	75.30	146.25
TOTAL			2,233.30	2,628.52	4,861.82
COST PER S.F.			1.12	1.31	2.43

*Not included in systems under 2000 S.F.

8.2-110	Wet Pipe Sprinkler Systems	MAT.	INST.	TOTAL
0520	Wet pipe sprinkler systems, steel, black, sch. 40 pipe			
0530	Light hazard, one floor, 500 S.F.	.70	1.33	2.03
0560	1000 S.F.	1.03	1.37	2.40
0580	2000 S.F.	1.12	1.31	2.43
0600	5000 S.F.	.55	.96	1.51
0620	10,000 S.F.	.37	.81	1.18

Figure 23.22a

FIRE PROTECTION — B8.2-110 — Wet Pipe Sprinkler Systems

8.2-110	Wet Pipe Sprinkler Systems	COST PER S.F. MAT.	INST.	TOTAL
0640	50,000 S.F.	.29	.73	1.02
0660	Each additional floor, 500 S.F.	.40	1.11	1.51
0680	1000 S.F.	.36	1.02	1.38
0700	2000 S.F.	.32	.90	1.22
0720	5000 S.F.	.23	.79	1.02
0740	10,000 S.F.	.23	.73	.96
0760	50,000 S.F.	.25	.69	.94
1000	Ordinary hazard, one floor, 500 S.F.	.79	1.46	2.25
1020	1000 S.F.	1.02	1.32	2.34
1040	2000 S.F.	1.19	1.42	2.61
1060	5000 S.F.	.63	1.04	1.67
1080	10,000 S.F.	.47	1.07	1.54
1100	50,000 S.F.	.45	1.04	1.49
1140	Each additional floor, 500 S.F.	.50	1.25	1.75
1160	1000 S.F.	.34	1	1.34
1180	2000 S.F.	.39	1	1.39
1200	5000 S.F.	.40	.96	1.36
1220	10,000 S.F.	.33	.99	1.32
1240	50,000 S.F.	.31	.93	1.24
1500	Extra hazard, one floor, 500 S.F.	2.74	2.43	5.17
1520	1000 S.F.	1.66	1.99	3.65
1540	2000 S.F.	1.23	1.85	3.08
1560	5000 S.F.	.84	1.66	2.50
1580	10,000 S.F.	.81	1.53	2.34
1600	50,000 S.F.	.83	1.45	2.28
1660	Each additional floor, 500 S.F.	.59	1.51	2.10
1680	1000 S.F.	.58	1.44	2.02
1700	2000 S.F.	.49	1.44	1.93
1720	5000 S.F.	.41	1.29	1.70
1740	10,000 S.F.	.46	1.17	1.63
1760	50,000 S.F.	.44	1.09	1.53
2020	Grooved steel, black sch. 40 pipe, light hazard, one floor, 2000 S.F.	1.20	1.15	2.35
2060	10,000 S.F.	.50	.73	1.23
2100	Each additional floor, 2000 S.F.	.40	.73	1.13
2150	10,000 S.F.	.27	.62	.89
2200	Ordinary hazard, one floor, 2000 S.F.	1.22	1.23	2.45
2250	10,000 S.F.	.50	.91	1.41
2300	Each additional floor, 2000 S.F.	.42	.82	1.24
2350	10,000 S.F.	.36	.83	1.19
2400	Extra hazard, one floor, 2000 S.F.	1.34	1.59	2.93
2450	10,000 S.F.	.73	1.20	1.93
2500	Each additional floor, 2000 S.F.	.61	1.19	1.80
2550	10,000 S.F.	.50	1.07	1.57
3050	Grooved steel black sch. 10 pipe, light hazard, one floor, 2000 S.F.	1.20	1.14	2.34
3100	10,000 S.F.	.43	.69	1.12
3150	Each additional floor, 2000 S.F.	.40	.72	1.12
3200	10,000 S.F.	.28	.61	.89
3250	Ordinary hazard, one floor, 2000 S.F.	1.22	1.22	2.44
3300	10,000 S.F.	.51	.89	1.40
3350	Each additional floor, 2000 S.F.	.42	.81	1.23
3400	10,000 S.F.	.37	.81	1.18
3450	Extra hazard, one floor, 2000 S.F.	1.34	1.58	2.92
3500	10,000 S.F.	.68	1.18	1.86
3550	Each additional floor, 2000 S.F.	.61	1.18	1.79
3600	10,000 S.F.	.49	1.06	1.55
4050	Copper tubing, type M, light hazard, one floor, 2000 S.F.	1.11	1.43	2.54
4100	10,000 S.F.	.39	.99	1.38
4150	Each additional floor, 2000 S.F.	.32	1.03	1.35

302

Figure 23.22b

FIRE PROTECTION — B8.2-110 — Wet Pipe Sprinkler Systems

8.2-110	Wet Pipe Sprinkler Systems	MAT.	INST.	TOTAL
4200	10,000 S.F.	.24	.91	1.15
4250	Ordinary hazard, one floor, 2000 S.F.	1.15	1.60	2.75
4300	10,000 S.F.	.45	1.20	1.65
4350	Each additional floor, 2000 S.F.	.36	1.10	1.46
4400	10,000 S.F.	.30	1.11	1.41
4450	Extra hazard, one floor, 2000 S.F.	1.25	2.04	3.29
4500	10,000 S.F.	.77	1.70	2.47
4550	Each additional floor, 2000 S.F.	.52	1.64	2.16
4600	10,000 S.F.	.48	1.52	2
5050	Copper tubing, type M, T-drill system, light hazard, one floor			
5060	2000 S.F.	1.11	1.41	2.52
5100	10,000 S.F.	.37	.90	1.27
5150	Each additional floor, 2000 S.F.	.32	1.01	1.33
5200	10,000 S.F.	.22	.82	1.04
5250	Ordinary hazard, one floor, 2000 S.F.	1.12	1.45	2.57
5300	10,000 S.F.	.43	1.14	1.57
5350	Each additional floor, 2000 S.F.	.32	1.03	1.35
5400	10,000 S.F.	.29	1.06	1.35
5450	Extra hazard, one floor, 2000 S.F.	1.16	1.76	2.92
5500	10,000 S.F.	.64	1.41	2.05
5550	Each additional floor, 2000 S.F.	.47	1.37	1.84
5600	10,000 S.F.	.35	1.23	1.58

Figure 23.22c

Means Forms

COST ANALYSIS

PROJECT	Fire Protection Systems Worksheet	SHEET NO.
		ESTIMATE NO.
ARCHITECT		DATE

TAKE OFF BY:	QUANTITIES BY:	PRICES BY:		EXTENSIONS BY:	CHECKED BY:	

Type of Component	Hazard or Class	Source/Dimensions Table Number	Qty		Unit Cost	Total Costs	Comments
Sprinkler	Light	B8.2-110-0620	18,900	SF	1.18	22302	First Floor
	Wet Pipe	B8.2-110-0740	37,800	SF	.96	36288	2nd & 3rd Flrs.
Standpipe	Class III	B8.2-310-1540	2	Ea.	2975	5950	2 Standpipes w/ A Total of
	Wet	B8.2-310-1560	84	LF	71.50	6006	104 Feet of Riser
Cabinets & Components	With Hose	B8.2 390 8400	6		655	3930	2 Ea. Floor
Subtotal						74476	
Quality Complexity		C8.2-303	5%			3724	
Total						78200	

Figure 23.23

FIRE PROTECTION — C8.2-300 — General

Table 8.2-301 Standpipe Systems

The basis for standpipe system design is National Fire Protection Association NFPA 14, however, the authority having jurisdiction should be consulted for special conditions, local requirements and approval.

Standpipe systems, properly designed and maintained, are an effective and valuable time saving aid for extinguishing fires, especially in the upper stories of tall buildings, the interior of large commercial or industrial malls, or other areas where construction features or access make the laying of temporary hose lines time consuming and/or hazardous. Standpipes are frequently installed with automatic sprinkler systems for maximum protection.

There are three general classes of service for standpipe systems:

Class I for use by fire departments and personnel with special training for heavy streams (2-1/2" hose connections).
Class II for use by building occupants until the arrival of the fire department (1-1/2" hose connector with hose).
Class III for use by either fire departments and trained personnel or by the building occupants (both 2-1/2" and 1-1/2" hose connections or one 2-1/2" hose valve with an easily removable 2-1/2" by 1-1/2" adapter).

Standpipe systems are also classified by the way water is supplied to the system. The four basic types are:

Type 1: Wet standpipe system having supply valve open and water pressure maintained at all times.
Type 2: Standpipe system so arranged through the use of approved devices as to admit water to the system automatically by opening a hose valve.
Type 3: Standpipe system arranged to admit water to the system through manual operation of approved remote control devices located at each hose station.
Type 4: Dry standpipe having no permanent water supply.

Table 8.2-302 NFPA 14 Basic Standpipe Design

Class	Design-Use	Pipe Size Minimums	Water Supply Minimums
Class I	2½" hose connection on each floor. All areas within 30' of nozzle with 100' of hose. Fire Department Trained Personnel	Height to 100', 4" dia. Heights above 100', 6" dia. (275' max. except with pressure regulators 400' max.)	For each standpipe riser 500 GPM flow. For common supply pipe allow 500 GPM for first standpipe plus 250 GPM for each additional standpipe (2500 GPM max. total). 30 min. duration 65 PSI at 500 GPM
Class II	1½" hose connection with hose on each floor. All areas within 30' of nozzle with 100' of hose. Occupant personnel	Height to 50', 2" dia. Height above 50', 2½" dia.	For each standpipe riser 100 GPM flow. For multiple riser common supply pipe 100 GPM. 30 min. duration, 65 PSI at 100 GPM
Class III	Both of above. Class I valved connections will meet Class III with addition of 2½" by 1½" adapter and 1½" hose.	Same as Class I	Same as Class I

Combined Systems

Combined systems are systems where the risers supply both automatic sprinklers and 2-1/2" hose connection outlets for fire department use. In such a system the sprinkler spacing pattern shall be in accordance with NFPA 13 while the risers and supply piping will be sized in accordance with NFPA 14. When the building is completely sprinklered the risers may be sized by hydraulic calculation. The minimum size riser for buildings not completely sprinklered is 6".

The minimum water supply of a completely sprinklered, light hazard, high-rise occupancy building will be 500 GPM while the supply required for other types of completely sprinklered high-rise buildings is 1000 GPM.

General System Requirements

1. Approved valves will be provided at the riser for controlling branch lines to hose outlets.
2. A hose valve will be provided at each outlet for attachment of hose.
3. Where pressure at any standpipe outlet exceeds 100 PSI a pressure reducer must be installed to limit the pressure to 100 PSI. Note that the pressure head due to gravity in 100' of riser is 43.4 PSI. This must be overcome by city pressure, fire pumps, or gravity tanks to provide adequate pressure at the top of the riser.
4. Each hose valve on a wet system having linen hose shall have an automatic drip connection to prevent valve leakage from entering the hose.
5. Each riser will have a valve to isolate it from the rest of the system.
6. One or more fire department connections as an auxiliary supply shall be provided for each Class I or Class III standpipe system. In buildings having two or more zones, a connection will be provided for each zone.
7. There will be no shutoff valve in the fire department connection, but a check valve will be located in the line before it joins the system.
8. All hose connections street side will be identified on a cast plate or fitting as to purpose.

Table 8.2-303 Quality/Complexity Adjustment for Sprinkler/Standpipe Systems

Economy installation, add	0 to 5%
Good quality, medium complexity, add	5 to 15%
Above average quality and complexity, add	15 to 25%

392

Figure 23.24

FIRE PROTECTION — B8.2-310 — Wet Standpipe Risers

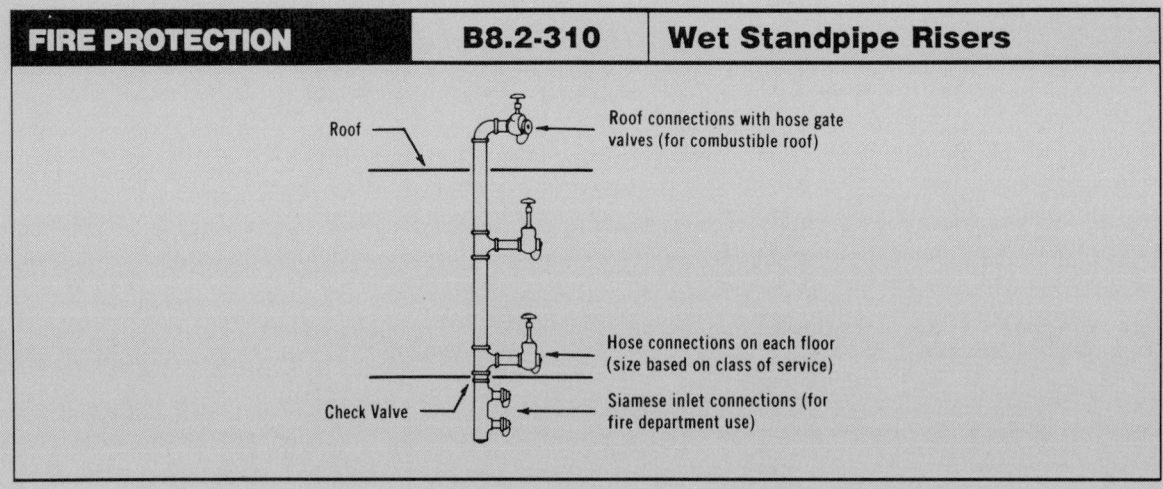

System Components	QUANTITY	UNIT	COST PER FLOOR		
			MAT.	INST.	TOTAL
SYSTEM 082-310-0560					
WET STANDPIPE RISER, CLASS I, STEEL, BLACK, SCH. 40 PIPE, 10' HEIGHT					
4" DIAMETER PIPE, ONE FLOOR					
Pipe, steel, black, schedule 40, threaded, 4" diam	20.000	L.F.	200.40	259.60	460
Pipe, Tee, malleable iron, black, 150 lb threaded, 4" pipe size	2.000	Ea.	77	243	320
Pipe, 90° elbow, malleable iron, black, 150 lb threaded 4" pipe size	1.000	Ea.	29.70	80.30	110
Pipe, nipple, steel, black, schedule 40, 2-½" pipe size x 3" long	2.000	Ea.	13.42	60.58	74
Fire valve, gate, 300 lb, brass w/handwheel, 2-½" pipe size	1.000	Ea.	85.80	39.20	125
Fire valve, pressure restricting, adj, rgh brs, 2-½" pipe size	1.000	Ea.	196.36	83.64	280
Valve, swing check, w/ball drip, CI w/brs ftngs, 4" pipe size	1.000	Ea.	113.30	166.70	280
Standpipe conn wall dble flush brs w/plugs & chains 2-½"x2-½"x4"	1.000	Ea.	245.30	99.70	345
Valve, swing check, bronze, 125 lb, regrinding disc, 2-½" pipe size	1.000	Ea.	85.80	34.20	120
Roof manifold, fire, w/valves & caps, horiz/vert brs 2-½"x2-½"x4"	1.000	Ea.	266.20	103.80	370
Fire, hydrolator, vent & drain, 2-½" pipe size	1.000	Ea.	88	42	130
Valve, gate, iron body 125 lb, OS&Y, threaded, 4" pipe size	1.000	Ea.	286	49	335
TOTAL			1,687.28	1,261.72	2,949

8.2-310	Wet Standpipe Risers, Class I	COST PER FLOOR		
		MAT.	INST.	TOTAL
0550	Wet standpipe risers, Class I, steel black sch. 40, 10' height			
0560	4" diameter pipe, one floor	1,675	1,250	2,925
0580	Additional floors	425	435	860
0600	6" diameter pipe, one floor	2,850	2,050	4,900
0620	Additional floors	790	615	1,405
0640	8" diameter pipe, one floor	3,925	2,475	6,400
0660	Additional floors	995	730	1,725

8.2-310	Wet Standpipe Risers, Class II	COST PER FLOOR		
		MAT.	INST.	TOTAL
1030	Wet standpipe risers, Class II, steel black sch. 40, 10' height			
1040	2" diameter pipe, one floor	605	455	1,060
1060	Additional floors	225	190	415
1080	2-½" diameter pipe, one floor	825	655	1,480
1100	Additional floors	245	220	465

Figure 23.25a

FIRE PROTECTION		B8.2-310	Wet Standpipe Risers			
8.2-310		Wet Standpipe Risers, Class III		COST PER FLOOR		
				MAT.	INST.	TOTAL
1530	Wet standpipe risers, Class III, steel black sch. 40, 10' height					
1540		4" diameter pipe, one floor		1,725	1,250	2,975
1560		Additional floors		350	365	715
1580		6" diameter pipe, one floor		2,875	2,050	4,925
1600		Additional floors		805	615	1,420
1620		8" diameter pipe, one floor		3,975	2,475	6,450
1640		Additional floors		1,025	730	1,755

Figure 23.25b

FIRE PROTECTION		B8.2-390	Standpipe Equipment		
8.2-390	Standpipe Equipment		COST EACH		
			MAT.	INST.	TOTAL
0100	Adapters, reducing, 1 piece, FxM, hexagon, cast brass, 2-½" x 1-½"		32		32
0200	Pin lug, 1-½" x 1"		10.30		10.30
0250	3" x 2-½"		30		30
0300	For polished chrome, add 75% mat.				
0400	Cabinets, D.S. glass in door, recessed, steel box, not equipped				
0500	Single extinguisher, steel door & frame		72	64	136
0550	Stainless steel door & frame		140	64	204
0600	Valve, 2-½" angle, steel door & frame		73	42	115
0650	Aluminum door & frame		79	41	120
0700	Stainless steel door & frame		145	41	186
0750	Hose rack assy, 2-½" x 1-½" valve & 100' hose, steel door & frame		115	86	201
0800	Aluminum door & frame		115	85	200
0850	Stainless steel door & frame		255	82	337
0900	Hose rack assy,& extinguisher,2-½"x1-½" valve & hose,steel door & frame		125	100	225
0950	Aluminum		125	100	225
1000	Stainless steel		270	100	370
1550	Compressor, air, dry pipe system, automatic, 200 gal., ⅓ H.P.		490	215	705
1600	520 gal., 1 H.P.		565	215	780
1650	Alarm, electric pressure switch (circuit closer)		58	10.95	68.95
2500	Couplings, hose, rocker lug, cast brass, 1-½"		13.20		13.20
2550	2-½"		22		22
3000	Escutcheon plate, for angle valves, polished brass, 1-½"		8.25		8.25
3050	2-½"		14.30		14.30
3500	Fire pump, electric, w/controller, fittings, relief valve				
3550	4" pump, 30 H.P., 500 G.P.M.		10,500	1,550	12,050
3600	5" pump, 40 H.P., 1000 G.P.M.		12,200	1,800	14,000
3650	5" pump, 100 H.P., 1000 G.P.M.		15,100	1,975	17,075
3700	For jockey pump system, add		1,375	250	1,625
5000	Hose, per linear foot, synthetic jacket, lined,				
5100	300 lb. test, 1-½" diameter		1.25		1.25
5150	2-½" diameter		1.72		1.72
5200	500 lb. test, 1-½" diameter		1.45		1.45
5250	2-½" diameter		2.34		2.34
5500	Nozzle, plain stream, polished brass, 1-½" x 10"		22		22
5550	2-½" x 15" x ¹³/₁₆" or 1-½"		62		62
5600	Heavy duty combination adjustable fog and straight stream w/handle 1-½"		215		215
5650	2-½" direct connection		275		275
6000	Rack, for 1-½" diameter hose 100 ft. long, steel		22	25	47
6050	Brass		38	25	63
6500	Reel, steel, for 50 ft. long 1-½" diameter hose		47	36	83
6550	For 75 ft. long 2-½" diameter hose		58	36	94
7050	Siamese, w/plugs & chains, polished brass, sidewalk, 4" x 2-½" x 2-½"		305	205	510
7100	6" x 2-½" x 2-½"		400	255	655
7200	Wall type, flush, 4" x 2-½" x 2-½"		245	100	345
7250	6" x 2-½" x 2-½"		365	105	470
7300	Projecting, 4" x 2-½" x 2-½"		225	98	323
7350	6" x 2-½" x 2-½"		265	110	375
7400	For chrome plate, add 15% mat.				
8000	Valves, angle, wheel handle, 300 Lb., rough brass, 1-½"		23	23	46
8050	2-½"		53	40	93
8100	Combination pressure restricting, 1-½"		34	23	57
8150	2-½"		76	39	115
8200	Pressure restricting, adjustable, satin brass, 1-½"		58	23	81
8250	2-½"		98	42	140
8300	Hydrolator, vent and drain, rough brass, 1-½"		38	23	61
8350	2-½"		88	42	130
8400	Cabinet assy, incls. 2-½" valve, adapter, rack, hose, nozzle & hydrolator		460	195	655

Figure 23.26

These systems may be modified to match conditions other than those shown in this example. For instance, if these two standpipes were fed from one common siamese inlet connection with interconnecting piping, the system cost could be modified by deducting two 4" x 2½" x 2½" standpipe connections (found in the unit price section). One 6" x 2½" x 2½" standpipe connection must then be added with the appropriate markup percentages. Finally, 6" interconnecting piping for the fire department connection to the two standpipe risers must be added. The pricing information for the additional piping and connections can be found in the unit price section of *Means Mechanical Cost Data*.

Heating and Cooling

The first step in preparing a systems estimate for heating a building is the same as for designing the heating system; determining the heat loss of the structure. In the design stage all of the necessary parameters are known: the types of materials and their coefficient of heat transmission for the building wall areas, and window, door, roof, floor, and slab.

Occupancy, geographic location, building use, and solar orientation all play a part in determining the heating or cooling load for a building. This can be a precise and time consuming procedure, as these loads are further refined by individual rooms and areas. In the preliminary stages of design most of these details are not known, nor is there sufficient time to complete an accurate heat loss study. Therefore, tables have been prepared to shorten the heat loss determination process. This can be done by combining many of the variables into a single numerical value to be multiplied by the building volume. A heat loss table can be found in *Means Mechanical Cost Data* and is reproduced in Figure 23.27.

This table does not take into consideration today's well insulated building envelope and results in a higher heating load estimation than the actual final design will indicate. However, it provides a quick and easy estimate adequate for the preliminary design stage.

The cubic content of the proposed three story office building is 793,800 C.F. The geographic location will warrant an outdoor design temperature of zero degrees Fahrenheit, and offices need to be heated to seventy degrees. With this information the table in Figure 23.27 can be used to estimate the heat loss.

The heat loss factor for a three story building to be heated to seventy degrees when the outside temperature is zero degrees is 4.3. The 4.3 factor is multiplied by the cubic foot area of the building (793,800 C.F.) denoting a heat loss of 3,413,340 British Thermal Units (B.T.U.). Working with figures in the millions can be difficult, therefore, a method to reduce the B.T.U. calculation into thousands has been designed (M.B.H.). Thus, the heat loss for the three story office building is 3,413 M.B.H.

Many factors influence the selection of the type of HVAC system for a building. Some of the considerations are initial cost, operating costs, maintenance, available space, fuel cost/availability, geographic location, proposed building use, and quality of construction. In addition to the above, local codes influence the number of air changes per hour for ventilation. At this stage of a preliminary estimate, many of the above factors may be known, In this case, a more precise selection of the systems can be made. For this estimating exercise, however, it will be assumed that building use (office), geographic location (Boston area), and quality of construction (above average) are the only known factors.

HEATING — C8.3-010 — Heat Transfer

Table 8.3-011 Factor for Determining Heat Loss for Various Types of Buildings

General: While the most accurate estimates of heating requirements would naturally be based on detailed information about the building being considered, it is possible to arrive at a reasonable approximation using the following procedure:

1. Calculate the cubic volume of the room or building.
2. Select the appropriate factor from Table 8.3-011. Note that the factors apply only to inside temperatures listed in the first column and to 0°F outside temperature.
3. If the building has bad north and west exposures, multiply the heat loss factor by 1.1
4. If the outside design temperature is other than 0°F, multiply the factor from Table 8.3-011 by the factor from Table 8.3-012
5. Multiply the cubic volume by the factor selected from Table 8.3-011. This will give the estimated BTUH heat loss which must be made up to maintain inside temperature.

Building Type	Conditions	Qualifications	Loss Factor*
Factories & Industrial Plants General Office Areas 70°F	One Story	Skylight in Roof	6.2
		No Skylight in Roof	5.7
	Multiple Story	Two Story	4.6
		Three Story	4.3
		Four Story	4.1
		Five Story	3.9
		Six Story	3.6
	All Walls Exposed	Flat Roof	6.9
		Heated Space Above	5.2
	One Long Warm Common Wall	Flat Roof	6.3
		Heated Space Above	4.7
	Warm Common Walls on Both Long Sides	Flat Roof	5.8
		Heated Space Above	4.1
Warehouses 60°F	All Walls Exposed	Skylights in Roof	5.5
		No Skylights in Roof	5.1
		Heated Space Above	4.0
	One Long Warm Common Wall	Skylight in Roof	5.0
		No Skylight in Roof	4.9
		Heated Space Above	3.4
	Warm Commom Walls on Both Long Sides	Skylight in Roof	4.7
		No Skylight in Roof	4.4
		Heated Space Above	3.0

*Note: This table tends to be conservative particularly for new buildings designed for minimum energy consumption.

Table 8.3-012 Outside Design Temperature Correction Factor (for Degrees Fahrenheit)

Outside Design Temperature	50	40	30	20	10	0	-10	-20	-30
Correction Factor	0.29	0.43	0.57	0.72	0.86	1.00	1.14	1.28	1.43

Figure 23.27

The budget estimate will be based on a heating and cooling system, with exhaust and makeup air ventilation. This includes a packaged chiller for cooling and a packaged boiler for heating utilizing one piping system (dual-temperature) feeding air handling units with ductwork distribution. This system provides either heating or cooling of the entire building. It cannot heat one area while cooling another.

Figure 23.28 is a page from the unit price section of *Means Mechanical Cost Data*, 1987, showing costs for cast iron, jacketed, hot water boilers. A boiler with a minimum output of 3,413 MBH is required. The smallest boiler that will meet this requirement is found on line 15.5-080-3400. It has an output of 3808 MBH. However, because the heat loss factor is inclined to be conservative, rather than go up 400 MBH to the large boiler, go down 150 MBH to the next smaller boiler of 3264 MBH shown on line 15.5-080-3380. This unit price cost is entered on the heating and air conditioning worksheet (shown in Figure 23.29). Unit price costs for the heating boiler are used because a price for the boiler only is needed. The related components required for a complete system will be shared with the cooling system that follows to complete the price for heating and cooling.

The cooling load is determined in a manner similar to the heating load except that it is based on the floor area rather than on building volume. Tables in *Means Mechanical Cost Data* list the cost per square foot for various types of occupancies for several kinds of air conditioning systems. One such table is shown in Figures 23.30a and 23.30b. Look up the occupancy, in this case for offices, and multiply the cost per square foot by the floor area (tabulated as closely as possible based on the gross floor area of the proposed building). In this example, line 8.4-120-4040 from Figure 23.30b is a cost for 60,000 S.F. offices. This category is chosen rather than the value for 40,000 S.F. offices (or interpolating between them) because the higher number of $6.71 per square foot allows a desirable conservative margin.

If the type of occupancy desired is not priced out in the systems tables, go to Figure 23.31. The lowest requirement for air conditioning on this chart is for apartment corridors with 550 S.F. per ton; the highest is for bars and taverns at 90 S.F. per ton of air conditioning. The occupancies priced out are spread between these two extremes. Pick the one desired that has been priced as an assembly. The result is entered on the heating and cooling worksheet shown in Figure 23.29.

The graphics and components listed for this chilled water system do not indicate a ductwork distribution system, but are based on many individual fan coil units. Using fewer units per floor and increasing their size or capacity, and relying on ductwork distribution would be approximately the same cost per square foot. The system table 8.4-120 does not include any provision for heating. As discussed earlier, the heating boiler selected from the unit price section must be added to the worksheet to round out a complete HVAC system. This price can then be compared to the prices for other systems from *Means Mechanical Cost Data* to select the most cost effective system.

080	0010	BOILERS, GAS FIRED Natural or propane, standard controls									080
	1000	Cast iron, with insulated jacket									
	2000	Steam, gross output, 81 MBH	Q-7	1.40	22.860	Ea.	895	510		1,405	1,725
	2020	102 MBH		1.30	24.620		1,030	550		1,580	1,925
	2040	122 MBH		1	32		1,130	710		1,840	2,275
	2060	163 MBH		.90	35.560		1,370	790		2,160	2,650
	2080	203 MBH		.90	35.560		1,560	790		2,350	2,875
	2100	240 MBH		.85	37.650		1,750	840		2,590	3,150
	2120	280 MBH		.80	40		2,140	890		3,030	3,650
	2140	320 MBH		.70	45.710		2,230	1,025		3,255	3,925
	2160	360 MBH		.65	49.230		2,520	1,100		3,620	4,375
	2180	400 MBH		.60	53.330		2,680	1,175		3,855	4,675
	2200	440 MBH		.55	58.180		2,860	1,300		4,160	5,025
	2220	544 MBH		.50	64		3,960	1,425		5,385	6,425
	2240	765 MBH		.45	71.110		5,390	1,575		6,965	8,225
	2260	892 MBH		.40	80		6,140	1,775		7,915	9,350
	2280	1275 MBH		.36	88.890		8,610	1,975		10,585	12,400
	2300	1530 MBH		.31	103		10,100	2,300		12,400	14,500
	2320	1875 MBH		.28	114		12,000	2,550		14,550	16,900
	2340	2170 MBH		.26	123		13,850	2,750		16,600	19,200
	2360	2675 MBH		.24	133		16,750	2,975		19,725	22,800
	2380	3060 MBH		.23	139		18,650	3,100		21,750	25,000
	2400	3570 MBH		.18	178		21,700	3,950		25,650	29,600
	2420	4207 MBH		.16	200		25,600	4,450		30,050	34,700
	2440	4720 MBH		.14	229		28,450	5,100		33,550	38,700
	2460	5660 MBH		.13	246		29,450	5,475		34,925	40,400
	2480	6100 MBH		.12	267		34,950	5,925		40,875	47,100
	2500	6390 MBH		.10	320		36,400	7,125		43,525	50,500
	2520	6680 MBH		.09	356		37,550	7,925		45,475	53,000
	2540	6970 MBH		.08	400		38,900	8,900		47,800	56,000
	3000	Hot water, gross output, 80 MBH		1.46	21.920		795	490		1,285	1,575
	3020	100 MBH		1.35	23.700		930	530		1,460	1,800
	3040	122 MBH		1.10	29.090		1,030	650		1,680	2,075
	3060	163 MBH		1	32		1,270	710		1,980	2,425

080	3080	203 MBH	Q-7	1	32	Ea.	1,460	710		2,170	2,650	080
	3100	240 MBH		.95	33.680		1,650	750		2,400	2,900	
	3120	280 MBH		.90	35.560		1,890	790		2,680	3,225	
	3140	320 MBH		.80	40		2,080	890		2,970	3,575	
	3160	360 MBH		.75	42.670		2,370	950		3,320	4,000	
	3180	400 MBH		.70	45.710		2,530	1,025		3,555	4,275	
	3200	440 MBH		.65	49.230		2,710	1,100		3,810	4,575	
	3220	544 MBH		.60	53.330		3,790	1,175		4,965	5,900	
	3240	765 MBH		.55	58.180		5,220	1,300		6,520	7,625	
	3260	1088 MBH		.50	64		6,720	1,425		8,145	9,475	
	3280	1275 MBH		.46	69.570		8,440	1,550		9,990	11,500	
	3300	1530 MBH		.42	76.190		9,940	1,700		11,640	13,400	
	3320	2000 MBH		.38	84.210		12,870	1,875		14,745	16,900	
	3340	2312 MBH		.36	88.890		14,400	1,975		16,375	18,700	
	3360	2856 MBH		.33	96.970		17,400	2,150		19,550	22,300	
	3380	3264 MBH		.30	107		19,400	2,375		21,775	24,800	
	3400	3808 MBH		.26	123		23,050	2,750		25,800	29,300	
	3420	4488 MBH		.22	145		26,850	3,250		30,100	34,300	
	3440	4720 MBH		.18	178		28,250	3,950		32,200	36,800	
	3460	5520 MBH		.14	229		31,300	5,100		36,400	41,800	
	3480	6100 MBH		.12	267		36,800	5,925		42,725	49,100	
	3500	6390 MBH		.10	320		38,150	7,125		45,275	52,500	
	3520	6680 MBH		.09	356		39,400	7,925		47,325	55,000	
	3540	6970 MBH		.08	400		40,750	8,900		49,650	58,000	

Figure 23.28

Means Forms
COST ANALYSIS

PROJECT: Heating & Air Conditioning Systems Worksheet
ARCHITECT:
SHEET NO.:
ESTIMATE NO.:
DATE:

TAKE OFF BY: QUANTITIES BY: PRICES BY: EXTENSIONS BY: CHECKED BY:

Equipment	Type	Table	Number	Qty.		Unit Cost	Total Cost	Comments
Heat Source Boiler Forced Hot Water	C.I. Gas	15.5-	080-3380	1	Ea.	248 00	248 00	Volume 18,900 SF x 14 = 264,600 x 3 793,800 x 4.3 3,413,340 BTUH
Pipe		with A/C						
Duct		with A/C						
Terminals		with A/C						
Cold Source - Chiller		B8.4-	120-4040	56,700	SF	6 71	380457	18,900 x 3 = 56,700 SF. By taking the higher number rather than interpoling gives an allowance for piping between boiler and chiller
Pipe		Included						
Duct		Included						See mechanical Cost Data Book, 8.4-120 for discussion of air handling units vs. duct distribution
Terminals		Included						
Subtotal							405257	
Quality Complexity		C8.4-008		10%			40526	
Total							445783	

Figure 23.29

AIR CONDITIONING — B8.4-120 — Chilled Water, Water Cooled

General: Water cooled chillers are available in the same sizes as air cooled units. They are also available in larger capacities.

Design Assumptions: The chilled water systems with water cooled condenser, include reciprocating hermetic compressors, water cooling tower, pumps, piping and expansion tanks and are based on a two pipe system. Chilled water piping is insulated. No ducts are included and fan-coil units are cooling only. Area distribution is through use of multiple fan coil units. Fewer but larger fan coil units with duct distribution would be approximately the same S.F. Cost (See C8.4-001 for methods of adding heat.) Water treatment and balancing are not included.

System Components	QUANTITY	UNIT	MAT.	INST.	TOTAL
SYSTEM 08.4-120-1320 PACKAGED CHILLER, WATER COOLED, WITH FAN COIL UNIT APARTMENT CORRIDORS, 4,000 S.F., 7.33 TON					
Fan coil air conditioner unit, cabinet mounted & filters, chilled water	2.000	Ea.	2,566.37	304.15	2,870.52
Water chiller, reciprocating, water cooled, 1 compressor semihermetic	1.000	Ea.	6,146.36	1,662.54	7,808.90
Cooling tower, draw thru single flow, belt drive	1.000	Ea.	470.07	65.02	535.09
Cooling tower pumps & piping	1.000	System	235.07	146.09	381.16
Chilled water unit coil connections	2.000	Ea.	712.54	1,227.46	1,940
Chilled water distribution piping	520.000	L.F.	4,264	9,776	14,040
TOTAL			14,394.41	13,181.26	27,575.67
COST PER S.F.			3.60	3.30	6.90

*Cooling requirements would lead to choosing multiple chillers

8.4-120	Chilled Water, Cooling Tower Systems	MAT.	INST.	TOTAL
1300	Packaged chiller, water cooled, with fan coil unit			
1320	Apartment corridors, 4,000 S.F., 7.33 ton	3.60	3.30	6.90
1360	6,000 S.F., 11.00 ton	3.07	2.80	5.87
1400	10,000 S.F., 18.33 ton	2.32	2.10	4.42
1440	20,000 S.F., 26.66 ton	1.82	1.58	3.40
1480	40,000 S.F., 73.33 ton	2.76	1.66	4.42
1520	60,000 S.F., 110.00 ton	2.73	1.72	4.45
1600	Banks and libraries, 4,000 S.F., 16.66 ton	5	3.70	8.70
1640	6,000 S.F., 25.00 ton	4.27	3.20	7.47
1680	10,000 S.F., 41.66 ton	3.57	2.42	5.99
1720	20,000 S.F., 83.33 ton	4.78	2.42	7.20
1760	40,000 S.F., 166.66 ton	4.45	2.90	7.35
1800	60,000 S.F., 250.00 ton	4.40	3.15	7.55
1880	Bars and taverns, 4,000 S.F., 44.33 ton	8.60	4.74	13.34
1920	6,000 S.F., 66.50 ton	10.05	4.93	14.98
1960	10,000 S.F., 110.83 ton	9.90	4.22	14.12
2000	20,000 S.F., 221.66 ton	8.95	4.55	13.50
2040	40,000 S.F., 440 ton*			
2080	60,000 S.F., 660 ton*			
2160	Bowling alleys, 4,000 S.F., 22.66 ton	6	4.01	10.01
2200	6,000 S.F., 34.00 ton	5.05	3.57	8.62
2240	10,000 S.F., 56.66 ton	4.45	2.68	7.13

Figure 23.30a

AIR CONDITIONING	B8.4-120	Chilled Water, Water Cooled			
8.4-120	Chilled Water, Cooling Tower Systems		COST PER S.F.		
			MAT.	INST.	TOTAL
2280	20,000 S.F., 113.33 ton		5.55	2.63	8.18
2320	40,000 S.F., 226.66 ton		5.15	3.12	8.27
2360	60,000 S.F., 340 ton*				
2440	Department stores, 4,000 S.F., 11.66 ton		4.34	3.52	7.86
2480	6,000 S.F., 17.50 ton		3.68	3	6.68
2520	10,000 S.F., 29.17 ton		2.75	2.22	4.97
2560	20,000 S.F., 58.33 ton		2.32	1.68	4
2600	40,000 S.F., 116.66 ton		3.33	1.81	5.14
2640	60,000 S.F., 175.00 ton		3.67	2.83	6.50
2720	Drug stores, 4,000 S.F., 26.66 ton		6.40	4.12	10.52
2760	6,000 S.F., 40.00 ton		5.70	3.60	9.30
2800	10,000 S.F., 66.66 ton		6.40	3.30	9.70
2840	20,000 S.F., 133.33 ton		6.20	2.86	9.06
2880	40,000 S.F., 266.67 ton		5.95	3.55	9.50
2920	60,000 S.F., 400 ton*				
3000	Factories, 4,000 S.F., 13.33 ton		4.27	3.50	7.77
3040	6,000 S.F., 20.00 ton		3.80	2.98	6.78
3080	10,000 S.F., 33.33 ton		3.02	2.30	5.32
3120	20,000 S.F., 66.66 ton		3.66	2.09	5.75
3160	40,000 S.F., 133.33 ton		3.64	1.92	5.56
3200	60,000 S.F., 200.00 ton		3.93	2.95	6.88
3280	Food supermarkets, 4,000 S.F., 11.33 ton		4.26	3.50	7.76
3320	6,000 S.F., 17.00 ton		3.33	2.90	6.23
3360	10,000 S.F., 28.33 ton		2.69	2.21	4.90
3400	20,000 S.F., 56.66 ton		2.35	1.69	4.04
3440	40,000 S.F., 113.33 ton		3.30	1.81	5.11
3480	60,000 S.F., 170.00 ton		3.64	2.82	6.46
3560	Medical centers, 4,000 S.F., 9.33 ton		3.65	3.21	6.86
3600	6,000 S.F., 14.00 ton		3.17	2.86	6.03
3640	10,000 S.F., 23.33 ton		2.54	2.11	4.65
3680	20,000 S.F., 46.66 ton		2.08	1.63	3.71
3720	40,000 S.F., 93.33 ton		3.01	1.74	4.75
3760	60,000 S.F., 140.00 ton		3.35	2.76	6.11
3840	Offices, 4,000 S.F., 12.66 ton		4.13	3.47	7.60
3880	6,000 S.F., 10.00 ton		3.73	3.05	6.78
3920	10,000 S.F., 31.66 ton		2.99	2.32	5.31
3960	20,000 S.F., 63.33 ton		3.60	2.09	5.69
4000	40,000 S.F., 126.66 ton		3.78	2.73	6.51
4040	60,000 S.F., 190.00 ton		3.80	2.91	6.71
4120	Restaurants, 4,000 S.F., 20.00 ton		5.35	3.72	9.07
4160	6,000 S.F., 30.00 ton		4.58	3.34	7.92
4200	10,000 S.F., 50.00 ton		4.08	2.56	6.64
4240	20,000 S.F., 100.00 ton		5.25	2.57	7.82
4280	40,000 S.F., 200.00 ton		4.62	2.92	7.54
4320	60,000 S.F., 300.00 ton		4.92	3.32	8.24
4400	Schools and colleges, 4,000 S.F., 15.33 ton		4.72	3.62	8.34
4440	6,000 S.F., 23.00 ton		4.03	3.13	7.16
4480	10,000 S.F., 38.33 ton		3.36	2.37	5.73
4520	20,000 S.F., 76.66 ton		4.58	2.38	6.96
4560	40,000 S.F., 153.33 ton		4.20	2.83	7.03
4600	60,000 S.F., 230.00 ton		4.09	3.02	7.11

Figure 23.30b

AIR CONDITIONING — C8.4-000 | General

Table 8.4-001

General: The purpose of air conditioning is to control the environment of a space so that comfort is provided for the occupants and/or conditions are suitable for the processes or equipment contained therein. The several items which should be evaluated to define system objectives are:

- Temperature Control
- Humidity Control
- Cleanliness
- Odor, smoke and fumes
- Ventilation

Efforts to control the above parameters must also include consideration of the degree or tolerance of variation, the noise level introduced, the velocity of air motion and the energy requirements to accomplish the desired results.

The variation in **temperature** and **humidity** is a function of the sensor and the controller. The controller reacts to a signal from the sensor and produces the appropriate suitable response in either the terminal unit, the conductor of the transporting medium (air, steam, chilled water, etc.), or the source (boiler, evaporating coils, etc.).

The **noise level** is a by-product of the energy supplied to moving components of the system. Those items which usually contribute the most noise are pumps, blowers, fans, compressors and diffusers. The level of noise can be partially controlled through use of vibration pads, isolators, proper sizing, shields, baffles and sound absorbing liners.

Some **air motion** is necessary to prevent stagnation and stratification. The maximum acceptable velocity varies with the degree of heating or cooling which is taking place. Most people feel air moving past them at velocities in excess of 25 FPM as an annoying draft, however, velocities up to 45 FPM may be acceptable in certain cases. Ventilation, expressed as air changes per hour and percentage of fresh air, is usually an item regulated by local codes.

Selection of the system to be used for a particular application is usually a trade-off. In some cases the building size, style, or room available for mechanical use limits the range of possibilities. Prime factors influencing the decision are first cost and total life (operating, maintenance and replacement costs). The accuracy with which each parameter is determined will be an important measure of the reliability of the decision and subsequent satisfactory operation of the installed system.

Heat delivery may be desired from an air conditioning system. Heating capability usually is added as follows: A gas fired burner or hot water/steam/electric coils may be added to the air handling unit directly and heat all air equally. For limited or localized heat requirements the water/steam/electric coils may be inserted into the duct branch supplying the cold areas. Gas fired duct furnaces are also available. **Note:** when water or steam coils are used the cost of piping and boiler must also be added. For a rough estimate use the cost per square foot of the appropriate sized hydronic system with unit heater. This will give the boiler, piping, and the unit heaters which would approximate the cost of the heating coils. The installed cost of electric and gas heaters, boilers and other heat related items on a unit basis may be located in Section 15.5 of *Means Mechanical Cost Data*.

Table 8.4-002 Air Conditioning Requirements
BTU's Per Hour Per S.F. of Floor Area and S.F. Per Ton of Air Conditioning

Type Building	BTU per S.F.	S.F. per Ton	Type Building	BTU per S.F.	S.F. per Ton	Type Building	BTU per S.F.	S.F. per Ton
Apartments, Individual	26	450	Dormitory, Rooms	40	300	Libraries	50	240
Corridors	22	550	Corridors	30	400	Low Rise Office, Exterior	38	320
Auditoriums & Theaters	666	18*	Dress Shops	43	280	Interior	33	360
Banks	50	240	Drug Stores	80	150	Medical Centers	28	425
Barber Shops	48	250	Factories	40	300	Motels	28	425
Bars & Taverns	133	90	High Rise Office-Ext. Rms.	46	263	Office (small suite)	43	280
Beauty Parlors	66	180	Interior Rooms	37	325	Post Office, Individual Office	42	285
Bowling Alleys	68	175	Hospitals, Core	43	280	Central Area	46	260
Churches	600	20*	Perimeter	46	260	Residences	20	600
Cocktail Lounges	68	175	Hotel, Guest Rooms	44	275	Restaurants	60	200
Computer Rooms	141	85	Public Spaces	55	220	Schools & Colleges	46	260
Dental Offices	52	230	Corridors	30	400	Shoe Stores	55	220
Dept. Stores, Basement	34	350	Industrial Plants, Offices	38	320	Shop'g. Ctrs., Super Markets	34	350
Main Floor	40	300	General Offices	34	350	Retail Stores	48	250
Upper Floor	30	400	Plant Areas	40	300	Specialty Shops	60	200

*Persons per ton   12,000 BTU = 1 ton of air conditioning

Figure 23.31

The Estimate Summary

To summarize the mechanical systems prices from separate subcontracting firms, or from one total mechanical contractor, add the three subtotals shown in Figure 23.32 together and round off to a manageable figure. In the event that the owner or designer has not informed the estimator of the type or quality of the system to be priced, the estimator should, at this time, qualify his proposal by stating the quality and the type of systems being priced.

These rounded off prices all include subcontractors' overhead and profit. At this time, a location factor adjustment, as discussed in Chapter 22, "How to Use *Means Mechanical Cost Data* (Figure 22.12), should be added. Any sales taxes and other adjustments (extra overhead, contingencies, etc.) that are required must also be applied. When these factors have been included, a *budget* price for the mechanical work is complete.

Means Forms

CONDENSED ESTIMATE SUMMARY

PROJECT: Office Building
LOCATION: Boston Area
ARCHITECT:
PRICES BY: JJM

TOTAL AREA/VOLUME: 56,700 S.F.
COST PER S.F./C.F.: 10.55
EXTENSIONS BY: RG

SHEET NO.:
ESTIMATE NO.:
DATE: 2-6-87
NO. OF STORIES: 3
CHECKED BY: RG

Systems Estimate	Total	Cost/S.F.
Plumbing	74000	1.30
Fire Protection	78000	1.38
HVAC	446000	7.87
Total Mechanical	598000	10.55

Figure 23.32

Chapter 24
UNIT PRICE ESTIMATING EXAMPLE

This chapter contains a sample mechanical estimate for a three story office building. To prepare the Unit Price Estimate, procedures and forms are utilized, and calculations made, as previously described in Part I of this book. The estimate that follows is prepared from the perspective of a prime mechanical contractor who has been invited by a familiar general contractor to bid on a regional office building for a nationwide firm. All costs and reference tables are from the annual *Means Mechanical Cost Data*. All forms are from *Means Forms for Building Construction Professionals*.

Project Description

The sample project is a three story office building with an open parking garage beneath the building. The structure is steel frame with curtain wall enclosure. Plans for the mechanical work are shown in Figures 24.1 through 24.5. These drawings have been reduced in size and are not to be scaled. The plans are for illustration only, and, while fairly representative, do not show all of the detail one would expect from a project of this magnitude. A full set of specifications would normally accompany a set of plans, spelling out the scope of work and level of quality of materials and equipment.

The quantities given on the following takeoff, summary, and cost analysis forms are representative of an actual takeoff of the various systems made from the original plans.

Some liberties have been taken in material selection, etc., to provide the reader with a variety of estimating procedures. For example, in the piping estimates, steel pipe, cast iron, copper tube, and polyvinyl chloride (PVC) have all been used. The estimate also includes a wide variety of joining methods, including threaded, butt weld, flanged, mechanical joint, lead and oakum, no hub, solvent weld, and (no lead) soft solder. This practice has been continued throughout the entire mechanical estimate to provide a diversified estimate while staying within the confines of practicality.

In keeping with accepted practices of the trade in many areas of this country, all building service piping, gas, water, storm, and sanitary systems either stop or begin at the appropriate meter or five feet outside the building foundation.

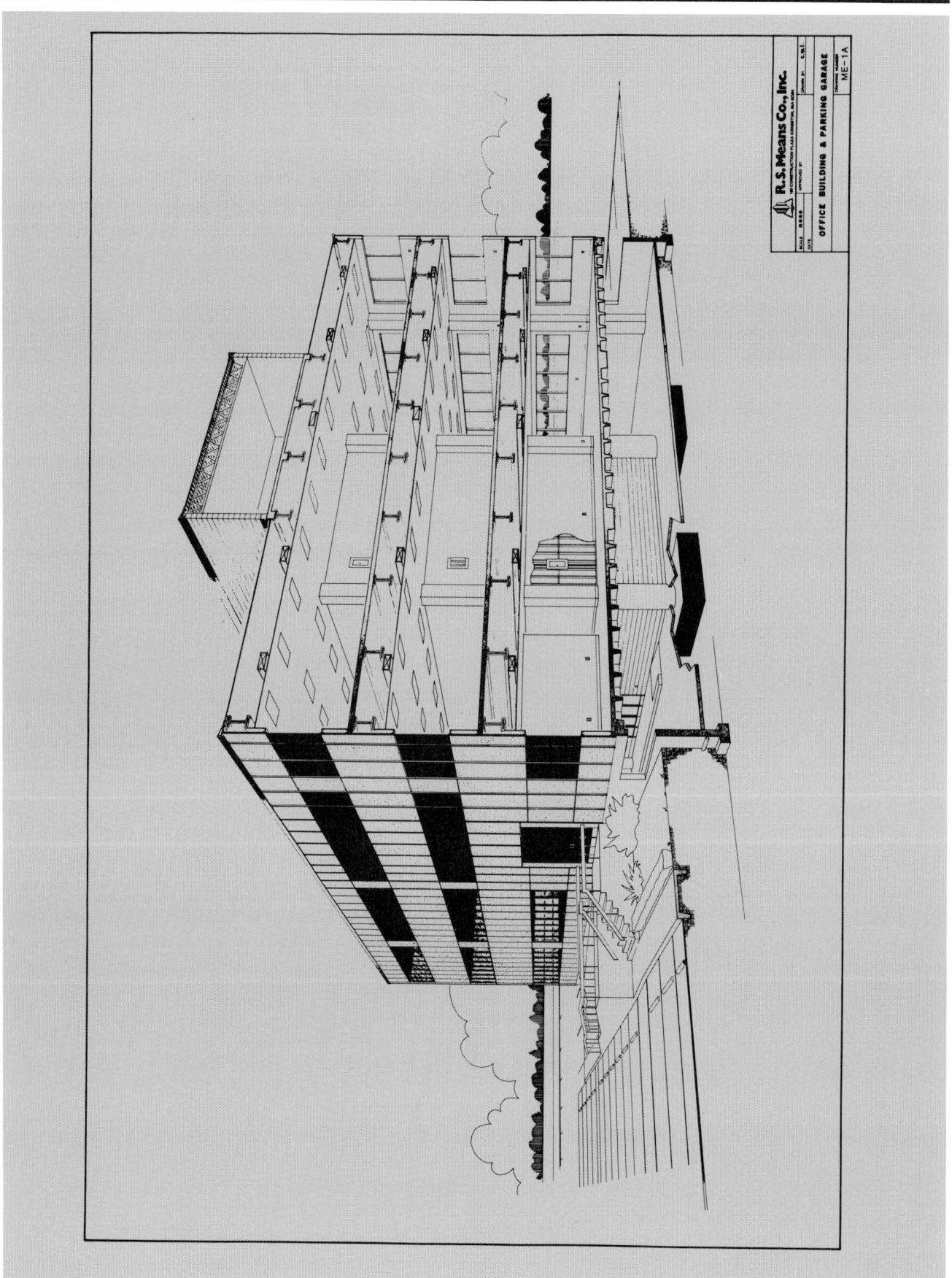

Figure 24.1

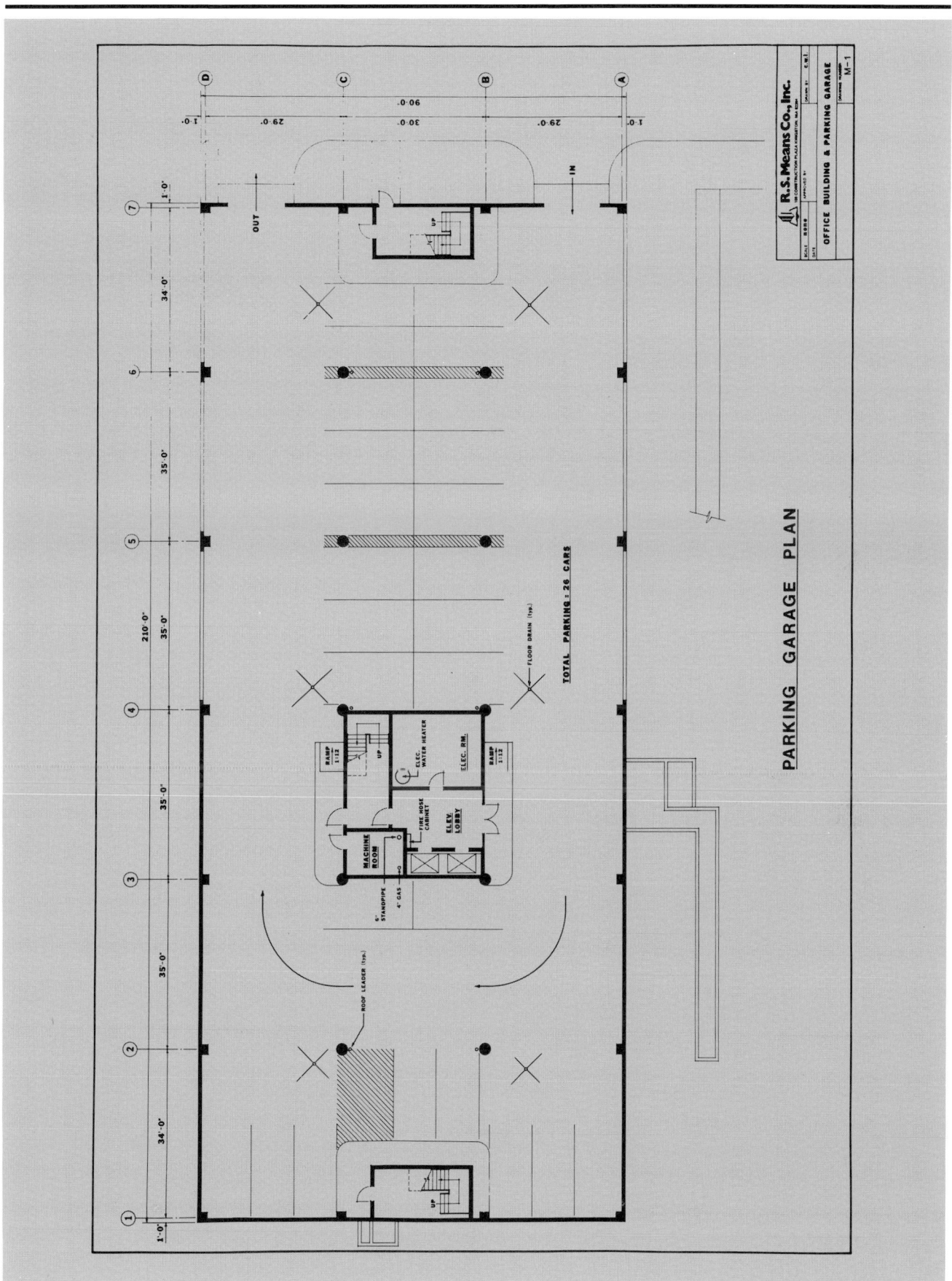

Figure 24.2

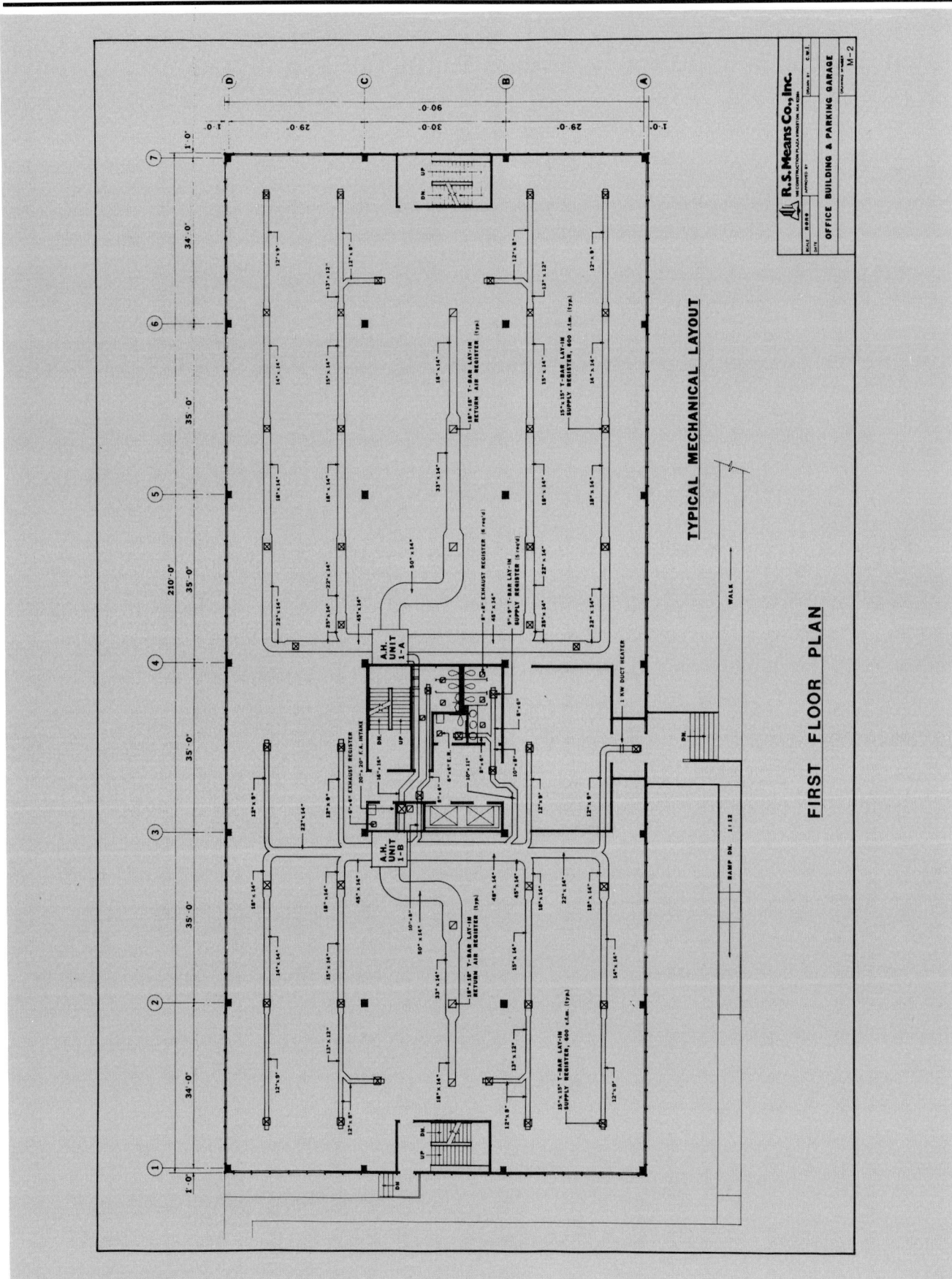

Figure 24.3

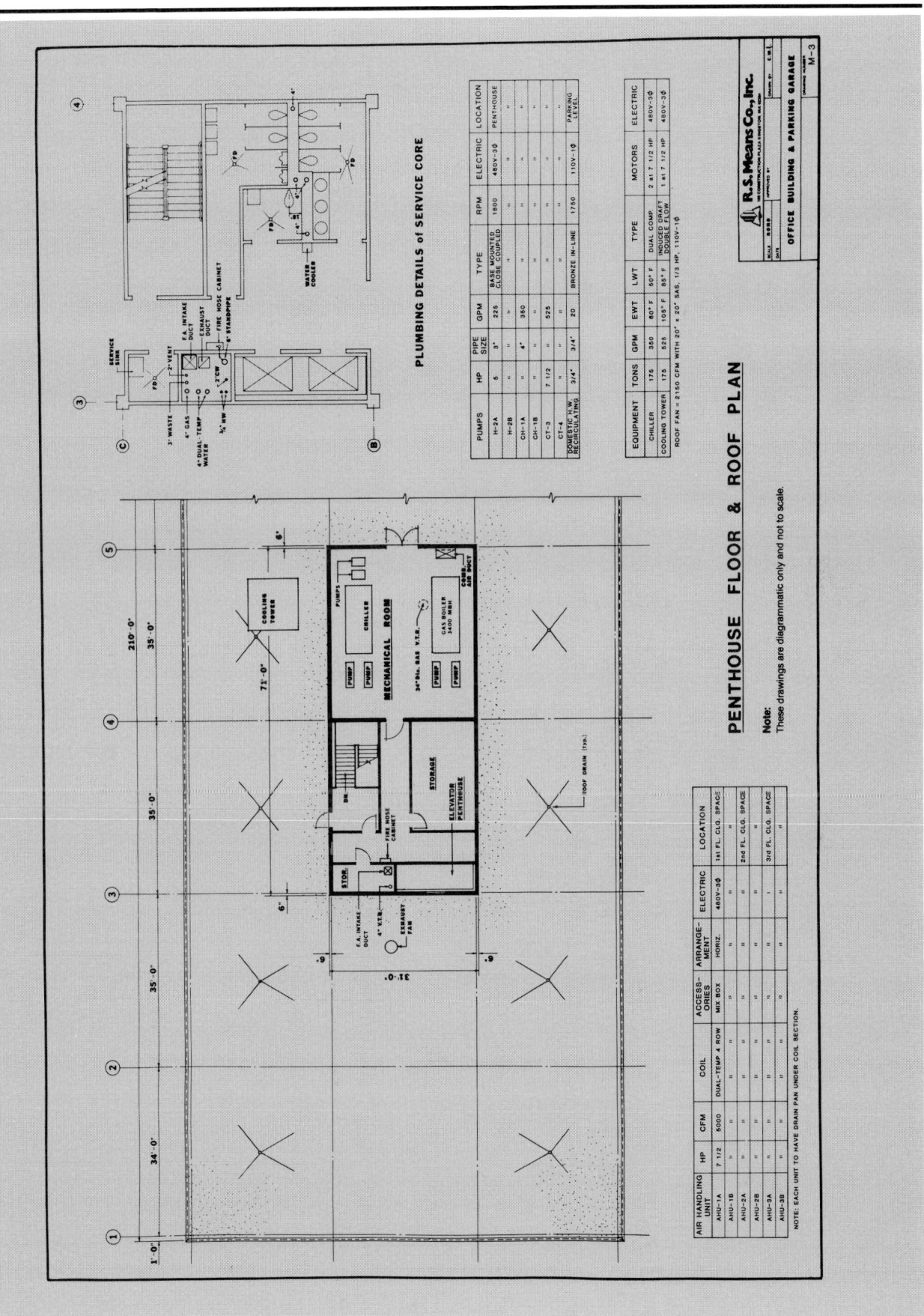

Figure 24.4

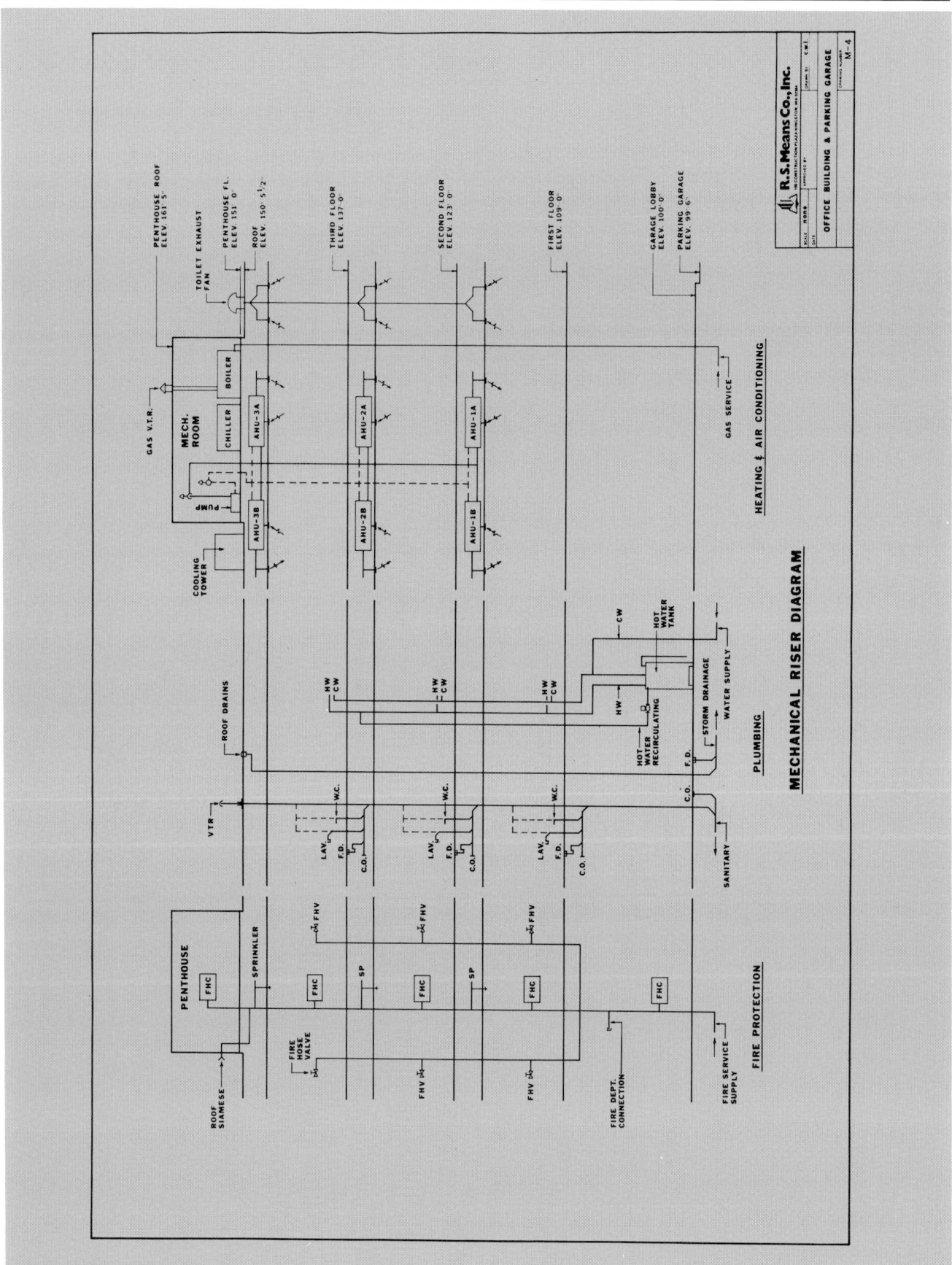

Figure 24.5

Getting Started

A good estimator must be able to visualize the proposed mechanical installation from source (i.e., water heaters or chillers) to end use (i.e., fixture). This process helps identify each component of the mechanical portion of a building. Before starting the takeoff, the estimator should follow some basic steps:

- Read the mechanical plans and specifications carefully (with special attention given to the General and Special Conditions).
- Examine the drawings for general content; type of structure, number of floors, etc.
- Clarify any unclear areas with the designer, making sure that the scope of work is understood. Addenda to the contract documents may be necessary to clarify certain items of work so that the responsibility for the performance of all work is defined.
- Immediately contact suppliers, manufacturers, and subsystem specialty contractors in order to get their quotations, and sub-bids.

Certain information should be taken from the complete architectural plans and specifications if the mechanical work is to be properly estimated. The following items should be given special attention:

- Temporary heat
- Temporary water or sanitary facilities
- Job completion dates (partial occupancy)
- Cleaning

Forms are a necessary tool in the preparation of estimates. This is especially true for the mechanical trades, which include a wide variety of items and materials. Careful measurements and a count of components cannot compensate for an oversight such as forgetting to include pipe insulation. A well-designed form acts as a checklist, a guide for standardization, and a permanent record. A typical preprinted form and checklist is shown in Figure 24.6.

Plumbing

As in any estimate, the first step in preparing the plumbing estimate is to visualize the scope of the job. This can be done by scanning the drawings and specifications. The next step is to make up a material "takeoff" sheet. Having read through the plumbing section of the specifications, list each of the item headings on the summary sheet. This will serve as a quantity checklist. Also include labor that will not be specified in the plans, such as cleaning, adjusting, testing, and balancing.

The easiest way to make a material "quantity count" or takeoff is by system, as most of the pipe components of a system will tend to be of the same material, class weight, and grade. A waste system, for example, would consist of pipe varying from 2" to 15" in diameter, but it would all be of service weight cast iron or DWV copper up to a specified size, and extra heavy cast iron for the larger sizes.

		1	2
	Labor w/Payroll Taxes, Welfare		
	Travel		
	Fixtures		
	Drains, Carriers, etc.		
	Sanitary - Interior		
	Storm - Interior		
	Water - Interior		
	Valves - Interior		
	Water Exterior - Street Connection		
	Water Exterior - Extension to Building		
	Sanitary & Storm Exterior - Street Connection		
	Sanitary & Storm Exterior - Extension to Building		
	Water Meter		
	Gas Service		
	Gas Interior		
	Accessories		
	Flashings for: Vents, Floor & Roof Drains, Pans		
	Hot Water Tanks: Automatic w/trim		
	Hot Water Tanks: Storage, w/stand and trim		
	Hot Water Circ. Pumps, Aquastats, Mag. Starters		
	Sump Pumps and/or Ejectors; Controls, Mag. Starters		
	Flue Piping		
	Rigging, Painting, Excavation and Backfill		
	Therm. Gauges, P.R.V. Controls, Specialties, etc.		
	Record Dwgs., Tags and Charts		
	Fire Extinguishers, Equipment, and/or Standpipes		
	Permits		
	Sleeves, Inserts and Hangers		
	Staging		
	Insulation		
	Acid Waste System		
	Sales Tax		
	Trucking and Cartage		

NAME OF JOB:
LOCATION:
ARCHITECT:
DUE DATE:
BID SENT TO:
ENGINEER:

Figure 24.6

Fixture takeoff usually involves nothing more than counting the various types, sizes, and styles, and entering them on a fixture form. It is important, however, that each fixture be fully identified. A common source of error is to overlook what does or does not come with the fixture (i.e., trim, carrier, flush valve). Equipment such as pumps, water heaters, water softeners, and all items not previously counted are also listed at this time. The order of proceeding is not as important as the development of a consistent method which will assist the estimator to improve in speed while minimizing the chances of overlooking any item or class of items. The various counts should then be totalled and transferred to the summary quantity sheet (Figure 24.7). Miscellaneous items which add to the plumbing contract should also be listed. Many of these would have been identified during the preliminary review of the plans and specifications. It is desirable to take off the fixtures and equipment before the piping for several reasons. The fixture and equipment lists can be given to suppliers for pricing while the estimator is performing the more arduous and time consuming piping takeoff. Taking off the fixtures and equipment first also gives the estimator a good perspective of the building and its systems. Coloring the fixture and equipment as the takeoff is made aids the estimator when he returns for the piping takeoff by acting as targets for piping networks.

Pipe runs for any type of system consist of straight section and fittings of various shapes, styles, and purposes. Depending on job size, separate forms may be used for pipe and fittings, or they may be combined on one sheet. The estimator should start at one end of each system and using a tape, ruler, or wheeled indicator set at the corresponding scale, should measure and record the straight lengths of each size of pipe. When a fitting is encountered, it should be recorded with a check in the appropriate column. The use of colored pencils will aid in marking runs that have been completed, or the termination points on the main line where measurements are stopped, so that a branch may be taken off. Any changes in material should be noted. This would only occur at a joint, therefore, the estimator should always verify that the piping material going into a joint is the same as that leaving the joint. The plumbing takeoff is shown in Figures 24.8a through 24.8f. Subcontractors (i.e., insulation, controls, etc.) should be notified of the availability of your plans after you have completed your takeoff.

Rounding off piping quantities to the nearest five or ten feet is good practice. In larger projects, rounding might be carried to the nearest hundred or thousand feet. This makes pricing and extending less prone to error. Rounding off pipe quantities is also practical because pipe is purchased in uniform lengths (i.e., rigid copper tubing and steel pipe is rounded off to 20' lengths, cast iron soil pipe to 5' or 10' lengths, etc.).

The plumbing totals are transferred to cost analysis forms for pricing and extending, as shown in Figures 24.9a through 24.9e. Units of cost are assigned to each item. These costs may be taken from historical data, or from a reliable reference, such as *Means Mechanical Cost Data*. Be sure to match the description of items to be priced with the historical data as closely as possible when obtaining unit costs. These costs are then extended (multiplied by the total quantity of each item) in order to obtain a total cost per item.

Means Forms

QUANTITY SHEET

PROJECT: Office Building
~~LOCATION~~ Plumbing
ARCHITECT:
TAKE OFF BY: MJM
EXTENSIONS BY: MJM
CHECKED BY:
SHEET NO.:
ESTIMATE NO.: 87-28
DATE: 3-17-87

DESCRIPTION	NO.	DIMENSIONS		UNIT		UNIT		UNIT		UNIT
Fixture Takeoff										
Water Heater 54 KW		80	Gal.	1	Ea.					
Circulating Pump - all bronze		3/4"		1	Ea.					
Water Closet, Wall Hung, w/Flush Valve				15	Ea.					
Water Closet, Floor Type w/Flush (Paraplegic)				3	Ea.					
Urinal, Stall Type				6	Ea.					
Lav. 19"x 16" Oval Vanity				9	Ea.					
Lav. 19"x 17" Wall Hung				6	Ea.					
Lav. 28"x 21" Wall Hung (Paraplegic)				3	Ea.					
Service Sinks 24"x 20" Wall Hung				3	Ea.					
Elec. Water Cooler, Wall Hung				6	Ea.					

Figure 24.7

Means Forms
PIPING SCHEDULE

JOB: Office Building
SYSTEM: Plumbing - Storm System   PVC/Cast Iron
PAGE: P1 of 6
DATE: 3-17-87
BY: JJM

	12	10	8	6	5	4	3 1/2	3	2 1/2	2	1 1/2	1 1/4	1	3/4	1/2
							PIPE DIAMETER IN INCHES								
PVC Sched 40 Pipe								720 180 200							
								PVC 3" Roof Drains (12)							
DWV 1/4 Bends								(1100)							
								(24)							
								4" Garage Floor Drains Galv. (6)							
Underground															
Cast Iron Pipe Service Weight				(220)		(120)									
Cleanout Tee C.I.						(12)									
San. Wye C.I.				(1)		(6)									

Figure 24.8a

Means Forms
PIPING SCHEDULE

JOB: Office Building
SYSTEM: Plumbing - Gas - Schedule #40 Blk. T&C
PAGE: P-2 of 6
DATE: 3-17-87
BY: JJM

	12	10	8	6	5	4	3 1/2	3	2 1/2	2	1 1/2	1 1/4	1	3/4	1/2
						PIPE DIAMETER IN INCHES							Vent		
Steel Pipe						60 100 160							20		
Blk mall Ells						7							4		
B.m. Tees						3									
Caps						3									
Gas Cocks						6									

Figure 24.8b

Means Forms
PIPING SCHEDULE

JOB: Office Building
SYSTEM: Plumbing - Water Pipe, Cold water   Type L
PAGE: P-3 of 6
DATE: 3-17-87
BY: JJM

	12	10	8	6	5	4	3 1/2	3	2 1/2	2	1 1/2	1 1/4	1	3/4	1/2
Copper Tubing										100 58 (160)			28 12 15 (60)	117 228 80 93 (460)	
Ells										|||| (6)			||||||| (18)	|||| |||| |||| |||| (26)	
Tees										|||| (10)				|||| |||| (12)	
Gate Valve Copper										||| (3)					
Water Meter										1				|||| (6)	
Hose Bibbs															
Back flow Preventer										1 Bldg.			1 Boiler Room		
Shock Absorber														|||| (4)	
Vacuum Breaker														|||| (4)	

PIPE DIAMETER IN INCHES

Figure 24.8c

Means Forms
PIPING SCHEDULE

JOB: Office Building
SYSTEM: Plumbing - Water Pipe, Hot Water Type L

PAGE: P.4 of 6
DATE: 3-17-87
BY: JJM

PIPE DIAMETER IN INCHES

	12	10	8	6	5	4	3 1/2	3	2 1/2	2	1 1/2	1 1/4	1	3/4	1/2
"L" Copper Tubing														66 22 5 20 /// 220'	
90° Ells														++++ ++++ ++ = 12	
Tees														//// ++++ ++++ /// (18)	
Relief Valve														—	
Unions														= (2)	
Temp. Reg./mixing Valve														(1) —	
Gate Valve														++++ - (6)	
Copper Check Valve														—	

Figure 24.8d

Means Forms
PIPING SCHEDULE

JOB: Office Building
SYSTEM: Plumbing - Sanitary & Vent
PAGE: P.5 of 6
DATE: 3-17-87
BY: JJM

	| PIPE DIAMETER IN INCHES												
	12	10	8	6	5	4	3 1/2	3	2 1/2	2	1 1/2	1 1/4	1	3/4	1/2
Serv. Weight B+S C.I. Pipe						52 35 ⑨⓪		13 27 ④⓪							
Comb. Y & ⅛ Bend						⑤		⑦							
Floor Cleanout (FCO)						③		-①							
¼ Bend								②							
No Hub C.I. Pipe						60 30 30 ①⑦⓪		30 10 20 ②①⓪							
Y (SanT)						③⓪		②⓪							
FCO						③									
Floor Drains						②									
Vent Flashing VTR						⑥		①②							
Wall cleanout (WCO)						⑥		②							
No Hubs ¼ Bends						⑥									

Figure 24.8e

Means Forms
PIPING SCHEDULE

JOB: Office Building
SYSTEM: Plumbing - Sanitary & Vent
PAGE: P-6 of 6
DATE: 3-17-87
BY: JJM

	12	10	8	6	5	4	3 1/2	3	2 1/2	2	1 1/2	1 1/4	1	3/4	1/2
No Hub 1/8 Bends						(11)		(4)							
No Hub Stainless Coupling						90		72							
Copper DWV										(300)					
San. Tee DWV										(84)					
Ells DWV										(49)					

Figure 24.8f

On the final sheet of the plumbing estimate, (Figure 24.9e) page subtotals are recorded and totalled to arrive at the "bare" cost for plumbing. These values must then be adjusted for sales tax (if applicable), overhead, and profit on material, labor, and equipment as discussed in Chapter 22, "How to Use *Means Mechanical Cost Data*". Finally, the marked up totals are adjusted by the appropriate location factor (Boston is used in the example) also as discussed in Chapter 22. This completes the plumbing estimate for the office building. The resultant price is what a general contractor would receive as a complete bid for the plumbing work.

Fire Protection

After having scanned the plans and specifications to visualize the scope and type of fire protection proposed for this project, estimate sheets are prepared, as shown in Figure 24.10a through 24.10c. When properly used, such forms will serve as a summary of the specifications.

The specifications should be studied, and all components of fire protection systems should be entered on the takeoff sheets (Figures 24.10a through 24.10 c). Material pricing and labor considerations should be listed.

The General Conditions and the Special Conditions should indicate the fire protection contractor's responsibilities regarding excavation and backfill, cutting and patching, masonry and concrete, painting, electrical work, temporary services, and any other items that may affect the estimate. Note on the estimate sheet the services that will be provided "by others". Indicate "labor only" items, such as cleaning, testing, flushing, application of decals, distribution of equipment to be stored in the cabinet, etc.

The material or quantity takeoff should be executed systematically, floor by floor. The fire protection estimator should be familiar with the architectural plans as this may be the only place that equipment such as extinguishers or other specialized equipment will be shown.

Hazardous areas requiring special considerations which may have been overlooked by the designer may be found on the architectural plans. The reflected ceiling plans and ductwork layout should also be studied to identify the degree of coordination required for the sprinkler head layout.

The piping takeoff for standpipes should be kept separate from the sprinkler distribution piping. As in other piping takeoffs, pipe should be listed by material and joining method. The sprinkler heads should be taken off first, floor by floor, by type and configuration.

Fire protection materials after the takeoff should be totalled and transferred to a cost analysis form for pricing and extending, as shown in Figures 24.11a through 24.11c. As with the plumbing estimate, bare costs are summarized, totalled, and marked up on the final page (Figure 24.11c) to arrive at the "bid" price for fire protection.

Means Forms
COST ANALYSIS

PROJECT: Office Building
CLASSIFICATION: Div. 15- Plumbing
SHEET NO. P1 of 5
LOCATION:
ARCHITECT:
ESTIMATE NO. 87-28
TAKE OFF BY: JJM
QUANTITIES BY: JJM
PRICES BY: MJM
EXTENSIONS BY: MJM
CHECKED BY: JJM
DATE: 3-19-87

DESCRIPTION	SOURCE/DIMENSIONS			QUANTITY	UNIT	MATERIAL UNIT COST	MATERIAL TOTAL	LABOR UNIT COST	LABOR TOTAL	EQUIPMENT UNIT COST	EQUIPMENT TOTAL	SUBCONTRACT UNIT COST	SUBCONTRACT TOTAL	TOTAL UNIT COST	TOTAL
Water Heater 36kw 80 Gal.	15.3	500	4300	1	Ea.	1870	1870	125	125						
Circulating Pump 3/4"	15.2	410	0640	1	Ea.	71	71	21	21						
Water Closet, white, wm/Flush w/c Rough In	15.2	800	3100	15	Ea.	235	3525	57	855						
w/c white, Fm/Flush Parapl.			3200	15		128.40	1926	160	2400						
w/c Rough In			3380	3		232	696	66	198						
		→	3400	3	→	89.36	268	185	555						
Urinal - Stall Type	15.2	680	5000	6	Ea.	370	2220	130	780						
Urinal - Rough In	→	→	6480	6	Ea.	68.61	412	165	990						
Lav. 19"x16" Oval Vanity, Mt.	15.2	320	0680	9	Ea.	102	918	52	468						
Lav. V/m Rough In			3580	9		59.75	538	170	1530						
Lav. 19"x17" Wall Mt.			4120	6		169	1014	41	246						
Lav. w/m Rough In			6960	6		111.10	667	200	1200						
Lav. 28"x21" Wall Mt. Parapl.			6210	3		209	627	47	141						
Lav. w/m Rough In	→	→	6960	3	→	111.10	333	200	600						
Service Sink 24"x20" w/m	15.2	600	7100	3	Ea.	270	810	110	330						
Service Sink w/m Rough In	→	→	8950	3	Ea.	151.68	455	255	765						
Water Cooler Wall Hung	15.3	450	0220	6	Ea.	350	2100	83	498						
Water Cooler w/H Rough In			9800	6		26.40	158	150	900						
Water Cooler-Stainless Cabinet	→		0640	6	→	38	228								
			Page P1 Subtotal				18836		12602						

Figure 24.9a

Means Forms
COST ANALYSIS

PROJECT: Office Building
LOCATION:
TAKE OFF BY: JJM
QUANTITIES BY: JJM
PRICES BY: MJM
EXTENSIONS BY: MJM
CLASSIFICATION: Div. 15 - Plumbing
ARCHITECT:
CHECKED BY: JJM
SHEET NO. P2 of 5
ESTIMATE NO. 87-28
DATE 3-19-87

DESCRIPTION		SOURCE/DIMENSIONS		QUANTITY	UNIT	MATERIAL UNIT COST	MATERIAL TOTAL	LABOR UNIT COST	LABOR TOTAL	EQUIPMENT UNIT COST	EQUIPMENT TOTAL	SUBCONTRACT UNIT COST	SUBCONTRACT TOTAL	TOTAL UNIT COST	TOTAL
PVC															
Roof Drain	3"	15.1	160 4780	12	Ea.	32	384	24	288						
Pipe	3"	→	490 4470	1100	LF	1.57	1727	6.25	6875						
90 Ells	3"	→	500 5080	24	Ea.	1.56	37	19.50	468						
Cast Iron Hub Type															
Floor Drain Galv.	4"	15.1	160 0440 / 0480	6	Ea.	225	1350	41	246						
Tee Clean Out	4"		100 0240	12		26.75	321	56	672						
San Y (Tee)	4"		380 0620	6		11.50	69	41	246						
	6"		380 0800	1		25	25	47	47						
Floor Cleanouts	3"		070 0120	1		36.75	37	23	23						
	4"		070 0140	3		50.25	151	31	93						
Comb. Y & 1/8 Bend	2"		380 1420	2		7.25	15	33	66						
	4"		1520	5		15.10	76	41	205						
90 Ell (1/4 Bend)	3"		0120	2		5.50	11	24	48						
Pipe	3"		370 2140	40	LF	3.73	149	6.90	276						
	4"		2160	210		4.99	1048	7.55	1586						
	6"	→	2200	220		9.23	2031	8.75	1925						
Cast Iron Hubless															
Pipe	3"	15.1	370 4140	210	LF	3.66	769	5.70	1197						
	4"		370 4160	170	LF	4.73	804	6.25	1063						
San Y (Tee)	3"		380 6520	20	Ea.	4.29	86								
	4"		380 6600	30		6.70	201								
Floor Cleanout	4"		070 0140	3		50.25	151	31	93						
Floor Drains	3"		160 2000 / 2120	12		41.75	501	28	336						
Wall Cleanout	4"	→	070 4100	6		44.50	267	18.40	110						
		Page P2 Subtotal					10210		15863						

Figure 24.9b

Means Forms — COST ANALYSIS

PROJECT: Office Building
LOCATION:
TAKE OFF BY: JJM
QUANTITIES BY: JJM
PRICES BY: JJM
EXTENSIONS BY: MJM
CLASSIFICATION: Div. 15 - Plumbing
ARCHITECT: MJM
SHEET NO. P3 of 5
ESTIMATE NO. 87-28
DATE 3-19-87
CHECKED BY: JJM

DESCRIPTION	SOURCE/DIMENSIONS			QUANTITY	UNIT	MATERIAL UNIT COST	MATERIAL TOTAL	LABOR UNIT COST	LABOR TOTAL	EQUIPMENT UNIT COST	EQUIPMENT TOTAL	SUBCONTRACT UNIT COST	SUBCONTRACT TOTAL	TOTAL UNIT COST	TOTAL
Cast Iron Hubless (Cont.)															
90 Ell (¼ Bend) 4"	15.1	380	6120	6	Ea.	5	30								
45 Ell (⅛ Bend) 3"			6260	4		2.73	11								
4"			6280	11		3.59	39								
St. St. Couplings 3"			8650	72		7.60	547	13.80	994						
4"			8660	90	→	8.80	792	15.05	1355						
Vent thru Roof Flash 3"	15.1	950	1450	2	Ea.	14.40	29	10.80	22						
4"	15.1	950	1460	2	Ea.	15.90	32	11.50	23						
Black Steel															
Pipe 1"	15.1	550	0580	20	LF	1.39	28	3.47	69						
4"		550	0650	160	LF	9.11	1458	9.20	1472						
90 Ell Blk mal 1"		560	5100	4	Ea.	1.33	5	14.15	57						
4"			5170	7		27	189	55	385						
Tee Blk mal 4"			5580	3		35	105	83	249						
Caps 4"			5780	3	→	21	63	16	48	(note: This is ½ Coupling Labor as there is no line number for caps)					
Gas Cocks 4"	15.1	930	7030	2	Ea.	330	660	110	220						
Copper (Sweat Joints)															
Type L Tube ¾"	15.1	400	2180	680	LF	1.03	700	4.18	2842						
1"			2200	60	LF	1.46	88	5.10	306						
2"			2260	160	LF	3.66	586	8	1280						
90 Ells ¾"		410	0120	38	Ea.	.39	15	9.70	369						
1"			0130	18	Ea.	.90	16	11.50	207						
2"			0160	6	Ea.	4.17	25	16.75	101						
	Page P3 Subtotal						5418		9999						

Figure 24.9c

Means Forms
COST ANALYSIS

PROJECT: Office Building
LOCATION:
TAKE OFF BY: JJM
QUANTITIES BY: JJM
PRICES BY: MJM
EXTENSIONS BY: MJM
CLASSIFICATION: Div. 15 - Plumbing
ARCHITECT:
SHEET NO. P4 of 5
ESTIMATE NO. 87-28
DATE 3-19-87
CHECKED BY: JJM

DESCRIPTION		SOURCE/DIMENSIONS		QUANTITY	UNIT	MATERIAL UNIT COST	MATERIAL TOTAL	LABOR UNIT COST	LABOR TOTAL	EQUIPMENT UNIT COST	EQUIPMENT TOTAL	SUBCONTRACT UNIT COST	SUBCONTRACT TOTAL	TOTAL UNIT COST	TOTAL
Copper (Sweat Joints, Cont.)															
Tee	3/4"	15.1	410 0500	18	Ea.	.66	12	15.35	276						
	2"		410 0540	10		8.05	80	26	260						
Shock Absorb.	3/4"		670 0500	4		29	116	15.35	61						
Vacuum Break	3/4"		760 1080	4		14.45	58	9.20	37						
Unions	3/4"		410 0900	2		1.93	4	10.20	20						
Gate Valve	3/4"		800 2940	18		14.90	268	9.20	166						
	2"		2980	3		43	129	16.75	50						
Check Valve	3/4"		1860	1		12.45	12	9.20	9						
Relief Valve	3/4"		5640	1		31	31	6.55	7						
Temp. Mix Valve	3/4"		8440	1		23	23	9.20	9						
DWV Copper (Sweat)															
DWV Tube	2"	15.1	400 4140	320	LF	2.49	797	7.65	2448						
San Tee	2"		410 2290	48	Ea.	5.35	257	26	1248						
90 Ell	2"		410 2070	24	Ea.	3.63	87	18.40	442						
Water Meter	2"	15.1	970 2360	1	Ea.	300	300	31	31						
Hose Bibbs	3/4"		220 5000	6		4.07	24	7.65	46						
Back Flow Prev.	1"		040 1120	1		130	130	13.15	13						
	2"		040 1160	1		325	325	26	26						
			Page P4 Subtotal				2653		5149						

Figure 24.9d

Means Forms
COST ANALYSIS

PROJECT: Office Building
LOCATION:
TAKE OFF BY: JJM **QUANTITIES BY:** JJM **PRICES BY:** MJM **EXTENSIONS BY:** MJM **CHECKED BY:** JJM
CLASSIFICATION: Div. 15 - Plumbing
ARCHITECT:
SHEET NO. P5 of 5
ESTIMATE NO. 87-28
DATE 3-19-87

DESCRIPTION	SOURCE/DIMENSIONS	QUANTITY	UNIT	MATERIAL UNIT COST	MATERIAL TOTAL	LABOR UNIT COST	LABOR TOTAL	EQUIPMENT UNIT COST	EQUIPMENT TOTAL	SUBCONTRACT UNIT COST	SUBCONTRACT TOTAL	TOTAL UNIT COST	TOTAL
Insulation for Hot/Cold Water													
Copper Tube													
Fiberglass w/ASJ 1"wall 3/4"	15.5 →650	680	LF							3.22	2190		
1"	→68.60	60	→							3.47	208		
2"	→68.90	160	→							4.07	651		
Miscellaneous													
Demonstration Tests		4	Hrs			23	92						
Sleeving		16				23	368						
Warranty		6				23	138						
Valves, Tags & Charts		4	→			23	92						
	Plumbing Page P5 Subtotal						690				3049		
Plumbing Subtotals P1					18836		12602						
P2					10210		15863						
P3					5418		9999						
P4					2653		5149						
P5							690				3049		
	Subtotal Total				37117		44303				3049		
Material Sales Tax				5%	1856								
Overhead & Profit markup				10%	38973	45.8%	20291			10%	305		110818
		Total			42870		64594				3354		
					×1.046		×1.106				×1.077		
City modifier for Boston					44842		71441				3612		119895
	Boston Total												

Figure 24.9e

Means Forms
PIPING SCHEDULE

JOB: Office Building
SYSTEM: Fire Protection Standpipe/Sprinkler Black Steel Sched. 40 T & C
PAGE: FP 1 of 3
DATE: 3-20-87
BY: JJM

							PIPE DIAMETER IN INCHES								
	12	10	8	6	5	4	3 1/2	3	2 1/2	2	1 1/2	1 1/4	1	3/4	1/2
Standpipe B.S. Pip Sch. 40			60	60		300			80						
90° Ells C.I. Screwed			3	1		4			16						
Tees C.I. Screwed			3	5-6x3 1/2 3-6x5		8									
F. Dept. Valves									13						
Reducers C.I.			2	6x3 1/2 x 2 1/2											
Roof Siamese				1											
OS & Y Gate Valve			1	3											
Swing Check Valve w/Ball Drip				1											
Wall Siamese				1											
Fire Hose											500				
Hose Adapter									5						
Hose Rack									5						
Hose Nozzle									5						

Figure 24.10a

Means Forms
PIPING SCHEDULE

JOB Office Building
SYSTEM Fire Protection Standpipe / Sprinkler

PAGE FP 2 of 3
DATE 3-20-87
BY JJM

	12	10	8	6	5	4	3 1/2	3	2 1/2	2	1 1/2	1 1/4	1	3/4	1/2	
Standpipe (cont.)																
Hydrolator									5							
Fire Hose Cabinet									5							
Sprink. Distribution																
Alarm Valve				1												
Alarm water mtr/Gong																size std. 1
Check Valve				1												
Flow Control Valve				1												
B.S. Pipe Grooved Sch. 10						50			300	960						
B.S. Pipe Grooved Sch. 40											1150	1150	1150			
Tee Grooved						10			36							

Figure 24.10b

Means Forms
PIPING SCHEDULE

JOB: Office Building
SYSTEM: Fire Protection
PAGE: FP 3 of 3
DATE: 3-20-87
BY: JJM

Standpipe/Sprinkler

	12	10	8	6	5	4	3 1/2	3	2 1/2	2	1 1/2	1 1/4	1	3/4	1/2
										2x1	1½x½	1¼x½			
Sprink. Distribution (cont.)															
mech. Joint Tees										86	80	40			
90° Ells C.I. Screwed													50		
Coupling Grooved Jt.						20			80	20					
90° Ells Grooved Jt.										10					
Sprinkler Heads Rec. Pendent															100 100 100 (300)
Tee C.I. Screwed													40		

PIPE DIAMETER IN INCHES

Figure 24.10c

Means Forms
COST ANALYSIS

PROJECT: Office Building
LOCATION: Boston, MA
TAKE OFF BY: JJM
QUANTITIES BY: JJM
PRICES BY: MJM
EXTENSIONS BY: MJM
CHECKED BY: JJM
CLASSIFICATION: Div. 15 - Fire Protection
ARCHITECT:
SHEET NO. FP 1 of 3
ESTIMATE NO. 87-16
DATE 3-20-87

DESCRIPTION	SOURCE/DIMENSIONS		QUANTITY	UNIT	MATERIAL UNIT COST	MATERIAL TOTAL	LABOR UNIT COST	LABOR TOTAL	EQUIPMENT UNIT COST	EQUIPMENT TOTAL	SUBCONTRACT UNIT COST	SUBCONTRACT TOTAL	TOTAL UNIT COST	TOTAL
Black Steel (thread & coupled)														
Pipe Sch. 40 2½"	15.1	550 0620	80	LF							14.65	1172		
4"		0650	300	→							23	6900		
6"		0670	60								51	3060		
8"		0680	60								65	3900		
90° Ells C.I. 2½"		560 0150	16	Ea.							44	704		
4"		0180	4								105	420		
6"		0200	1								165	165		
8"		0210	3								240	730		
Tees C.I. 4"		0620	8								155	1240		
6" x 2½"		0640	5								265	1325		
6" x 5"		0640	3								265	795		
8"		0650	3	→							415	1245		
Reducers C.I. 8"		0310	2								240	480		
F.D. Brz. Valves 2½"	15.4	500 0090	13	Ea.							93	1209		
Roof Siamese 6"x2½"x2½"	15.4	500 6060	1								405	405		
OS&Y Gate Valve 6"	15.1	820 3700	3								565	1695		
8"	15.1	820 3720	1								865	865		
Swing Check Valve w/Ball 6"	15.4	600 6540	1								485	485		
Wall Siamese 6"	15.4	300 7370	1	→							640	640		
Fire Hose 1½"	15.4	300 2260	500	LF							1.25	625		
" Adapter 2½"		11000	5	Ea.							19.80	99		
" Rack 1½"		2640	5								47	235		
" Nozzle 1½"		6700	5								22	110		
" Coupling 1½"		1410	5								13.20	66		
" Rack Nipple 1½"	→	2820	5	→							17.90	90		
								Page FP1 Subtotal				28650		

Figure 24.11a

Means Forms
COST ANALYSIS

PROJECT: Office Building
LOCATION: Boston, MA
TAKE OFF BY: JJM
QUANTITIES BY: JJM
PRICES BY: MJM
CLASSIFICATION: Div. 15 - Fire Protection
ARCHITECT:
EXTENSIONS BY: MJM
CHECKED BY: JJM
SHEET NO. FP 2 of 3
ESTIMATE NO. 87-16
DATE 3-20-87

DESCRIPTION		SOURCE/DIMENSIONS			QUANTITY	UNIT	MATERIAL UNIT COST	MATERIAL TOTAL	LABOR UNIT COST	LABOR TOTAL	EQUIPMENT UNIT COST	EQUIPMENT TOTAL	SUBCONTRACT UNIT COST	SUBCONTRACT TOTAL	TOTAL UNIT COST	TOTAL
Hydrolator	1½"	15.4	506	4200	5	Ea.							61	305		
Fire Hose Cabinet		15.4	100	4200	5	Ea.							200	1000		
Alarm Valve	6"	15.4	600	6300	1	Ea.							855	855		
Wtr. Alarm Mtr./Gong				1220	1								175	175		
Check Valve	6"			6840	1								500	500		
Flow Control Valve	6"	→		8860	1	→							1900	1900		
Blk Stl Grooved																
Pipe Sch. 10	2"	15.1	610	0550	960	LF							10.05	9648		
	2½"			0580	300								12.60	3780		
	4"			0590	50								17.25	863		
Pipe Sch. 40	1"			1050	1150								6	6900		
	1¼"			1060	570								6.85	3905		
	1½"	→		1070	1150	→							7.80	8970		
Tee Grvd.	2½"			4780	30	Ea							34	1020		
	4"			4800	10								63	630		
90° Ells Grvd.	2"			4070	10								18.35	184		
Coupling Grvd.	2"			4990	20								13.65	273		
	2½"			5000	80								16	1280		
	4"	→		5030	20	→							26	520		
Tee mech Jt.	1¼" x ½"	15.1	570	9510	40	Ea.							35	1400		
	1½" x 1"			9560	80								37	2960		
	2" x 1"	→		9590	80	→							43	3440		
										Page FP2 Subtotal					50508		

Figure 24.11b

Means Forms
COST ANALYSIS

PROJECT Office Building
LOCATION Boston, MA
TAKE OFF BY JJM
QUANTITIES BY JJM
PRICES BY MJM
EXTENSIONS BY MJM
CLASSIFICATION Div. 15-Fire Protection
ARCHITECT
SHEET NO. FP 3 of 3
ESTIMATE NO. 87-16
DATE 3-20-87
CHECKED BY JJM

DESCRIPTION	SOURCE/DIMENSIONS		QUANTITY	UNIT	MATERIAL UNIT COST	MATERIAL TOTAL	LABOR UNIT COST	LABOR TOTAL	EQUIPMENT UNIT COST	EQUIPMENT TOTAL	SUBCONTRACT UNIT COST	SUBCONTRACT TOTAL	TOTAL UNIT COST	TOTAL
90° Ell SCR CI 1"	15.1	560 0110	50	Ea.							22	1100		
Tee SCR CI 1"	15.1	560 0550	40	Ea.							36	1440		
Sprinkler Heads	15.4	600 4830	300	Ea.							39	11700		
												14240		
					Page FP 3 Subtotal										
Fire Protection Subtotals														
FP 1												28650		
FP 2												50508		
FP 3												14240		
							Subtotal Total					93398		
Profit markup											10%	9340		
									Total			102738		102738
City modifier for Boston							Boston Total				x 1.077			
												110649		110649

Figure 24.11c

Heating, Ventilating, and Air Conditioning

Visualizing the air conditioning system and its varied and numerous components requires more than just scanning the plans and specifications. An in-depth perusal is required to understand the operation, distribution, and control of what appears to be and can be a complex installation. A detailed form should be utilized for estimating HVAC systems, which may be three or more pages long.

The General and Special Conditions define the scope of the project and should delegate financial responsibility for such pertinent items as excavation and backfill, masonry and concrete, cutting and patching, temporary heat (and operation), power and control wiring, hoisting, rigging, etc. When these essential items have been noted on the estimate either as "by others" or "to be priced", the equipment takeoff can begin.

The estimator should begin a plan by plan takeoff of the equipment, starting either at the roof or at the basement. Each floor should be listed separately. Starting at the roof provides a takeoff point at a less cluttered location. The complexity increases as the estimator moves down gradually floor by floor. The starting point is a matter of personal preference, however, it should be consistent throughout the takeoff. Marking off each system on the plans with a different colored pencil will help assure that every item has been accounted for. Consistency is the key to accuracy.

After the equipment is taken off and grouped together for pricing, price requests can be mailed out to equipment suppliers (see Figure 24.12). This may be done by postcard. This point is too soon, however, to notify subcontractors, as they will tie up the plans while the estimator is trying to complete the ductwork and piping takeoffs.

Equipment schedules are often included on the drawings rather than listed in the specifications. Typical details, such as trap assemblies, pump connections, coil connections, and duct assemblies, are also a great asset when they are provided with the plans. From the equipment totals many of these details will give the material quantities instantly thereby avoiding an item by item takeoff while going through the floor plans.

The piping takeoff for the HVAC estimate is similar to those previously covered in plumbing and fire protection. Welding fittings for the larger pipe sizes will be more predominant in the heating and cooling systems. The HVAC takeoff is shown in Figures 24.13a through 24.13e.

Means Forms
QUANTITY SHEET

PROJECT: Office Building
LOCATION: HVAC
ARCHITECT:
TAKE OFF BY: MJM
EXTENSIONS BY:
SHEET NO.
ESTIMATE NO. 87-32
DATE: 3-23-87
CHECKED BY: JJM

DESCRIPTION	NO.	DIMENSIONS		UNIT		UNIT		UNIT		UNIT
Boiler 3500 mBH Gas C.I.		Water		1 Ea.						
ASME Expansion Tank 79 Gal.		Heating		1						
" " " 30 "		Chill Wtr.		1						
Hot Water Pumps		5 HP		2						
Chill Water Pumps		5 HP		2						
Condensed Water Pumps		7½ HP		2 ✓						
Packaged Chiller		175 Ton		1 Ea.						
Air Handling Units		30 Ton		6 Ea.						
Electric Duct Coil		1 Kw		1 Ea.						
Cooling Tower - Double Flow		175 Ton		1 Ea.						
Roof Exhaust Fan } Aluminum w/Curb & Sal }		2200 CFM		1 Ea.						

Figure 24.12

Ductwork is scaled off the plans much like the piping. It is then converted to pounds of metal for takeoff and pricing. The takeoff for ductwork is shown in Figures 24.14a through 24.14d.

The tables for calculating the weight of duct from the linear measure and for obtaining the surface area of ductwork are shown in Figures 18.3 and 18.4

Insulation follows the same takeoff procedures as for piping and ductwork, converting the fittings to linear feet for pricing as previously discussed. Figure 18.3 and 18.4 indicate the method of obtaining the duct surface area in square feet for insulation purposes.

The pipe, valve, fitting, duct work, and equipment takeoffs are combined and the totals transferred to cost analysis sheets for pricing (see Figures 24.15a through 24.15g). Any subcontractor totals are also included on the cost analysis sheets when appropriate. All page totals are summarized on the last sheet (Figure 24.15g). Miscellaneous and subcontractor items are added and totalled to apply the final markups. Appropriate sales taxes, markups, and modifiers (as discussed earlier), are then added to complete the HVAC estimate.

Alternate Pricing Method

An alternate pricing method for the HVAC takeoff of the proposed project is illustrated in Figures 24.16a through 24.16d. The alternate method involves giving a list of material to the suppliers for lump sum quotations. This method is used when the estimator has the luxury of time and the advantage of reliable suppliers with pricing capabilities. It can save countless hours of pricing and extending, and is a free service provided by suppliers.

The estimator can break these lump sums down into pricing units for future reference. For example, bevel end pipe and threaded and coupled (T&C) pipe are priced separately. From the T&C total, the estimator can derive a certain percentage for screwed fittings and nipples. The estimator should retain costs from previous jobs, as they are good references for these percentages. Often, a pattern will emerge for valve or pipe hanger percentages based on pipe costs and job type. Labor is estimated from the lump sums instead of by a line item basis (shown in Figure 24.16c. Figure 24.16d is the estimate summary sheet.

This alternative method is based on the office building estimated earlier in this chapter, using only the heating and cooling system as an example. This pricing method could also be used for plumbing and/or fire protection depending on the estimator's preference.

Means Forms
PIPING SCHEDULE

JOB: Office Building
SYSTEM: HVAC — Hot water and Chilled water Piping (weld 2½" and up)
PAGE: HVAC 1 of 5
DATE: 3-24-87
BY: JJM

	12	10	8	6	5	4	3 1/2	3	2 1/2	2	1 1/2	1 1/4	1	3/4	1/2	
Blk Steel Pipe Sched.40						240 20 20 18		40 14 14	40 14 11	40 20 20 20			20			
Ells						28				36			4			Auto Air Vent — 10
Tees						17		2	2	45						Gauges — 20
Reducers						4										Thermometers — 15
Unions						8		2	2	21						Drain Cocks — 19
Gate Valves						4		2	2	13		2	1			
Check Valves						4							1			
Balancing Cocks						4										
Strainers						8										
Flexible Connectors																3/4" nozzle welds — 24
Weld Neck Flanges						32										
Bolts, Nuts, Gaskets						48										

PIPE DIAMETER IN INCHES

Figure 24.13a

Means Forms
PIPING SCHEDULE

JOB: Office Building
SYSTEM: Condenser Water Piping — All Welded
PAGE: HVAC 2 of 5
DATE: 3-24-87
BY: JM

	12	10	8	6	5	4	3 1/2	3	2 1/2	2	1 1/2	1 1/4	1	3/4	1/2		
Blk Steel Pipe Sched. 40					80 (10, 30, 42)												
Ells					11											Gauges	4
Tees					5												
Gate Valves					2											Thermometers	2
Check Valves					2												
Balancing Cocks					2											Drain Cocks	2
Strainers					2												
Flexible Connectors					4											3/4" Nozzle Welds	8
Weld Neck Flanges					11												
Bolts, Nuts, & Gaskets					18												

PIPE DIAMETER IN INCHES

Figure 24.13b

Means Forms
PIPING SCHEDULE

JOB: Office Building
SYSTEM: HVAC Condensate Drains
PVC (Socket Joint)

PAGE: HVAC 3 of 5
DATE: 3-24-87
BY: JJM

	12	10	8	6	5	4	3 1/2	3	2 1/2	2	1 1/2	1 1/4	1	3/4	1/2
PVC Pipe #40 DWV											56 / 120 / 18 (200)				
Sanitary Tee PVC											24				
Quarter Bends PVC											2				
Adapters PVC											6				
P Traps PVC											6				

PIPE DIAMETER IN INCHES

Figure 24.13c

Means Forms
PIPING SCHEDULE

JOB: Office Building
SYSTEM: HVAC - Cooling Tower Drain System
PAGE: HVAC 4 of 5
DATE: 3-24-87
BY: JJM

	12	10	8	6	5	4	3 1/2	3	2 1/2	2	1 1/2	1 1/4	1	3/4	1/2
Cool Tower Drain Galv. Steel Sch40 T&C										20 ++++					
90 Ells Cast Iron Galv.										(5)					
Tees Cast Iron Galv.										1					
Galv. Union										1					
Gate Valve										1					

PIPE DIAMETER IN INCHES

Figure 24.13d

Means Forms
PIPING SCHEDULE

JOB: Office Building
SYSTEM: HVAC - Boiler/Chiller/Cooling Tower   Cold Water Make Up

PAGE HVAC 5 of 5
DATE 3-25-87
BY JJM

	12	10	8	6	5	4	3 1/2	3	2 1/2	2	1 1/2	1 1/4	1	3/4	1/2
Make Up L Copper													40		
Ells Copper													12		
Tees Copper													4		
Gate Valve Copper													4		
Globe Valve Copper													1		
Check Valve Copper													3		
Reduce Pressure Valve-Copper													1		

PIPE DIAMETER IN INCHES

Figure 24.13e

Means Forms

DUCTWORK SCHEDULE

PAGE Sm 1 OF 4
DATE 3-26-87
BY EEC

JOB Office Building
SYSTEM noted: Duct   HVAC
MATERIAL Galv. Steel
DRAWING NO. M-2

DUCT SIZE HEIGHT	WIDTH	LINING	INSUL.	GAUGE	LENGTH	TOTAL LENGTH	LBS./FOOT	TOTAL POUNDS
Supply Duct					AHU-1A Same layout for 2A & 3A			
14 ×	65	no	2"	20	4	4	22.9	92
14 ×	45			22	6, 34	40	14.8	592
14 ×	25			24	20, 8, 18	46	8.4	386
14 ×	22				50, 16, 16, 50	132	7.8	1030
14 ×	18				24, 24, 24, 24	96	6.9	662
14 ×	15				23, 23	46	6.2	285
14 ×	14				23, 23	46	6.0	276
12 ×	13				4, 4	8	5.4	43
12 ×	9				24, 24	48	4.5	216
12 ×	8	↓	↓	↓	20, 10, 10, 20	60	4.3	258
							Total	3840#
Supply Duct					AHU-1B, Same layout for 2B & 3B			
14 ×	65	no	2"	20	4	4	22.9	92
14 ×	48			22	16	16	15.5	248
14 ×	45			22	12, 10	22	14.8	326
14 ×	22			24	16, 16	32	7.8	250
14 ×	18				12, 12, 12, 12	48	6.9	331
14 ×	14				24, 24, 24, 24	96	6.0	576
12 ×	13				12, 12	24	5.4	130
12 ×	9				20, 20, 24, 24, 20, 20	128	4.5	576
12 ×	8				16, 16, 8, 8	48	4.3	206
10 ×	11				12	12	4.5	54
10 ×	8				8, 12	20	3.9	78
8 ×	6	↓	↓	↓	16, 18	34	3.0	102
							Total	2969#
							1 Floor Total	6809#
					Surface Area for Insulation (Top Floor Only)			
				20	92 + 92 = 184/1.656 =	111		
				22	592 + 248 + 326 = 1166/1.406 =	829		
				24	5459/1.156 =	4722		
					1 Floor Total	5662 SF		

Figure 24.14a

Means Forms

DUCTWORK SCHEDULE

JOB: Office Building
SYSTEM: Noted: Duct   HVAC
MATERIAL: Galv. Steel

PAGE: Sm 2 of 4
DATE: 3-26-87
BY: EEC
DRAWING NO.: m-2

DUCT SIZE HEIGHT	DUCT SIZE WIDTH	LINING	INSUL.	GAUGE	LENGTH	TOTAL LENGTH	LBS./FOOT	TOTAL POUNDS
Return Air					AHU-1A, Same layout for 2A & 3A			
30	30	no	no	24	8	8	12.9	103
14	50	↓	↓	22	32	32	16.0	512
14	33	↓	↓	22	24	24	11.8	283
14	18	↓	↓	24	22	22	6.9	152
							Total	1050#
Return Air					AHU-1B, Same layout for 2B & 3B			
30	30	no	no	24	6	6	12.9	77
14	50	↓	↓	22	16	16	16.0	256
14	33	↓	↓	22	16	16	11.8	189
14	18	↓	↓	24	18	18	6.9	124
							Total	646#
							1 Floor Total	1696#
Fresh Air Intake (Insulated)								
20	42	no	2"	22	8	8	15.5	124
20	32	↓	↓	22	15	15	13.0	195
20	20	↓	↓	24	15	15	10.0	150
16	16	↓	↓	24	108	108	3.4	367
							Total	836#
					Insulation (FAI)			
				22	124 + 195 = 319/1.406 =	227 SF		
				24	150 + 367 = 517/1.156 =	447 SF		
					Total	674 SF		

Figure 24.14b

Means Forms

DUCTWORK SCHEDULE

JOB: Office Building
SYSTEM: noted: Duct   HVAC
MATERIAL: Galv. Steel

PAGE: Sm 3 of 4
DATE: 3-26-87
BY: EEC
DRAWING NO.: m-2

DUCT SIZE HEIGHT	DUCT SIZE WIDTH	LINING	INSUL.	GAUGE	LENGTH	TOTAL LENGTH	LBS./FOOT	TOTAL POUNDS
Toilet Exhaust								
18	18	no	no	24	8	8	7.8	62
15	15				15	15	6.5	98
12	12				15	15	5.2	78
8	8				30	30	3.4	102
8	8				30	30	3.0	90
6	6				36	36	2.6	94
6	4	↓	↓		78	78	2.2	172
							Total	696#
Mechanical Room (FAI)								
24	48	no	no	22	5	5	18	90
12	48	"	"	22	6	6	15	90
							Total	180#

Figure 24.14c

Means Forms
QUANTITY SHEET

PROJECT: Office Building
LOCATION:
TAKE OFF BY: EEC
ARCHITECT:
EXTENSIONS BY: MJM
SHEET NO. SM 4 of 4
ESTIMATE NO.
DATE: 3-26-87
CHECKED BY: JJM

DESCRIPTION	NO.	DIMENSIONS		Galv. Steel	# UNIT	Duct Insul.	SF UNIT	Ea.	UNIT	LF	UNIT
Duct											
Supply Top Floor				6809	lbs.	5662	SF				
" Floor 1 & 2				13618	lbs.						
Return Top Floor				1696	lbs.						
" Floor 1 & 2				3628	lbs.						
Fresh Air Intake				836	lbs.	674	SF				
Toilet Exhaust				696	lbs.						
Mechanical Room (FAI)				180	lbs.						
Total				27463	lbs.	6336	SF				
Diffusers											
T-Bar Lay-In		15	15					120	Ea.		
" "		9	9					15	Ea.		
Return Air Register											
T-Bar Lay-In		18	18					18	Ea.		
Flex Connector										180	LF
Fresh Air Intake - Hood/Curb		20	42					1	Ea.		
" " " Damper						6	SF				
Volume Dampers		16	16					6	Ea.		
Toilet Exhaust Register		6	6					6	Ea.		
" "		8	8					18	Ea.		
Fire Damper		12	12					1	Ea.		
" "		15	15					1	Ea.		
" " Access Door								2	Ea.		
Toilet Exh. Fan Rooftop	2200	CFM (5/min. Air Change)									
Elevator Relief Wall Lvr		18	12					1	Ea.		
" " Trans Grille		18	12					1	Ea.		
Mech. Room Wall Louver		48	48					1	Ea.		
" " Damper		48	24					1	Ea.		
" " "		48	12					1	Ea.		
Boiler Flue with Boiler											

Figure 24.14d

Means Forms
COST ANALYSIS

PROJECT: Office Building
LOCATION:
TAKE OFF BY: EEC
QUANTITIES BY: EEC
PRICES BY: MJM
CLASSIFICATION: Div. 15 - HVAC
ARCHITECT:
EXTENSIONS BY: MJM
CHECKED BY: JJM
SHEET NO. HVAC 1 of 7
ESTIMATE NO. 87-32
DATE: 3-27-87

DESCRIPTION	SOURCE/DIMENSIONS			QUANTITY	UNIT	MATERIAL UNIT COST	MATERIAL TOTAL	LABOR UNIT COST	LABOR TOTAL	EQUIPMENT UNIT COST	EQUIPMENT TOTAL	SUBCONTRACT UNIT COST	SUBCONTRACT TOTAL	TOTAL UNIT COST	TOTAL
Sheet Metal															
Total Duct	15.7	640	0580	27,500	lbs.	.29	7975	1.72	47300						
Diffusers															
T-Bar 15"x15"	15.7	520	2060	120	Ea.	96.25	11550	16.80	2016						
9" x 9"	15.7	520	2020	15	Ea.	57.75	866	13.20	198						
Return Air Register															
T-Bar 18" x 18"	15.7	520	2080	18	Ea.	98.50	1773	18.50	333						
Flex Connector	15.7	610	2000	180	LF	1.18	212	1.85	333						
FAI Hood/Curb 20"x 42"	15.7	960	5680	1	Ea.	730	730	83	83						
FAI Damper 20" x 42"	15.7	960	7600	6	SF	39.20	235	3.33	20						
Volume Dampers 16" x 16"	15.7	610	8180	6	Ea.	30	180	12.30	74						
Toilet Exhaust Reg. 6"x 6"	15.7	580	5040	6	Ea.	10.40	62	7.70	46						
8" x 8"	15.7	580	5100	18	Ea.	12.30	221	9.75	176						
Fire Damper 12" x 12"	15.7	610	3180	1	Ea.	20.75	21	9.25	9						
15" x 15"			3240	1	Ea.	23.50	24	10.25	10						
" Access Door	↓		1020	2	Ea.	14.65	29	16.80	34						
Toilet Exhaust Fan 2200CFM	15.7	700	7180	1	Ea.	1200	1200	105	105						
Elev. Relf. Wall Lvr. 18"x 12"		790	3200	1.5	SF	26.25	39	6.60	10						
" Transfer Grile 18"x12"	↓	550	5160	1	Ea.	32.75	33	9.25	9						
Page HVAC 1 Subtotal							25150		50756			For Sheetmetal			

Figure 24.15a

Means Forms
COST ANALYSIS

PROJECT: Office Building
LOCATION:
TAKE OFF BY: EEC
QUANTITIES BY: EEC
PRICES BY:
CLASSIFICATION: Div. 15 - HVAC
ARCHITECT:
EXTENSIONS BY: MJM
CHECKED BY: JJM
SHEET NO. HVAC 2 of 7
ESTIMATE NO. 87-32
DATE 3-27-87

DESCRIPTION	SOURCE/DIMENSIONS		QUANTITY	UNIT	MATERIAL UNIT COST	MATERIAL TOTAL	LABOR UNIT COST	LABOR TOTAL	EQUIPMENT UNIT COST	EQUIPMENT TOTAL	SUBCONTRACT UNIT COST	SUBCONTRACT TOTAL	TOTAL UNIT COST	TOTAL
Sheet metal (cont.)														
Mech. Rm. Wall Lvr 48"x48"	15.7	790 3200	16	SF	26.25	420	6.60	106						
" Relief Damp 48"x24"	→	3500	8		13.12	105	1.98	16						
" Manual " 48"x12"	→	3500	4	→	13.12	52	1.98	8						
Page HVAC2 Subtotal						577		130						
Sheet metal Subtotals														
HVAC 1						25150		50756						
HVAC 2						577		130						
Subtotals Total						25727		50886						
Material Sales Tax					5%	1286								
Overhead & Profit					10%	2573 47.4%	24130							
Total						29586		75006						104592

Figure 24.15b

Means Forms
COST ANALYSIS

PROJECT: Office Building
LOCATION:
TAKE OFF BY: MJM
QUANTITIES BY: MJM
PRICES BY: MJM
EXTENSIONS BY: MJM
CHECKED BY: JJM
CLASSIFICATION: Div. 15 - HVAC
ARCHITECT:
SHEET NO. HVAC 3 of 7
ESTIMATE NO. 87-32
DATE: 3-27-87

Piping

DESCRIPTION	SOURCE/DIMENSIONS			QUANTITY	UNIT	MATERIAL UNIT COST	MATERIAL TOTAL	LABOR UNIT COST	LABOR TOTAL	EQUIPMENT UNIT COST	EQUIPMENT TOTAL	SUBCONTRACT UNIT COST	SUBCONTRACT TOTAL	TOTAL UNIT COST	TOTAL
Blk Stl. Pipe #40 T&C 1"	15.1	550	0580	20	LF	1.39	28	3.47	69						
" 2"			0610	220		2.76	607	5.20	1144						
BE 2½"			2080	40		3.79	152	7.05	282	.40	16				
3"			2090	40		4.72	189	7.70	308	.44	18				
4"			2110	240		6.88	1651	8.95	2148	.51	122				
5"			2120	80	→	17.18	1374	10.35	828	.59	47				
Elbow Cast Iron 90° 1"	15.1	500	0110	4	Ea.	1.53	6	14.15	57						
" 2"		560	0140	39		4.61	180	18.40	718						
weld 4"		570	3130	26		21	546	66	1716	3.80	99				
" 5"		570	3140	11	→	59	649	105	1155	3.80	42				
Tee Cast Iron 2"	15.1	560	0580	42	Ea.	6.25	263	30	1260						
" 2½"		570	3420	2		31	62	66	132	3.80	8				
" weld 3"			3430	2		34	68	83	166	4.75	10				
" 4"			3440	17		41	697	110	1870	6.35	108				
" 5"			3450	5	→	64	420	170	850	6.35	32				
Union Blk mall. 1"	15.1	560	7050	1	Ea.	3.77	4	15.35	15						
" 2"	15.1	560	7080	21	Ea.	8.05	169	19.50	410						
Gate Valve, Bronze 1"	15.1	800	3450	1	Ea.	16.40	16	9.70	10						
" 2"	15.1	800	3480	12	Ea.	39	468	16.75	201						
		Page HVAC 3 Subtotal					7549		13339		502				

Figure 24.15c

Means Forms
COST ANALYSIS

PROJECT: Office Building
LOCATION:
TAKE OFF BY: MJM
QUANTITIES BY: MJM
PRICES BY: MJM
EXTENSIONS BY: MJM
CLASSIFICATION: Div. 15-HVAC
ARCHITECT:
SHEET NO. HVAC 4 of 7
ESTIMATE NO. 87-32
DATE: 3-27-87
CHECKED BY: JJM

DESCRIPTION		SOURCE/DIMENSIONS		QUANTITY	UNIT	MATERIAL UNIT COST	MATERIAL TOTAL	LABOR UNIT COST	LABOR TOTAL	EQUIPMENT UNIT COST	EQUIPMENT TOTAL	SUBCONTRACT UNIT COST	SUBCONTRACT TOTAL	TOTAL UNIT COST	TOTAL
Piping (Cont.)															
Reducer Welded	2½"	15.1	570 3493	2	Ea.	19.95	40	37	74	2.11	4				
	3"		3494	2		21	42	41	82	2.38	5				
	4"		3495	4		28	112	55	220	3.17	13				
Weld Neck Flange	4"	15.1	570 6500	58	Ea.	30	1740	33	1914	1.90	110				
	5"	15.1	570 6510	20	Ea.	38	760	41	820	2.38	48				
Flanged Fittings															
OS&Y Gate Valve	4"	15.1	820 3680	8	Ea.	180	1440	110	880						
	5"	15.1	820 3690	2		285	570	150	300						
Strainer	4"	15.6	750 1060	4		156.75	627	110	440						
	5"	15.6	750 1080	2		247	494	155	310						
Iron Body Check	4"	15.1	820 6670	4		105	420	66	264						
	5"	15.1	820 6680	2		145	290	86	172						
Flex Connect	4"	15.6	200 0540	8		305	2440	42	336						
	5"	15.6	200 0560	4		340	1360	48	192						
Lube Plug Valve	4"	15.1	930 7030	4		330	1320	110	440						
	5"		930 7040	2		565	1130	170	340						
Bolt & Gasket Set	4"		570 0670	58		8.35	484	6.80	394						
	5"		570 0680	20		12.50	250	7.10	142						
Air Control Valve	4"	15.6	080 0120	1		710	710	110	110						
3-Way Temp.Cont.Valve	2"	15.7	400 7400	6	Ea.	191	1146	19.60	118						
Thread-O-Let	¾"	15.1	570 5633	32		4.57	146	15.75	504	.90	29				
Automatic Air Vent	¾"	15.6	040 1130	8		15.65	125	5.85	47						
Thermometers	¾"	15.7	400 4500	17		24	408	5.85	99						
		Page HVAC 4 Subtotal					16054		8178		209				

Figure 24.15d

Means Forms
COST ANALYSIS

PROJECT: Office Building
LOCATION:
TAKE OFF BY: MJM
QUANTITIES BY: MJM
PRICES BY: MJM
EXTENSIONS BY: MJM
CLASSIFICATION: Div. 15 - HVAC
ARCHITECT:
SHEET NO. HVAC 5 of 7
ESTIMATE NO. 87-32
DATE 3-27-87
CHECKED BY JJM

DESCRIPTION		SOURCE/DIMENSIONS		QUANTITY	UNIT	MATERIAL UNIT COST	MATERIAL TOTAL	LABOR UNIT COST	LABOR TOTAL	EQUIPMENT UNIT COST	EQUIPMENT TOTAL	SUBCONTRACT UNIT COST	SUBCONTRACT TOTAL	TOTAL UNIT COST	TOTAL
Piping (Cont.)															
Bronze Gate Valve Scr.	1½"	15.1	800 3470	2	Ea.	.28	56	14.15	28						
Gauges	¾"	15.7	400 2300	24		12	288	5.85	140						
Drain Valve (Hose Bibb)	¾"	15.1	220 5000	21	→	4.07	85	7.65	161						
PVC Piping DWV															
Pipe Sch. 40	1½"	15.1	490 4420	200	LF	.83	166	5.10	1020						
Quarter Bend	1½"		500 5060	2	Ea.	.40	1	11.50	23						
San Tee	1½"		500 5270	24		.75	18	18.40	442						
P Traps	1½"		730 6910	6		1.45	9	10.20	61						
Adapt PVC/THD	1½"	→	500 8520	6	→	.98	6	10.20	61						
Galv. Stl. Piping															
Pipe Sch. 40	2"	15.1	550 1360	20	LF	3.05	61	5.20	104						
90° C.I. Ells	2"		560 0800	5	Ea.	7.05	35	18.40	92						
Tee C.I.	2"		1180	1		8.50	9	30	30						
Union	2"	→	7080 7350	1		8.45	8	14.50	20						
Gate Valve	2"	→	820 1700	1	→		120	16.75	17						
Copper (Sweat)															
"L" Tube	1"	15.1	400 2200	40	LF	1.46	58	5.10	204						
90° Ells	1"		410 0130	12	Ea.	.90	11	11.50	138						
Tees	1"		410 0510	4		2.44	10	18.40	74						
Gate Valve	1"	→	800 2950	4		18.10	72	9.70	39						
Globe Valve	1"	→	800 4970	1	→	30	30	9.70	10						
		Page HVAC 5 Subtotal					1043		2664						

Figure 24.15e

Means Forms
COST ANALYSIS

PROJECT: Office Building
LOCATION:
TAKE OFF BY: MJM
QUANTITIES BY: MJM
PRICES BY: MJM
CLASSIFICATION: Div. 15 - HVAC
ARCHITECT:
EXTENSIONS BY: MJM
CHECKED BY: JJM
SHEET NO. HVAC 6 of 7
ESTIMATE NO. 87-32
DATE 3-27-87

DESCRIPTION	SOURCE/DIMENSIONS			QUANTITY	UNIT	MATERIAL UNIT COST	MATERIAL TOTAL	LABOR UNIT COST	LABOR TOTAL	EQUIPMENT UNIT COST	EQUIPMENT TOTAL	SUBCONTRACT UNIT COST	SUBCONTRACT TOTAL	TOTAL UNIT COST	TOTAL
Copper (Sweat)(Cont.)															
Check Valve 1"	15.1	800	1870	3	Ea.	16.30	49	9.70	29						
Reduce Press. Valve 1"	15.1	800	6960	1	Ea.	78	78	9.70	10						
Heating Equipment															
Boiler Flue															
Dbl. Wall Straight 24∅	15.5	920	0340	12	LF	80	960	16.15	194						
Roof Flashing Collar 24∅			1170	1	Ea.	133	133	43	43						
Dbl. Wall Tee 24∅			1330	1		390	390	43	43						
Tee Cap 24∅			1590	1		26	26	25	25						
Rain Cap & Screen 24∅	→		1880	1	↓	365	365	26	26						
Boiler 3500 MBH Gas															
C.I. water	15.5	080	3400	1	Ea.	23050	23050	2750	2750						
Electric Duct Htr. 1 KW		290	1020	1		229	229	35	35						
Expansion Tank ASME 79G	→	890	3060	1		1210	1210	67	67						
Hot Water Pump 5 HP	15.2	410	4300	2	↓	1020	2040	185	310						
Air Conditioning															
Chiller 175 Ton	15.7	980	1600	1	Ea.	61500	61500	1900	11900						
Fan Coil Units 30 Ton		730	0260	6	Ea.	3960	23760	870	5220						
Cooling Tower 175 Ton	→	490	1900	175	Ton	39.25	6869	4.14	725						
Condens. Water Pump 7.5 HP	15.2	410	4420	2	Ea.	1240	2480	205	410						
Chilled Water Pump 5 HP	15.2	410	4410	2		1050	2100	205	410						
Expansion Tank ASME 30G	15.5	890	3020	1	↓	770	770	43	46						
Page HVAC 6 Subtotal							126009		22299						

Figure 24.15f

Means Forms
COST ANALYSIS

PROJECT: Office Building
LOCATION:
TAKE OFF BY: MJM
QUANTITIES BY: MJM
PRICES BY:
CLASSIFICATION: Div. 15 - HVAC
ARCHITECT:
EXTENSIONS BY: MJM
CHECKED BY: JJM
SHEET NO. HVAC 7 of 7
ESTIMATE NO. 87-32
DATE 3-27-87

DESCRIPTION	SOURCE/DIMENSIONS		QUANTITY	UNIT	MATERIAL UNIT COST	MATERIAL TOTAL	LABOR UNIT COST	LABOR TOTAL	EQUIPMENT UNIT COST	EQUIPMENT TOTAL	SUBCONTRACT UNIT COST	SUBCONTRACT TOTAL	TOTAL
HVAC Subtotals													
HVAC 1						SHEET		METAL					
2						SHEET		METAL					
3						7549		13339		502			
4						16054		8198		209			
5						1043		2664					
6						126009		22299					
Miscellaneous													
Record Drawings			8	Hrs	23.30		23.30	186					
Operating Instructions			8		23.30		23.30	186					
Maintenance Manuals			4		23.30		23.30	93					
Cleaning System			16		23.30		23.30	373					
Subcontracts													
Insulation HVAC	15.5	650				-QUOTED-						14477	
Balancing Air	15.7	160				-QUOTED-						5113	
water		170				-QUOTED-						1422	
Temp. Control Electronic		460				-QUOTED-						30512	
Rigging	1.5	200				-QUOTED-						1132	
Piping Subtotal						150655		47338		711		52656	
Sales Tax					5%	7533							
Handling, Overhead & Profit markup					10%	15066	45.8%	21681	10%	71	10%	5266	
Piping Total						173254		69019		782		57922	
Sheet metal Total						29586		75006					
Combined Total						202840		144025		782		57922	
City Modifier for Boston						x 1.046		x 1.106		x 1.046		x 1.077	
Boston Total						212171		159292		818		62382	434663

Figure 24.15g

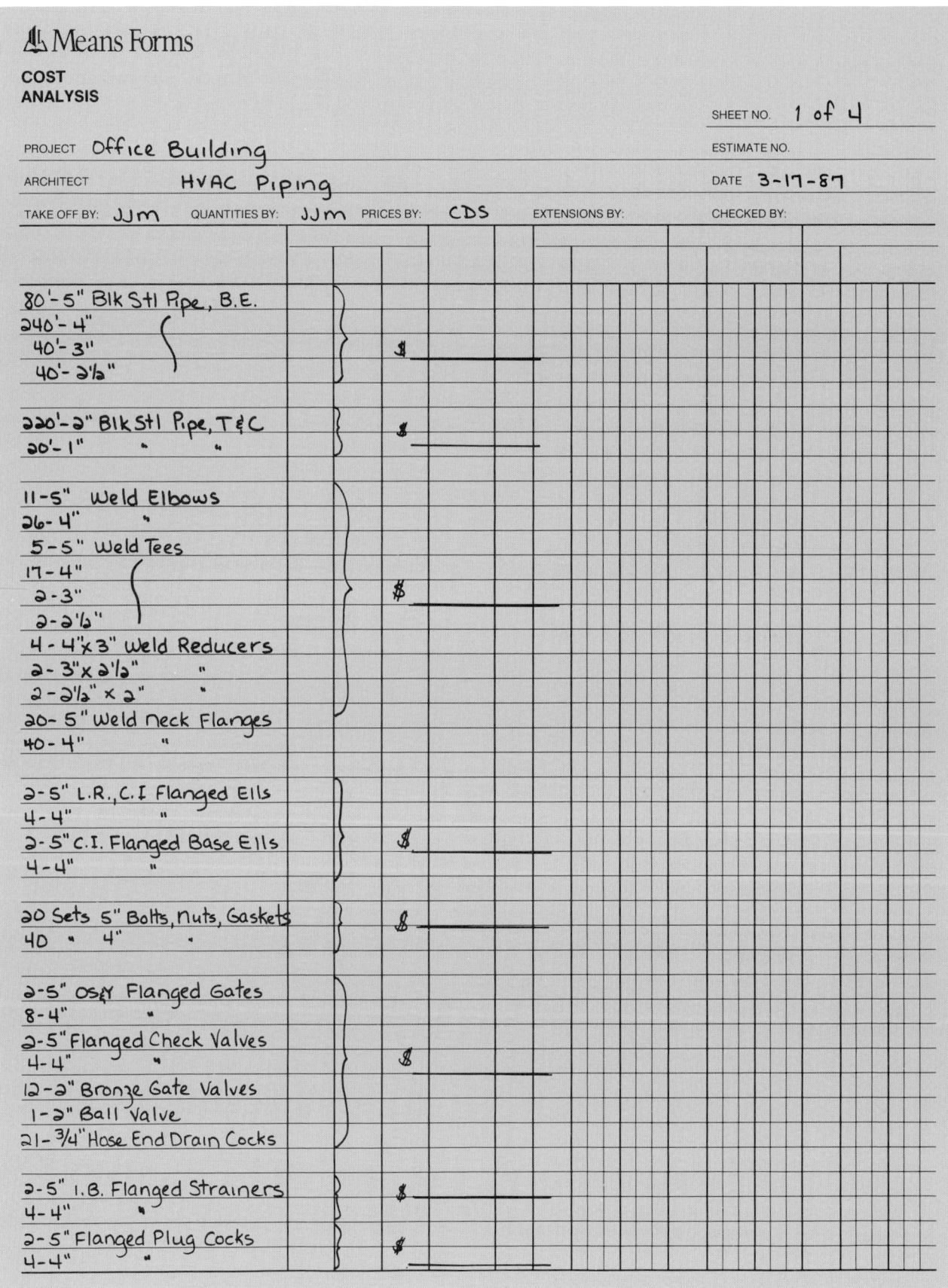

Figure 24.16a

Means Forms

TELEPHONE QUOTATION

Page 2 of 4
DATE 3-18-87
TIME 1:15 p.m.

PROJECT: Office Building
FIRM QUOTING: C.D.S. Piping Supply
PHONE: (617) 747-1270
ADDRESS:
BY: D.B.
ITEM QUOTED: Dual Temp. System
RECEIVED BY:

WORK INCLUDED		AMOUNT OF QUOTATION
Bevel End Steel Pipe	$ 3186	
T & C Steel Pipe	$ 635	
Weld Fittings & Flanges	$ 4502	
Cast Iron Flanged Elbows	$ 895	
Bolts, Nuts, and Gaskets	$ 584	
Valves	$ 3830	
Strainers	$ 1121	
Lubricated Plug Cocks	$ 2450	

DELIVERY TIME: Stock Items

TOTAL BID

DOES QUOTATION INCLUDE THE FOLLOWING: — If ☐ NO is checked, determine the following:

STATE & LOCAL SALES TAXES	☐ YES	☑ NO	MATERIAL VALUE
DELIVERY TO THE JOB SITE	☑ YES	☐ NO	WEIGHT
COMPLETE INSTALLATION	☐ YES	☐ NO	QUANTITY
COMPLETE SECTION AS PER PLANS & SPECIFICATIONS	☐ YES	☐ NO	DESCRIBE BELOW

EXCLUSIONS AND QUALIFICATIONS

Note: Substitute Butterfly Valves for Lubricated Plugs Deduct $1960

ADDENDA ACKNOWLEDGEMENT

TOTAL ADJUSTMENTS

ADJUSTED TOTAL BID

Figure 24.16b

Means Forms

CONDENSED ESTIMATE SUMMARY

SHEET NO. 3 of 4
PROJECT: Office Building
ESTIMATE NO.
LOCATION: Boston
TOTAL AREA/VOLUME:
DATE: 3-25-87
ARCHITECT: Means
COST PER S.F./C.F.:
NO. OF STORIES:
PRICES BY: JJM
EXTENSIONS BY: JJM
CHECKED BY: MJM

	LABOR	DAYS (PAIR)
1	Cast Iron Boiler Packaged Work with Riggers	4
2	Expansion Tanks and Miscellaneous Trim	3
1	Packaged Chiller Work with Riggers	5
6	Air Handling Units Work with Riggers	6
1	Cooling Tower Work with Riggers	5
6	Base Mounted Pumps	8
	Piping	35
	Welding	25
	Sleeves, Hangers and Inserts	3
	Supports, Anchors and Guides	3
	Testing	5
	Supervision	6
	Record Drawings	3
	Operating Instructions & Manuals	2
	Valve Tags, Charts, Arrows	2
	Clean Systems	2
	Gauges, Thermometers, Air Vents	2
	Call Backs	2
		121
	Lost Time 5%	6
	Total	127

127 Days × 375 = $47,625

Figure 24.16c

Means Forms

CONDENSED ESTIMATE SUMMARY

SHEET NO. 4 of 4

PROJECT: Office Building
LOCATION: Boston, MA
ARCHITECT: R.S. Means
PRICES BY: JJM
EXTENSIONS BY: JJM
DATE: 3-27-87
NO. OF STORIES: 3

Qty	Description	Amount
1	Boiler C.I. Gas 3500 MBH	23050
2	Expansion Tanks and Misc. Specialties	2770
	Boiler Flue – Prefab	1875
1	Chiller 175 Ton	
6	Air Handling Units	86900
1	Duct Coil Electric	
1	Roof Exhaust Fan	
1	Cooling Tower	6870
6	Pumps	6620
27,600#	Ductwork	105000
	Black Steel Pipe 635 T&C 3185 B.E.	3820
	Fittings Cast Iron 100% 635 Weld 5136	5770
	Galv. Pipe	60
	Galv. Fittings	50
	Bolts, Nuts, & Gaskets	785
	PVC Pipe	165
	PVC Fittings	35
	Copper Tube Type L	60
	Copper Fittings	20
	Pipe Hangers & Supports	180
	Miscellaneous Steel, Supports, Anchors, Guides	500
	Valves 2½ and up flanged	6290
	Air Vents	125
	Gauges and Thermometers, 17	695
12	Flexible Pump Connectors	3800
	Strainers	1120
	Trucking and Cartage	200
	Subcontractors	
	Balancing	6535
	Rigging	1130
	Insulation	14500
	Temperature Control	30500
	Labor 127 days/pair @ 375/day	47625
	Cost	**357050**

Figure 24.16d

Means Forms
COST ANALYSIS

PROJECT Office Building (Quote Preparation By Subcontractor) **SHEET NO.**

LOCATION **CLASSIFICATION** Div. 15 - Balancing **ESTIMATE NO.**

TAKE OFF BY **ARCHITECT** **DATE** 3-25-87

QUANTITIES BY **PRICES BY** **CHECKED BY** JJF

DESCRIPTION	SOURCE/DIMENSIONS			QUANTITY	UNIT	MATERIAL UNIT COST	MATERIAL TOTAL	LABOR UNIT COST	LABOR TOTAL	EQUIP. UNIT COST	EQUIP. TOTAL	SUBCONTRACT UNIT COST	SUBCONTRACT TOTAL	UNIT COST	TOTAL
Balancing Air															
Roof Exhaust Fan	15.7	160	1400	1	Ea.							109	109		
Air Supply Diff.	→	→	3000	135	→							32.70	4415		
Air Return Reg.	→	→	3000	18	→							32.70	589		
Air Total														5113		
Balancing Water																
Packaged Chiller	15.7	190	0200	1	Ea.							231	231		
Cooling Tower	→	→	0500	1	→							177	177		
Fan Coil Units	→	→	0600	6	→							49	294		
Pumps	→	→	1000	6	→							120	720		
Water Total														1422		
															6535

Figure 24.17a

Means Forms

TELEPHONE QUOTATION

DATE 3-27-87
TIME 10:45 a.m.

PROJECT Office Building
FIRM QUOTING Wind & Water Balance Co., Inc.
PHONE (543) 548-5980
ADDRESS
BY Jack Frost
ITEM QUOTED Independent HVAC testing, balance, adjust.
RECEIVED BY Marcia

WORK INCLUDED TBA

Description	Amount of Quotation
Air balancing of roof exhaust fan, air handling units, register, diffusers and tower fan	$ 5113
Water balance of pumps, hot and chilled water systems including the flow through the chiller evaporator and condenser	$ 1422
Price is based on sending the original balancing report to the architect plus a marked up set of the contract drawings will be left with the owner	
TOTAL BID	$ 6535

DELIVERY TIME

DOES QUOTATION INCLUDE THE FOLLOWING:

			If ☐ NO is checked, determine the following:
STATE & LOCAL SALES TAXES	☑ YES	☐ NO	MATERIAL VALUE
DELIVERY TO THE JOB SITE	☐ YES	☐ NO	WEIGHT
COMPLETE INSTALLATION	☐ YES	☐ NO	QUANTITY
COMPLETE SECTION AS PER PLANS & SPECIFICATIONS	☑ YES	☐ NO	DESCRIBE BELOW

EXCLUSIONS AND QUALIFICATIONS

"Entire system must be in and operating before we are able to do this work. All areas must be available to our technicians. HVAC contractor is to supply us with a complete set of plans."

ADDENDA ACKNOWLEDGEMENT

TOTAL ADJUSTMENTS
ADJUSTED TOTAL BID

Figure 24.18a

Means Forms
COST ANALYSIS
(Quote Preparation By Subcontractor)

PROJECT: Office Building
CLASSIFICATION: Div. 15 - Crane
SHEET NO.
LOCATION:
ARCHITECT:
ESTIMATE NO.
TAKE OFF BY: JFG
QUANTITIES BY: JFG
PRICES BY:
EXTENSIONS BY: JFG
DATE: 3-26-87
CHECKED BY: WHG

DESCRIPTION	SOURCE/DIMENSIONS			QUANTITY	UNIT	MATERIAL UNIT COST	MATERIAL TOTAL	LABOR UNIT COST	LABOR TOTAL	EQUIPMENT UNIT COST	EQUIPMENT TOTAL	SUBCONTRACT UNIT COST	SUBCONTRACT TOTAL	TOTAL UNIT COST	TOTAL
Crane Service															
Truck Crane - 1 Day	1.5	200	1800	1	Day					565	565				
Equip. Operating Cost	1.5	200	1800	8	Hrs.					14.25	114				
Equipment Operator				1	Day					248	248				
Equipment Oiler				1	Day					205	205				
Total											1132				1132

Figure 24.17b

Means Forms

TELEPHONE QUOTATION

DATE 3-26-87
PROJECT Office Building
TIME 2:30 p.m.
FIRM QUOTING Sky High Co., Inc.
PHONE (543) 789-1234
ADDRESS
BY Bill
ITEM QUOTED Rigging Service
RECEIVED BY Marcia

WORK INCLUDED	AMOUNT OF QUOTATION
Supply 30 ton crane for 1 day with oiler and operator	
To place max. weight 4 tons a max. distance 45' in from edge of 52' high building	$ 1132
TOTAL BID	$ 1132

DELIVERY TIME

DOES QUOTATION INCLUDE THE FOLLOWING: If NO is checked, determine the following:

STATE & LOCAL SALES TAXES	☑ YES	☐ NO	MATERIAL VALUE
DELIVERY TO THE JOB SITE	☑ YES	☐ NO	WEIGHT
COMPLETE INSTALLATION	☑ YES	☐ NO	QUANTITY
COMPLETE SECTION AS PER PLANS & SPECIFICATIONS	☐ YES	☐ NO	DESCRIBE BELOW

EXCLUSIONS AND QUALIFICATIONS

Quote based on items to be set in place being delivered to our yard, freight allowed!

ADDENDA ACKNOWLEDGEMENT

TOTAL ADJUSTMENTS
ADJUSTED TOTAL BID

Figure 24.18b

Means Forms
COST ANALYSIS

PROJECT: Office Building
LOCATION:
TAKE OFF BY: MEA
QUANTITIES BY: MEA
PRICES BY: MEA
EXTENSIONS BY: MEA
CLASSIFICATION: Div. 15 - Insulation
ARCHITECT:
SHEET NO.:
ESTIMATE NO.:
DATE: 3-26-87
CHECKED BY: JM

Quote Preparation By Subcontractor

DESCRIPTION	SOURCE/DIMENSIONS		QUANTITY	UNIT	MATERIAL UNIT COST	MATERIAL TOTAL	LABOR UNIT COST	LABOR TOTAL	EQUIPMENT UNIT COST	EQUIPMENT TOTAL	SUBCONTRACT UNIT COST	SUBCONTRACT TOTAL	TOTAL UNIT COST	TOTAL
Plumbing, Copper Tube														
Fiberglass w/ASJ 1" wall 3/4"	15.5	650 6840	680	LF	1.05	714	1.38	938						
1"	↓	6860	60		1.20	72	1.44	86						
2"	↓	6890	160	↓	1.56	250	1.58	253						
Plumbing Total						1036		1277						
Sales Tax					5%	52								
Overhead & Profit markup					10%	104	49.1%	627						
Total						1192		1904						3096
Sheet metal														
Blanket 1½"	15.5	650 3170	6336	SF	.22	1394	1.04	6589						
Sales Tax					5%	70								
Overhead & Profit Markup					10%	139	49.1%	3235						
Total						1603		9824						11427
HVAC Piping														
Fiberglass w/ASJ 1"wall 2"	15.5	650 6900	210	LF	1.69	355	1.66	349						
2½"	↓	6910	30		1.89	57	1.74	52						
3"		6920	30		2.10	63	1.84	55						
4"	↓	6940	240	↓	2.74	658	2.21	530						
HVAC Piping Subtotal						1133		986						
Add for Valves and Flanges					10%	113		99						
Sales Tax					5%	61								
Overhead & Profit					10%	125	49.1%	533						
Total						1432		1618						3050

Figure 24.17c

Means Forms
TELEPHONE QUOTATION

DATE 3-27-87
TIME 10:30 a.m.

PROJECT: Office Building
FIRM QUOTING: Eastern Allstate Insulation
PHONE: (466) 747-4321
ADDRESS:
BY: Marty Armstrong
ITEM QUOTED: Insulation
RECEIVED BY: Marcia

WORK INCLUDED	AMOUNT OF QUOTATION
Plumbing Insulation	$ 3096
HVAC Insulation	14477
TOTAL BID	$ 17573

DOES QUOTATION INCLUDE THE FOLLOWING: If ☐ NO is checked, determine the following:

STATE & LOCAL SALES TAXES	☑ YES	☐ NO	MATERIAL VALUE
DELIVERY TO THE JOB SITE	☑ YES	☐ NO	WEIGHT
COMPLETE INSTALLATION	☑ YES	☐ NO	QUANTITY
COMPLETE SECTION AS PER PLANS & SPECIFICATIONS	☑ YES	☐ NO	DESCRIBE BELOW

EXCLUSIONS AND QUALIFICATIONS

Complete per plans and specs

ADDENDA ACKNOWLEDGEMENT: None

TOTAL ADJUSTMENTS
ADJUSTED TOTAL BID

Figure 24.18c

Means Forms
COST ANALYSIS
(Quote Preparation by Subcontractor)

PROJECT: Office Building
CLASSIFICATION: Div. 15 - Temp. Control
SHEET NO.:
LOCATION:
ARCHITECT: SLP
ESTIMATE NO:
TAKE OFF BY: JVC
QUANTITIES BY: JA
PRICES BY:
EXTENSIONS BY: JA
DATE: 3-27-87
CHECKED BY: JBG

DESCRIPTION	SOURCE/DIMENSIONS		QUANTITY	UNIT	MATERIAL UNIT COST	MATERIAL TOTAL	LABOR UNIT COST	LABOR TOTAL	EQUIPMENT UNIT COST	EQUIPMENT TOTAL	SUBCONTRACT UNIT COST	SUBCONTRACT TOTAL	TOTAL UNIT COST	TOTAL
Control Systems														
Basic Control of Units	15.7	430 0220	1	Ea.							2690	2690		
Add for over 20 tons		0260	1	Ea.							672	672		
Heating Coils		0320	6	Ea.							2010	12060		
Cooling Coils		0520	6	Ea.							775	4650		
Cooling Tower		0620	1	Ea.							1980	1980		
Boiler Room Air		3000	1	Ea.							980	980		
Program Optimizer	→	4040	1	Ea.							3500	3500		
Subtotal												26532		
Electronic Control	15.7	460 0020									15%	3980		
Total												30512		30512

Figure 24.17d

Means Forms

TELEPHONE QUOTATION

	DATE 3-27-87
PROJECT Office Building	TIME 11:15 a.m.
FIRM QUOTING Barber Honeyshaw Controls, Inc.	PHONE (543) 762-5861
ADDRESS	BY JVC
ITEM QUOTED Electronic controls	RECEIVED BY marcia

WORK INCLUDED	AMOUNT OF QUOTATION
Install electronic control system for HVAC equipment.	
System to include controls tie in for the chiller, boiler, boiler room, air, heating/cooling coils with specified program optimizer.	$ 30512
DELIVERY TIME — TOTAL BID	$ 30512

DOES QUOTATION INCLUDE THE FOLLOWING: If ☐ NO is checked, determine the following:

STATE & LOCAL SALES TAXES	☑ YES	☐ NO	MATERIAL VALUE
DELIVERY TO THE JOB SITE	☑ YES	☐ NO	WEIGHT
COMPLETE INSTALLATION	☑ YES	☐ NO	QUANTITY
COMPLETE SECTION AS PER PLANS & SPECIFICATIONS	☑ YES	☐ NO	DESCRIBE BELOW

EXCLUSIONS AND QUALIFICATIONS

Job installed per plans & specs complete

Dampers to be sized by Tin Knockers.

Valve bodies to be installed by pipefitters

ADDENDA ACKNOWLEDGEMENT — TOTAL ADJUSTMENTS	
ADJUSTED TOTAL BID	

Figure 24.18*d*

Summarizing the Unit Price Estimate

Subcontractors: The prime mechanical contractor usually subcontracts a portion of the mechanical work. These subcontractors will employ estimators who specialize in their individual fields. Typical mechanical subcontractors are Air and Water Balancing, Hoisting and Rigging, Insulation, Sheet Metal, and Temperature Control. In this example, the prime mechanical contractor will accomplish the sheet metal work, and the other four trades will be subcontracted. Competitive bids will be solicited for the four subcontracted trades. Figures 24.17a through 24.17d are representative subcontractor's estimates for the proposed project, which are either mailed or telephoned to the prime mechanical contractor.

In a competitive situation, the subcontractors will usually submit their bids by telephone at the last possible minute to prevent any "bid shopping". Figures 24.18a through 24.18d are typical subcontractor quotes received by telephone and recorded by the prime mechanical contractor.

Overhead and Profit: Since fire protection, plumbing, and sheet metal contractors do not usually have to purchase large quantities of expensive equipment, their overhead markups are based primarily on labor. In this case the markup for labor makes up a relatively large percentage of the total cost, and, therefore, the overhead markup percentage will likely be low in relation to total project cost. A contractor whose work involves large equipment purchases and several subcontractors, on the other hand, must include overhead percentages for these items in addition to the labor percentage. Published industry guides such as *Means Mechanical Cost Data* provide useful overhead percentages. However, such books should be used primarily as a reference. Every mechanical task has unique labor and material requirements, and markups for overhead and profit should ultimately be based on the contractor's individual situation.

The Recap Sheet: It is generally good practice to list the cost totals from each segment of the work onto one sheet of paper, often called a "Recap" (shown in Figure 24.19). This procedure will allow the estimator to review all of the segments of work and to add them into a total contractor's cost.

The Project Schedule: When the work for all sections is priced, the estimator should complete the project schedule so that time-related costs in the Project Overhead Summary can be determined. When preparing the schedule, the estimator must visualize the entire construction process in order to determine the correct sequence of work. Certain tasks must be completed before others are begun. Different trades will work simultaneously. Material deliveries will also affect scheduling. All such variables must be incorporated into the Project Schedule. An example of a Project Schedule is shown in Figure 24.20. The man-hour figures, which have been calculated for each section, are used to assist with scheduling. The estimator must be careful not only to use the man-hours for each section independently, but to coordinate each section with related work.

Means Forms

CONDENSED ESTIMATE SUMMARY

PROJECT: Office Building
LOCATION: Boston, MA
ARCHITECT: R.S. Means
PRICES BY: JJM

TOTAL AREA/VOLUME: 56,700 SF
COST PER S.F./C.F.:
EXTENSIONS BY: JJM

SHEET NO.:
ESTIMATE NO.:
DATE: 3-27-87
NO. OF STORIES: 3
CHECKED BY: RG

Unit Price		Round Off
Plumbing	116283	116280
Fire Protection	110649	110650
HVAC	434664	434665

Figure 24.19

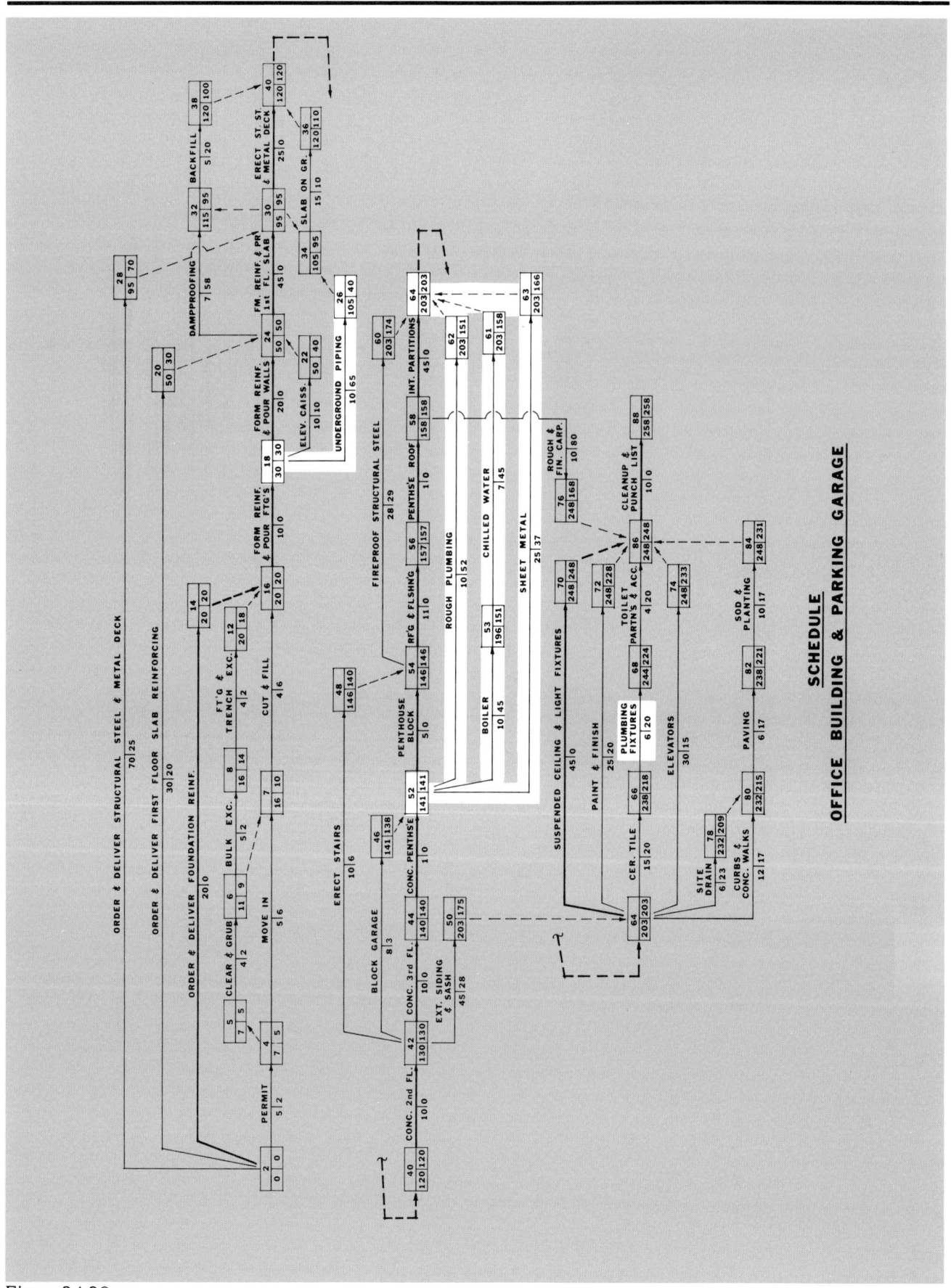

Figure 24.20

The schedule shows that the project will last approximately one year. The duration of the mechanical work will be about six months. Time dependent items, such as equipment rental and superintendent costs, can be analyzed on a Project Overhead Summary form (Figures 24.21a and 24.21b). Some items, such as permits and insurance, are dependent on total job costs. The total direct costs for the project from the Estimate Summary can be used to estimate these costs. These items should be analyzed individually if possible. Alternatively, percentages can be added to material and labor to cover these items. Since the example estimate has been prepared using *Means Mechanical Cost Data*, the percentage method has been used.

The Bottom Line: The estimator is now able to complete the Estimate Summary form as shown in Figure 24.19. Appropriate contingency, sales tax, and overhead and profit costs must be added to the direct costs of the project. The overhead and profit percentage for labor obtained from Figure 22.3 (see Chapter 22) has already been added to all labor costs. Wherever possible, contractors should determine appropriate mark-ups for their own companies as discussed earlier in this section and in Part I. Finally, Means prices represent national averages, and should be adjusted with the city cost index for your locality as shown in Chapter 22, Figure 22.12.

Since all totals for our estimate of the office building project have already been marked up to cover these items, the Estimate Summary shown here lists only the previously calculated totals for each portion of the project. Either applying mark-ups individually or or at the final summary is acceptable, depending on the estimator's preference.

The prime mechanical contractor may also wish to or need to add a final markup for profit if the cost of administering subcontracted items has not been covered previously. Market conditions and project overhead will dictate the amount of final mark-up, if any, to be applied. The complete mechanical estimate is now ready for submittal to the general contractor.

Means Forms

PROJECT OVERHEAD SUMMARY

PROJECT
LOCATION　　　　　　　　　　　ARCHITECT
QUANTITIES BY:　　　　PRICES BY:　　　　EXTENSIONS BY:　　　　CHECKED BY:
SHEET NO.
ESTIMATE NO.
DATE

DESCRIPTION	QUANTITY	UNIT	MATERIAL/EQUIPMENT UNIT	MATERIAL/EQUIPMENT TOTAL	LABOR UNIT	LABOR TOTAL	TOTAL COST UNIT	TOTAL COST TOTAL
Job Organization: Superintendent								
Project Manager								
Timekeeper & Material Clerk								
Clerical								
Safety, Watchman & First Aid								
Travel Expense: Superintendent								
Project Manager								
Engineering: Layout								
Inspection/Quantities								
Drawings								
CPM Schedule								
Testing: Soil								
Materials								
Structural								
Equipment: Cranes								
Concrete Pump, Conveyor, Etc.								
Elevators, Hoists								
Freight & Hauling								
Loading, Unloading, Erecting, Etc.								
Maintenance								
Pumping								
Scaffolding								
Small Power Equipment/Tools								
Field Offices: Job Office								
Architect/Owner's Office								
Temporary Telephones								
Utilities								
Temporary Toilets								
Storage Areas & Sheds								
Temporary Utilities: Heat								
Light & Power								
Water								
PAGE TOTALS								

Figure 24.21a

Means Forms

DESCRIPTION	QUANTITY	UNIT	MATERIAL/EQUIPMENT UNIT	MATERIAL/EQUIPMENT TOTAL	LABOR UNIT	LABOR TOTAL	TOTAL COST UNIT	TOTAL COST TOTAL
Totals Brought Forward								
Winter Protection: Temp. Heat/Protection								
Snow Plowing								
Thawing Materials								
Temporary Roads								
Signs & Barricades: Site Sign								
Temporary Fences								
Temporary Stairs, Ladders & Floors								
Photographs								
Clean Up								
Dumpster								
Final Clean Up								
Punch List								
Permits: Building								
Misc.								
Insurance: Builders Risk								
Owner's Protective Liability								
Umbrella								
Unemployment Ins. & Social Security								
Taxes								
City Sales Tax								
State Sales Tax								
Bonds								
Performance								
Material & Equipment								
Main Office Expense								
Special Items								
TOTALS:								

Figure 24.21b

APPENDIX

Table of Contents

Appendix A Spec-Aid	354
Appendix B Reference Tables	356
Appendix C Glossary of Terms	359
Appendix D Abbreviations	364
Appendix E Graphic Symbols	367
Appendix F Index	371

Appendix A

⚓ Means Forms
SPEC-AID
DATE _____

DIVISION 15: MECHANICAL

PROJECT _____ LOCATION _____

Building Drainage: Design Rainfall _____ ☐ Roof Drains _____ ☐ Court Drains _____
   ☐ Floor Drains _____ ☐ Yard Drains _____ ☐ Lawn Drains _____ ☐ Balcony Drains _____
   ☐ Area Drains _____ ☐ Sump Drains _____ Shower Drains _____ ☐ _____
   ☐ Drain Piping: Size _____ Describe _____
   ☐ Drain Gates _____ ☐ Clean Outs _____ ☐ Grease Traps _____

Sanitary System: ☐ No ☐ Yes ☐ Site Main _____ ☐ Manholes _____
   ☐ Sump Pumps _____ ☐ Bilge Pumps _____ ☐ Ejectors _____
   ☐ Soils, Stacks _____ ☐ Wastes, Vents _____ ☐ _____

Domestic Cold Water: ☐ No ☐ Water Meters _____ ☐ Law Sprinkler Connection _____
   ☐ Water Softening _____ ☐ Water Filtering _____
   ☐ Boiler Feed Water _____ ☐ Conditioning Apparatus _____
   ☐ Standpipe System _____ ☐ Hose Bibbs _____
   ☐ Pressure Tank _____ ☐ Booster Pumps _____
   ☐ Reducing Valves _____ ☐ _____

Domestic Hot Water: ☐ No ☐ Electric ☐ Gas ☐ Oil ☐ Solar _____
   ☐ Boiler _____ ☐ Conditioner _____ ☐ Fixture Connections _____
   ☐ Storage Tanks _____ Capacity _____
   ☐ Pumps _____

Piping: ☐ No ☐ Yes Material _____
   ☐ Air Chambers _____ ☐ Escutcheons _____ ☐ Expansion Joints _____
   ☐ Shock Absorbers _____ ☐ Hangers _____
   ☐ Valves _____ ☐ Paint _____

Special Piping: ☐ No ☐ Compressed Air _____ ☐ Vacuum _____
   ☐ Oxygen _____ ☐ Nitrous Oxygen _____
   ☐ Carbon Dioxide _____ ☐ Process Piping _____

Insulation Cold: ☐ No ☐ Yes Material _____ Jacket _____
   Hot: ☐ No ☐ Yes Material _____ Jacket _____

Fixtures Bathtub: ☐ No ☐ C.I. ☐ Steel ☐ Fiberglass ☐ _____ Color _____
   ☐ Curtain ☐ Rod ☐ Enclosure ☐ Wall Shower _____
   Drinking Fountain: ☐ No ☐ Yes ☐ Wall Hung ☐ Pedestal _____
   Hose Bibb: ☐ No ☐ Yes Describe _____
   Lavatory: ☐ No ☐ China ☐ C.I. ☐ Steel ☐ _____ Color _____
   ☐ Wall Hung ☐ Legs ☐ Acid Resisting _____
   Shower: ☐ No ☐ Individual ☐ Group ☐ Heads ☐ _____ Size _____
   Compartment: ☐ No ☐ Metal ☐ Stone ☐ Fiberglass ☐ _____ ☐ Door ☐ Curtain
   Receptor: ☐ No ☐ Plastic ☐ Metal ☐ Terrazzo ☐ _____
   Sinks: ☐ No ☐ Kitchen _____ ☐ Janitor _____
   ☐ Laundry _____ ☐ Pantry _____
   ☐ _____
   Urinals: ☐ No ☐ Floor Mounted ☐ Wall Hung _____
   Screens: ☐ No ☐ Floor Mounted ☐ Wall Hung _____
   Wash Centers: ☐ No ☐ Yes Describe _____
   Wash Fountains: ☐ No ☐ Floor Mounted ☐ Wall Hung _____ Size _____
   Describe _____
   Water Closets: ☐ No ☐ Floor Mounted ☐ Wall Hung Color _____
   Describe _____
   Water Coolers: ☐ No ☐ Floor Mounted ☐ Wall Hung _____ Capacity _____ gph.
   ☐ Water Supply ☐ Bottle ☐ Hot ☐ Compartment _____
   Other Fixtures: _____

Appendix A

Means Forms
SPEC-AID
DIVISION 15: MECHANICAL

Fire Protection: ☐ Carbon Dioxide System _____ ☐ Standpipe _____
   ☐ Sprinkler System   ☐ Wet _____ ☐ Dry _____ Spacing _____
   ☐ Fire Department Connection _____ ☐ Building Alarm _____
   ☐ Hose Cabinets _____ ☐ Hose Racks _____
   ☐ Roof Manifold _____ ☐ Compressed Air Supply _____
   ☐ Hydrants _____ ☐ _____

Special Plumbing _____

Gas Supply System: ☐ No   ☐ Natural Gas   ☐ Manufactured Gas _____
   Pipe: Schedule _____ Fittings _____
   Shutoffs: _____ Master Control Valve: _____
   Insulation: _____ Paint: _____

Oil Supply System: ☐ No   ☐ Tanks   ☐ Above Ground   ☐ Below Ground _____
   ☐ Steel   ☐ Plastic   ☐ _____ Capacity _____

Heating Plant: ☐ No   ☐ Electric   ☐ Gas   ☐ Oil   ☐ Solar _____
   ☐ Boilers _____ ☐ Pumps _____
   ☐ PRV Stations _____ ☐ Piping _____
   ☐ Heat Pumps _____

Cooling Plant: ☐ No   ☐ Yes _____ Tons _____
   Chillers: ☐ Steam   ☐ Water   ☐ Air _____
   Condenser—Compressor   ☐ Air   ☐ Water _____
   Pumps _____ Cooling Towers _____

System Type: _____
   ☐ Single Zone _____ ☐ Multi-Zone _____
   ☐ All Air _____ ☐ Terminal Reheat _____
   ☐ Double Duct _____ ☐ Radiant Panels _____
   ☐ Fan Coil _____ ☐ Unit Ventilators _____
   ☐ Perimeter Radiation _____ ☐ _____

Air Handling Units: Area Served _____ Number _____
   Total CFM _____ % Outside Air _____
   Cooling, Tons _____ Heating, MBH _____
   Filtration _____ Supply Fans _____
   Economizer _____

Fans: ☐ No   ☐ Return   ☐ Exhaust   ☐ _____
   Describe _____

Distribution: Ductwork _____ Material _____
   Terminals: ☐ Diffusers _____ ☐ Registers _____
      ☐ Grilles _____ ☐ Hoods _____
   Volume Dampers: _____
   Terminal Boxes: ☐ High Velocity _____ ☐ With Coil _____
      ☐ Double Duct _____ ☐ _____

Coils: _____
   ☐ Preheat _____ ☐ Reheat _____
   ☐ Cooling _____ ☐ _____

Piping: See Previous Page _____
Insulation: Cold: ☐ No   ☐ Yes   Material _____ Jacket _____
   ☐ Hot: ☐ No   ☐ Yes   Material _____ Jacket _____
   Automatic Temperature Controls: _____
   Air & Hydronic Balancing: _____

Special HVAC: _____

Appendix B

Useful Formulas for the Mechanical Trades

Rectangle
$A = W \times L$

Parallelogram
$A = H \times L$

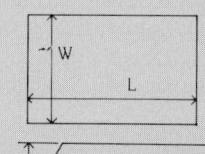

Trapezoid
$A = H \times \dfrac{L_1 + L_2}{2}$

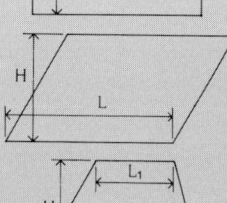

Triangle
$A = \dfrac{W \times H}{2}$

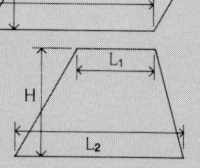

Circle
$A = 3.142 \times R \times R$

$C = 3.142 \times D$

$R = \dfrac{D}{2}$

$D = 2 \times R$

Sector of Circle
$A = \dfrac{3.142 \times R \times R \times \alpha}{360}$

$L = .01745 \times R \times \alpha$

$\alpha = \dfrac{L}{.01745 \times R}$

$R = \dfrac{L}{.01745 \times \alpha}$

Ellipse
$A = 3.142 \times A \times B$

$C = 6.283 \times \sqrt{\dfrac{A^2 + B^2}{2}}$

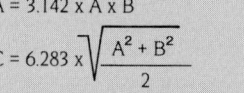

Rectangular Solid
$A_1 = 2[W \times L + L \times H + H \times W]$

$V = W \times L \times H$

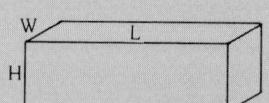

Cone
$A_1 = 3.142 \times R \times S + 3.142 \times R \times R$

$V = 1.047 \times R \times R \times H$

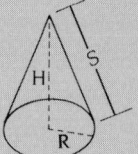

Cylinder
$A_1 = 6.283 \times R \times H + 6.283 \times R \times R$

$V = 3.142 \times R \times R \times H$

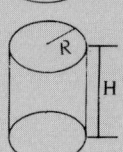

Elliptical Tanks
$V = 3.142 \times A \times B \times H$

$A_1 = 6.283 \times \sqrt{\dfrac{A^2 + B^2}{2}} \times H + 6.283 \times A \times B$

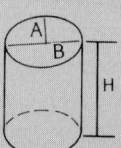

Sphere
$A_1 = 12.56 \times R \times R$

$V = 4.188 \times R \times R \times R$

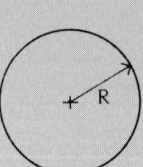

For above containers:

Capacity in gallons = $\dfrac{V}{231}$ when V is in cubic inches.

Capacity in gallons = $7.48 \times V$ when V is in cubic feet.

A = Area   C = Circumference   V = Volume

Appendix B

Conversion Tables

Heat and Power		Volume and Weight	
1 Btu =	778.3 ft.-lb. 0.2930 Int. whr. 252.0 I.T. calorie	1 cu. ft. =	7.481 gal. 1728 cu. in.
1000 LT. calories 1 LT. Kilocalorie =	3.968 Btu 3088 ft.-lb. 1.1628 Int. whr.	1 gallon (U.S.A.) =	231 cu. in. 0.1337 cu. ft.
		1 gal. (British or Imperial) =	277.42 cu. in.
1 horsepower =	0.7455 Int. kw 42.40 Btu per minute 33,000 ft.-lb. per minute 550 ft.-lb. per second	1 cu. ft. water at 60 F =	62.37 lb.
		1 cu. ft. water at 212 F =	59.83 lb.
		1 gal. water at 60 F =	8.338 lb.
		1 gal. water at 212 F =	7.998 lb.
1 boiler horsepower =	33,475 Btu per hour 9.809 Int. kw		
1 Int. watthour =	2656 ft.-lb. 3.413 Btu 3600 Int. joules 860 LT. calories		
1 Int. kilowatt (1000 watts) =	1.341 hp 56.88 Btu per minute 44,267 ft.-lb. per minute		
1 Int. kilowatthour =	3,413 Btu 3,517 lb. water evaporated from and at 212 F		
1 ton refrigeration =	12,000 Btu per hour 200 Btu per minute		
Latent heat of ice (fusion) =	143.4 Btu per pound		

Appendix B

Conversion Tables

Pressure		Metric	
1 lb. per square inch =	144 lb. per square foot 2.0360 in. mercury at 32 F 2.0422 in. mercury at 62 F 2.309 ft. water at 62 F 27.71 in. water at 62 F	1 cm. = 1 in. = 1 m = 1 ft. = 1 sq. cm. =	0.3937 in. = 0.0328 ft. 2.540 cm. 3.281 ft. 0.3048 m. 0.1550 sq. in.
1 oz. per square inch =	0.1276 in. mercury at 62 F 1.732 in. water at 62 F	1 sq. in. = 1 sq. m. =	6.452 sq. cm. 10.76 sq. ft.
1 atmosphere =	14.696 lb. per square inch 2116 lb. per square foot 33.94 ft. water at 62 F 30.01 in. mercury at 62 F 29.921 in. mercury at 32 F	1 sq. ft. = 1 cu. cm. = 1 cu. in. = 1 cu. m. = 1 cu. ft. =	0.09290 sq. m. 0.06102 cu. in. 16.39 cu. cm. 35.31 cu. ft. 0.02832 cu. m.
1 in. water at 62 F =	0.03609 lb. per squre inch 0.5774 oz. per square inch 5.197 lb. per square foot	1 liter = 1 kg. = 1 lb. =	1000 cu. cm. = 0.2642 gal. 2.205 lb. (avdp) 0.4536 kg.
1 ft. water at 62 F =	0.4330 lb. per square inch 62.37 lb. per square foot	1 metric ton =	2205 lb. (avdp)
1 in. mercury at 62 F =	0.4897 lb. per square inch 7.835 oz. per square inch 1.131 ft. water at 62 F 13.57 in. water at 62 F		
1 in. mercury at 32 F =	0.49115 lb. per square inch		

Glossary of Piping Terms

ANSI
American National Standards Institute.

API
American Petroleum Institute.

ASHRAE
American Society of Heating, Refrigerating, and Air Conditioning Engineers.

ASME
American Society of Mechanical Engineers.

ASTM
American Society for Testing Materials.

AWWA
American Water Works Association.

Backing Ring
A metal ring used during the welding process. Its purpose is to prevent melted metal from entering a pipe when making a butt-weld joint, and to provide a uniform gap for welding. Often referred to as "chill rings."

BE
Bevelled end. (The end of a pipe or fitting prepared for welding.)

Bell and Spigot Joint
The most commonly used joint in cast iron soil pipe. Each length is made with an enlarged bell at one end into which the spigot end of another piece is inserted. The joint is then made tight by lead and oakum or a rubber ring caulked into the bell around the spigot.

Black Steel Pipe
Steel pipe that has not been galvanized.

Blank Flange
A flange in which the bolt holes are not drilled.

Blind Flange
A flange used to close off the end of a pipe

Branch
The outlet or inlet of a fitting that is not in line with the run, and takes off at an angle to the run (e.g., tees, wyes, crosses, laterals, etc.).

Building Sewer
The pipe running from the outside wall of the building drain to the public sewer.

Building Storm Drain
A drain for carrying rain, surface water, condensate, etc., to the building drain or sewer.

Bull Head Tee
A tee with the branch larger than the run.

Butt Weld Joint
A welded pipe joint made with the ends of the two pipes butting each other.

Butt Weld Pipe
Pipe welded along the seam and not lapped.

Carbon Steel Pipe
Steel pipe which owes its properties chiefly to the carbon which it contains.

Companion Flange
A pipe flange which connects with another flange or with a flanged valve from fitting. This flange differs with a flange that is an integral part of a fitting or valve.

Cooling Tower
A device for cooling water by evaporation. A natural draft cooling tower is one where the air flow through the tower is due to its natural chimney effect. A mechanical draft tower employs fans to force or induce a draft.

Couplings
Fittings for joining two pieces of pipe.

Appendix C

Drainage System
The piping system in a building up to where it discharges to the sewer system.

Elbow
A fitting that makes an angle in a pipe run. The angle is 90 degrees unless another angle is specified.

Equivalent Length
The resistance of a duct or pipe elbow, valve, damper, orifice, bend, fitting, or other obstruction to flow expressed in the number of feet of straight duct or pipe of the same diameter which would have the same resistance.

ERW
Electric resistance weld. (A method of welding pipe in the manufacturing process.)

Expansion Joint
A joint whose primary purpose is to absorb the longitudinal expansion and contraction in the line due to temperature changes.

Expansion Loop
A large radius loop in a pipe line which absorbs the longitudinal expansion and contraction in the line due to temperature changes.

Flange
A ring-shaped plate at right angles to the end of the pipe. It is provided with holes for bolts to allow fastening of the pipe to a similar flange.

Flat Face
Pipe flanges which have the entire face of the flange faced straight across and which use a full face gasket. These are commonly employed for pressures less than 125 pounds.

Galvanizing
Coating iron or steel surfaces with a protective layer of zinc.

Gate Valve
A valve utilizing a gate, usually wedge-shaped, which allows fluid flow when the gate is lifted from the seat. Gate valves have less resistance to flow than globe valves, and should always be used fully open or fully closed.

Globe Valve
A valve with a rounded body utilizing a manually raised or lowered disc which, when closed, seats so as to prevent fluid flow. Globe valves are ideal for throttling in a semi-closed position.

Header
A large pipe or drum into which each of a group of boilers, chillers, or pumps are connected. (see Manifold)

Horizontal Branch
In plumbing, the horizontal line from the fixture drain to the waste stack.

Hot Water Heating System
One in which hot water is the heating medium. Flow is either gravity or forced circulation.

ID
Inside diameter.

IDHA
International District Heating Association.

Insulation
Thermal insulation is a material used for covering pipes, ducts, vessels, etc., to effect a reduction of heat loss or gain.

Lapped Joint
A lapped joint, like a van stone joint, is a type of pipe joint made using loose flanges on lengths of pipe. The ends of this pipe are lapped over to give a bearing surface for a gasket or metal to metal joint.

Appendix C

Lap Weld Pipe
Pipe made by welding along a scarfed longitudinal seam in which one part is overlapped by another.

LCL
Less carload lot. (Pipe is either ordered from the mill by the carload, or LCL.)

Lead Joint
A joint made by pouring molten lead into the space between a bell and spigot, and making the joint tight by caulking.

Malleable Iron
Cast iron which has been heat treated to reduce its brittleness.

Manifold
A fitting with several branch outlets.

Mill Length
Also known as "random length". The usual run-of-the-mill pipe is 16 to 20 feet in length. Line pipe for power plant or oil field use is often made in double random lengths of 30 to 35 feet.

Nipple
A piece of pipe less than 12 inches long and threaded on both ends. Pipe over 12 inches long is regarded as a cut measure.

Nominal
Name given to standard pipe size designations through 12 inches nominal O.D. For example, 2" nominal is 2-3/8" O.D.

OD
Outside diameter.

O.S. & Y.
Outside screw and yoke. A valve configuration where the valve stem, having exposed external threads supported by a yoke, indicates the open or closed position of the valve.

PE
Plain end. (Used to describe the ends of pipe which are shipped from the mill with unfinished ends. These ends may eventually be threaded, beveled, or grooved in the field.)

Pipe
A hollow cylinder or tube for conveyance of a fluid.

Plug Valve
A valve containing a tapered plug through which a hole is drilled so that fluid can flow through when the holes line up with the inlet and the outlet, but when the plug is rotated 90 degrees, the flow is stopped.

Plumbing Fixtures
Devices which receive water, liquid, or water-borne wastes, and discharge the wastes into a drainage system.

Plumbing System
Arrangements of pipes, fixtures, fittings, valves, and traps, in a building which supply water and remove liquid borne wastes, including storm water.

Potable Water
Water suitable for human consumption.

PSI
Pounds per square inch.

Reducer
A pipe coupling with a larger size at one end than the other. The larger size is designated first. Reducers are threaded, flanged, welded, etc. Reducing couplings are available in either eccentric or concentric configurations.

Riser
A vertical pipe extending one or more floors.

Roof Drain
A fitting which collects water on the roof surface and discharges it into the leader.

Schedule Number
Schedule numbers are American Standards Association designations for classifying the strength of pipe. Schedule 40 is the most common form of steel pipe used in the mechanical trades.

Screwed Joint
A pipe joint consisting of threaded male and female parts joined together.

Seamless Pipe
Pipe or tube formed by piercing a billet of steel and then rolling, rather than having welded seams.

Service Pipe
A pipe connecting water or gas mains into a building from the street.

Slip-on Flange
A flange slipped onto the end of a pipe and then welded in place.

Socket Weld
A pipe joint made by use of a socket weld fitting which has a female end or socket for insertion of the pipe to be welded.

Stainless Steel
An alloy steel having unusual corrosion-resisting properties, usually imparted by nickel and chromium.

Storm Sewer
A sewer carrying surface or storm water from roofs or exterior surfaces of a building.

Street Elbow
An elbow with a male thread on one end, and a female thread on the other end.

Swing Joint
An arrangement of screwed fittings to allow for movement in a pipe line.

Swivel Joint
A special pipe fitting designed to be pressure tight under continuous or intermittent movement of the equipment to which it is connected.

TBE
Thread both ends. Term used when specifying or ordering cut measures of pipe.

T & C
Threaded and coupled; an ordering designation for threaded pipe.

Tee
A pipe fitting that has a side port at right angles to the run.

TOE
Thread one end. Term used when specifying or ordering cut measures of pipe.

Union
A fitting used to join pipes. It commonly consists of three pieces. Unions are extensively used, because they allow dismantling and reassembling of piping assemblies with ease and without distorting the assembly.

Van Stone
A type of joint made by using loose flanges on lengths of pipe whose ends are lapped over to give a bearing surface for the flange.

Vents
Vents are used to permit air to escape from hydronic systems, condensate receivers, fuel oil storage tanks, as a breather line for gas regulators, etc.

Appendix C

Vent Stack
A vertical vent pipe which provides air circulation to and from the drainage system.

Vent System
Piping which provides a flow of air to or from a drainage system to protect trap seals from siphonage or back pressure.

Welding Fittings
Wrought steel elbows, tees, reducers, saddles, and the like, beveled for butt welding to pipe. Forged fittings with hubs or with ends counter-bored for fillet welding to pipe are used for small pipe sizes and high pressures.

Welding Neck Flange
A flange with a long neck beveled for butt welding to pipe.

Wye
A pipe fitting with a side outlet that is any angle other than 90 degrees to the main run or axis.

Appendix D

A	Area Square Feet; Ampere	Calc	Calculated	D.H.	Double Hung
ABS	Acrylonitrile Butadiene Styrene; Asbestos Bonded Steel	Cap.	Capacity	DHW	Domestic Hot Water
		Carp.	Carpenter	Diag.	Diagonal
A.C.	Alternating Current; Air Conditioning; Asbestos Cement	C.B.	Circuit Breaker	Diam.	Diameter
		C.C.F.	Hundred Cubic Feet	Distrib.	Distribution
		cd	Candela	Dk.	Deck
A.C.I.	American Concrete Institute	cd/sf	Candela per Square Foot	D.L.	Dead Load; Diesel
Addit.	Additional	CD	Grade of Plywood Face & Back	Do.	Ditto
Adj.	Adjustable	CDX	Plywood, grade C&D, exterior glue	Dp.	Depth
af	Audio-frenquency	Cefi.	Cement Finisher	D.P.S.T.	Double Pole, Single Throw
A.G.A.	American Gas Association	Cem.	Cement	Dr.	Driver
Agg.	Aggregate	CF	Hundred Feet	Drink.	Drinking
A.H.	Ampere Hours	C.F.	Cubic Feet	D.S.	Double Strength
A hr	Ampere-hour	CFM	Cubic Feet per Minute	D.S.A.	Double Strength A Grade
A.I.A.	American Institute of Architects	c.g.	Center of Gravity	D.S.B.	Double Strength B Grade
AIC	Ampere Interrupting Capacity	CHW	Commercial Hot Water	Dty.	Duty
Allow.	Allowance	C.I.	Cast Iron	DWV	Drain Waste Vent
alt.	Altitude	C.I.P.	Cast in Place	DX	Deluxe White, Direct Expansion
Alum.	Aluminum	Circ.	Circuit	dyn	Dyne
a.m.	ante meridiem	C.L.	Carload Lot	e	Eccentricity
Amp.	Ampere	Clab.	Common Laborer	E	Equipment Only; East
Approx.	Approximate	C.L.F.	Hundred Linear Feet	Ea.	Each
Apt.	Apartment	CLF	Current Limiting Fuse	Econ.	Economy
Asb.	Asbestos	CLP	Cross Linked Polyethylene	EDP	Electronic Data Processing
A.S.B.C.	American Standard Building Code	cm	Centimeter	E.D.R.	Equiv. Direct Radiation
Asbe.	Asbestos Worker	CMP	Corr. Metal Pipe	Eq.	Equation
A.S.H.R.A.E.	American Society of Heating, Refrig. & AC Engineers	C.M.U.	Concrete Masonry Unit	Elec.	Electrician; Electrical
		Col.	Column	Elev.	Elevator; Elevating
A.S.M.E.	American Society of Mechanical Engineers	CO_2	Carbon Dioxide	EMT	Electrical Metallic Conduit; Thin Wall Conduit
		Comb.	Combination		
A.S.T.M.	American Society for Testing and Materials	Compr.	Compressor	Eng.	Engine
		Conc.	Concrete	EPDM	Ethylene Propylene Diene Monomer
Attchmt.	Attachment	Cont.	Continuous; Continued		
Avg.	Average			Eqhv.	Equip. Oper., heavy
Bbl.	Barrel	Corr.	Corrugated	Eqlt.	Equip. Oper., light
B.&B.	Grade B and Better; Balled & Burlapped	Cos	Cosine	Eqmd.	Equip. Oper., medium
		Cot	Cotangent	Eqmm.	Equip. Oper., Master Mechanic
B.&S.	Bell and Spigot	Cov.	Cover	Eqol.	Equip. Oper., oilers
B.&W.	Black and White	CPA	Control Point Adjustment	Equip.	Equipment
b.c.c.	Body-centered Cubic	Cplg.	Coupling	ERW	Electric Resistance Welded
B.F.	Board Feet	C.P.M.	Critical Path Method	Est.	Estimated
Bg. Cem.	Bag of Cement	CPVC	Chlorinated Polyvinyl Chloride	esu	Electrostatic Units
BHP	Brake Horse Power	C. Pr.	Hundred Pair	E.W.	Each Way
B.I.	Black Iron	CRC	Cold Rolled Channel	EWT	Entering Water Temperature
Bit.; Bitum.	Bituminous	Creos.	Creosote	Excav.	Excavation
		Crpt.	Carpet & Linoleum Layer	Exp.	Expansion
Bk.	Backed	CRT	Cathode-ray Tube	Ext.	Exterior
Bkrs.	Breakers	CS	Carbon Steel	Extru.	Extrusion
Bldg.	Building	Csc	Cosecant	f.	Fiber stress
Blk.	Block	C.S.F.	Hundred Square Feet	F	Fahrenheit; Female; Fill
Bm.	Beam	C.S.I.	Construction Specification Institute	Fab.	Fabricated
Boil.	Boilermaker			FBGS	Fiberglass
B.P.M.	Blows per Minute	C.T.	Current Transformer	F.C.	Footcandles
BR	Bedroom	CTS	Copper Tube Size	f.c.c.	Face-centered Cubic
Brg.	Bearing	Cu	Cubic	f'c	Compressive Stress in Concrete; Extreme Compressive Stress
Brhe.	Bricklayer Helper	Cu. Ft.	Cubic Foot		
Bric.	Bricklayer	cw	Continuous Wave	F.E.	Front End
Brk.	Brick	C.W.	Cool White	FEP	Fluorinated Ethylene Propylene (Teflon)
Brng.	Bearing	Cwt.	100 Pounds		
Brs.	Brass	C.W.X.	Cool White Deluxe	F.G.	Flat Grain
Brz.	Bronze	C.Y.	Cubic Yard (27 cubic feet)	F.H.A.	Federal Housing Administration
Bsn.	Basin	C.Y./Hr.	Cubic Yard per Hour	Fig.	Figure
Btr.	Better	Cyl.	Cylinder	Fin.	Finished
BTU	British Thermal Unit	d	Penny (nail size)	Fixt.	Fixture
BTUH	BTU per Hour	D	Deep; Depth; Discharge	Fl. Oz.	Fluid Ounces
BX	Interlocked Armored Cable	Dis.; Disch.	Discharge	Flr.	Floor
c	Conductivity			F.M.	Frequency Modulation; Factory Mutual
C	Hundred; Centigrade	Db.	Decibel		
		Dbl.	Double	Fmg.	Framing
C/C	Center to Center	DC	Direct Current	Fndtn.	Foundation
Cab.	Cabinet	Demob.	Demobilization	Fori.	Foreman, inside
Cair.	Air Tool Laborer	d.f.u.	Drainage Fixture Units	Foro.	Foreman, outside

Appendix D

Fount.	Fountain	I.P.S.	Iron Pipe Size	M.C.M.	Thousand Circular Mils
FPM	Feet per Minute	I.P.T.	Iron Pipe Threaded	M.C.P.	Motor Circuit Protector
FPT	Female Pipe Thread	J	Joule	MD	Medium Duty
Fr.	Frame	J.I.C.	Joint Industrial Council	M.D.O.	Medium Density Overlaid
F.R.	Fire Rating	K.	Thousand; Thousand Pounds	Med.	Medium
FRK	Foil Reinforced Kraft	K.D.A.T.	Kiln Dried After Treatment	MF	Thousand Feet
FRP	Fiberglass Reinforced Plastic	kg	Kilogram	M.F.B.M.	Thousand Feet Board Measure
FS	Forged Steel	kG	Kilogauss	Mfg.	Manufacturing
FSC	Cast Body; Cast Switch Box	kgf	Kilogram force	Mfrs.	Manufacturers
Ft.	Foot; Feet	kHz	Kilohertz	mg	Milligram
Ftng.	Fitting	Kip.	1000 Pounds	MGD	Million Gallons per Day
Ftg.	Footing	KJ	Kiljoule	MGPH	Thousand Gallons per Hour
Ft. Lb.	Foot Pound	K.L.	Effective Length Factor	MH	Manhole; Metal Halide; Man Hour
Furn.	Furniture	Km	Kilometer	MHz	Megahertz
FVNR	Full Voltage Non Reversing	K.L.F.	Kips per Linear Foot	Mi.	Mile
FXM	Female by Male	K.S.F.	Kips per Square Foot	MI	Malleable Iron; Mineral Insulated
Fy.	Minimum Yield Stress of Steel	K.S.I.	Kips per Square Inch	mm	Millimeter
g	Gram	K.V.	Kilo Volt	Mill.	Millwright
G	Gauss	K.V.A.	Kilo Volt Ampere	Min.	Minimum
Ga.	Gauge	K.V.A.R.	Kilovar (Reactance)	Misc.	Miscellaneous
Gal.	Gallon	KW	Kilo Watt	ml	Milliliter
Gal./Min.	Gallon Per Minute	KWh	Kilowatt-hour	M.L.F.	Thousand Linear Feet
Galv.	Galvanized	L	Labor Only; Length; Long	Mo.	Month
Gen.	General	Lab.	Labor	Mobil.	Mobilization
Glaz.	Glazier	lat	Latitude	Mog.	Mogul Base
GPD	Gallons per Day	Lath.	Lather	MPH	Miles per Hour
GPH	Gallons per Hour	Lav.	Lavatory	MPT	Male Pipe Thread
GPM	Gallons per Minute	lb.; #	Pound	MRT	Mile Round Trip
GR	Grade	L.B.	Load Bearing; L Conduit Body	ms	millisecond
Gran.	Granular	L. & E.	Labor & Equipment	M.S.F.	Thousand Square Feet
Grnd.	Ground	lb./hr.	Pounds per Hour	Mstz.	Mosaic & Terrazzo Worker
H	High; High Strength Bar Joist; Henry	lb./L.F.	Pounds per Linear Foot	M.S.Y.	Thousand Square Yards
		lbf/sq in.	Pound-force per Square Inch	Mtd.	Mounted
H.C.	High Capacity	L.C.L.	Less than Carload Lot	Mthe.	Mosaic & Terrazzo Helper
H.D.	Heavy Duty; High Density	Ld.	Load	Mtng.	Mounting
H.D.O.	High Density Overlaid	L.F.	Linear Foot	Mult.	Multi; Multiply
Hdr.	Header	Lg.	Long; Length; Large	MVAR	Million Volt Amp Reactance
Hdwe.	Hardware	L. & H.	Light and Heat	MV	Megavolt
Help.	Helper average	L.H.	Long Span High Strength Bar Joist	MW	Megawatt
HEPA	High Efficiency Particulate Air Filter	L.J.	Long Span Standard Strength Bar Joist	MXM	Male by Male
Hg	Mercury			MYD	Thousand yards
H.O.	High Output	L.L.	Live Load	N	Natural; North
Horiz.	Horizontal	L.L.D.	Lamp Lumen Depreciation	nA	nanoampere
H.P.	Horsepower; High Pressure	lm	Lumen	NA	Not Available; Not Applicable
H.P.F.	High Power Factor	lm/sf	Lumen per Square Foot	N.B.C.	National Building Code
Hr.	Hour	lm/W	Lumen Per Watt	NC	Normally Closed
Hrs./Day	Hours Per Day	L.O.A.	Length Over All	N.E.M.A.	National Electrical Manufacturers Association
HSC	High Short Circuit	log	Logarithm		
Ht.	Height	L.P.	Liquefied Petroleum; Low Pressure	NEHB	Bolted Circuit Breaker to 600V.
Htg.	Heating			N.L.B.	Non-Load-Bearing
Htrs.	Heaters	L.P.F.	Low Power Factor	nm	nanometer
HVAC	Heating, Ventilating & Air Conditioning	Lt.	Light	No.	Number
		Lt. Ga.	Light Gauge	NO	Normally Open
Hvy.	Heavy	L.T.L.	Less than Truckload Lot	N.O.C.	Not Otherwise Classified
HW	Hot Water	Lt. Wt.	Lightweight	Nose.	Nosing
Hyd.; Hydr.	Hydraulic	L.V.	Low Voltage	N.P.T.	National Pipe Thread
		M	Thousand; Material; Male; Light Wall Copper	NQOB	Bolted Circuit Breaker to 240V.
Hz.	Hertz (cycles)			N.R.C.	Noise Reduction Coefficient
I.	Moment of Inertia	m/hr	Manhour	N.R.S.	Non Rising Stem
I.C.	Interrupting Capacity	mA	Milliampere	ns	nanosecond
ID	Inside Diameter	Mach.	Machine	nW	nanowatt
I.D.	Inside Dimension; Identification	Mag. Str.	Magnetic Starter	OB	Opposing Blade
		Maint.	Maintenance	OC	On Center
I.F.	Inside Frosted	Marb.	Marble Setter	OD	Outside Diameter
I.M.C.	Intermediate Metal Conduit	Mat.	Material	O.D.	Outside Dimension
In.	Inch	Mat'l.	Material	ODS	Overhead Distribution System
Incan.	Incandescent	Max.	Maximum	O & P	Overhead and Profit
Incl.	Included; Including	MBF	Thousand Board Feet	Oper.	Operator
Int.	Interior	MBH	Thousand BTU's per hr.	Opng.	Opening
Inst.	Installation	M.C.F.	Thousand Cubic Feet	Orna.	Ornamental
Insul.	Insulation	M.C.F.M.	Thousand Cubic Feet per Minute	O.S.&Y.	Outside Screw and Yoke
I.P.	Iron Pipe			Ovhd	Overhead

Appendix D

Oz.	Ounce	S.	Suction; Single Entrance; South	T.S.	Trigger Start
P.	Pole; Applied Load; Projection			Tr.	Trade
p.	Page	Scaf.	Scaffold	Transf.	Transformer
Pape.	Paperhanger	Sch.; Sched.	Schedule	Trhv.	Truck Driver, Heavy
PAR	Weatherproof Reflector			Trlr.	Trailer
Pc.	Piece	S.C.R.	Modular Brick	Trlt.	Truck Driver, Light
P.C.	Portland Cement; Power Connector	S.D.R.	Standard Dimension Ratio	TV	Television
		S.E.	Surfaced Edge	T.W.	Thermoplastic Water Resistant Wire
P.C.F.	Pounds per Cubic Foot	S.E.R.; S.E.U.	Service Entrance Cable	UCI	Uniform Construction Index
P.E.	Professional Engineer; Porcelain Enamel; Polyethylene; Plain End	S.F.	Square Foot	UF	Underground Feeder
		S.F.C.A.	Square Foot Contact Area	U.H.F.	Ultra High Frequency
Perf.	Perforated	S.F.G.	Square Foot of Ground	U.L.	Underwriters Laboratory
Ph.	Phase	S.F. Hor.	Square Foot Horizontal	Unfin.	Unfinished
P.I.	Pressure Injected	S.F.R.	Square Feet of Radiation	URD	Underground Residential Distribution
Pile.	Pile Driver	S.F.Shlf.	Square Foot of Shelf		
Pkg.	Package	S4S	Surface 4 Sides	V	Volt
Pl.	Plate	Shee.	Sheet Metal Worker	VA	Volt/amp
Plah.	Plasterer Helper	Sin.	Sine	V.A.T.	Vinyl Asbestos Tile
Plas.	Plasterer	Skwk.	Skilled Worker	VAV	Variable Air Volume
Pluh.	Plumbers Helper	SL	Saran Lined	Vent.	Ventilating
Plum.	Plumber	S.L.	Slimline	Vert.	Vertical
Ply.	Plywood	Sldr.	Solder	V.G.	Vertical Grain
p.m.	Post Meridiem	S.N.	Solid Neutral	V.H.F.	Very High Frequency
Pord.	Painter, Ordinary	S.P.	Static Pressure; Single Pole; Self Propelled	VHO	Very High Output
pp	Pages			Vib.	Vibrating
PP; PPL	Polypropylene	Spri.	Sprinkler Installer	V.L.F.	Vertical Linear Foot
P.P.M.	Parts per Million	Sq.	Square; 100 square feet	Vol.	Volume
Pr.	Pair	S.P.D.T.	Single Pole, Double Throw	W	Wire; Watt; Wide; West
Prefab.	Prefabricated	S.P.S.T.	Single Pole, Single Throw	w/	With
Prefin.	Prefinished	SPT	Standard Pipe Thread	W.C.	Water Column; Water Closet
Prop.	Propelled	Sq. Hd.	Square Head	W.F.	Wide Flange
PSF; psf	Pounds per Square Foot	S.S.	Single Strength; Stainless Steel	W.G.	Water Gauge
PSI; psi	Pounds per Square Inch	S.S.B.	Single Strength B Grade	Wldg.	Welding
PSIG	Pounds per Square Inch Gauge	Sswk.	Structural Steel Worker	Wrck.	Wrecker
PSP	Plastic Sewer Pipe	Sswl.	Structural Steel Welder	W.S.P.	Water, Steam, Petroleum
Pspr.	Painter, Spray	St.; Stl.	Steel	WT, Wt.	Weight
Psst.	Painter, Structural Steel	S.T.C.	Sound Transmission Coefficient	WWF	Welded Wire Fabric
P.T.	Potential Transformer	Std.	Standard	XFMR	Transformer
P. & T.	Pressure & Temperature	STP	Standard Temperature & Pressure	XHD	Extra Heavy Duty
Ptd.	Painted	Stpi.	Steamfitter, Pipefitter	Y	Wye
Ptns.	Partitions	Str.	Strength; Starter; Straight	yd	Yard
Pu	Ultimate Load	Strd.	Stranded	yr	Year
PVC	Polyvinyl Chloride	Struct.	Structural	Δ	Delta
Pvmt.	Pavement	Sty.	Story	%	Percent
Pwr.	Power	Subj.	Subject	†	Approximately
Q	Quantity Heat Flow	Subs.	Subcontractors	∅	Phase
Quan.; Qty.	Quantity	Surf.	Surface	@	At
Q.C.	Quick Coupling	Sw.	Switch	#	Pound; Number
r	Radius of Gyration	Swbd.	Switchboard	<	Less Than
R	Resistance	S.Y.	Square Yard	>	Greater Than
R.C.P.	Reinforced Concrete Pipe	Syn.	Synthetic		
Rect.	Rectangle	Sys.	System		
Reg.	Regular	t.	Thickness		
Reinf.	Reinforced	T	Temperature; Ton		
Req'd.	Required	Tan	Tangent		
Resi	Residential	T.C.	Terra Cotta		
Rgh.	Rough	T.D.	Temperature Difference		
R.H.W.	Rubber, Heat & Water Resistant; Residential Hot Water	TFE	Tetrafluoroethylene (Teflon)		
		T. & G.	Tongue & Groove; Tar & Gravel		
rms	Root Mean Square				
Rnd.	Round	Th.; Thk.	Thick		
Rodm.	Rodman	Thn.	Thin		
Rofc.	Roofer, Composition	Thrded	Threaded		
Rofp.	Roofer, Precast	Tilf.	Tile Layer Floor		
Rohe.	Roofer Helpers (Composition)	Tilh.	Tile Layer Helper		
Rots.	Roofer, Tile & Slate	THW.	Insulated Strand Wire		
R.O.W.	Right of Way	THWN; THHN	Nylon Jacketed Wire		
RPM	Revolutions per Minute				
R.R.	Direct Burial Feeder Conduit	T.L.	Truckload		
R.S.	Rapid Start	Tot.	Total		
RT	Round Trip				

Appendix E

Plumbing Fixture Symbols

Category	Type		Type	Category
Baths	Corner			Dishwasher
	Recessed		Single Basin	Sinks
	Angle		Twin Basin	
	Whirlpool		Single Drainboard	
			Double Drainboard	
Showers	Stall		Floor	Drinking Fountains
	Corner Stall		Recessed	
			Semi-Recessed	
	Wall Gang		Single	Laundry Trays
Water Closets	Tank		Double	
	Flush Valve		Wall	Service Sinks
	Bidet		Floor	
Urinals	Wall		Circular	Wash Fountains
	Stall		Semi-Circular	
	Trough		Heater	Hot Water
Lavatories	Counter		Tank	
	Wall		Gas	Separators
	Corner		Oil	
	Pedestal			

Appendix E

Piping Symbols

Valves, Fittings & Specialties			Valves, Fittings & Specialties (Cont.)
Gate		Pipe Pitch Up or Down (Up/Dn)	
Globe		Expansion Joint	
Check		Expansion Loop	
Butterfly		Flexible Connection	
Solenoid		Thermostat	
Lock Shield		Thermostatic Trap	
2-Way Automatic Control		Float and Thermostatic Trap	
3-Way Automatic Control		Thermometer	
Gas Cock		Pressure Gauge	
Plug Cock		Flow Switch	
Flanged Joint		Pressure Switch	
Union		Pressure Reducing Valve	
Cap		Humidistat	
Strainer		Aquastat	
Concentric Reducer		Air Vent	
Eccentric Reducer		Meter	
Pipe Guide		Elbow	
Pipe Anchor		Tee	
Elbow Looking Up			
Elbow Looking Down			
Flow Direction			

Appendix E

Piping Symbols (Cont.)

Plumbing	Symbol		Symbol	HVAC (Cont.)
Floor Drain	▢		—FOG—	Fuel Oil Gauge Line
Indirect Waste	—W—		—o—PD—o—	Pump Discharge
Storm Drain	—SD—		— — — —	Low Pressure Condensate Return
Combination Waste & Vent	—CWV—		—LPS—	Low Pressure Steam
Acid Waste	—AW—		—MPS—	Medium Pressure Steam
Acid Vent	— — AV — —		—HPS—	High Pressure Steam
Cold Water	—CW—		—BD—	Boiler Blow-Down
Hot Water	—HW—		—F—	Fire Protection Water Supply
Drinking Water Supply	—DWS—			
Drinking Water Return	—DWR—		—WSP—	Wet Standpipe
Gas-Low Pressure	—G—		—DSP—	Dry Standpipe
Gas-Medium Pressure	—MG—		—CSP—	Combination Standpipe
Compressed Air	—A—		—SP—	Automatic Fire Sprinkler
Vacuum	—V—		—o——o—	Upright Fire Sprinkler Heads
Vacuum Cleaning	—VC—			
Oxygen	—O—		—●——●—	Pendent Fire Sprinkler Heads
Liquid Oxygen	—LOX—			
Liquid Petroleum Gas	—LPG—		(hydrant symbol)	Fire Hydrant
Hot Water Heating Supply (HVAC)	—HWS—		(wall connection symbol)	Wall Fire Dept. Connection
Hot Water Heating Return	—HWR—		(sidewalk connection symbol)	Sidewalk Fire Dept. Connection
Chilled Water Supply	—CHWS—		FHR ○-|||||||	Fire Hose Rack
Chilled Water Return	—CHWR—			
Drain Line	—D—		FHC (surface mounted)	Surface Mounted Fire Hose Cabinet
City Water	—CW—			
Fuel Oil Supply	—FOS—		FHC (recessed)	Recessed Fire Hose Cabinet
Fuel Oil Return	—FOR—			
Fuel Oil Vent	—FOV—			

HVAC — Fire Protection

369

Appendix E

HVAC Ductwork Symbols

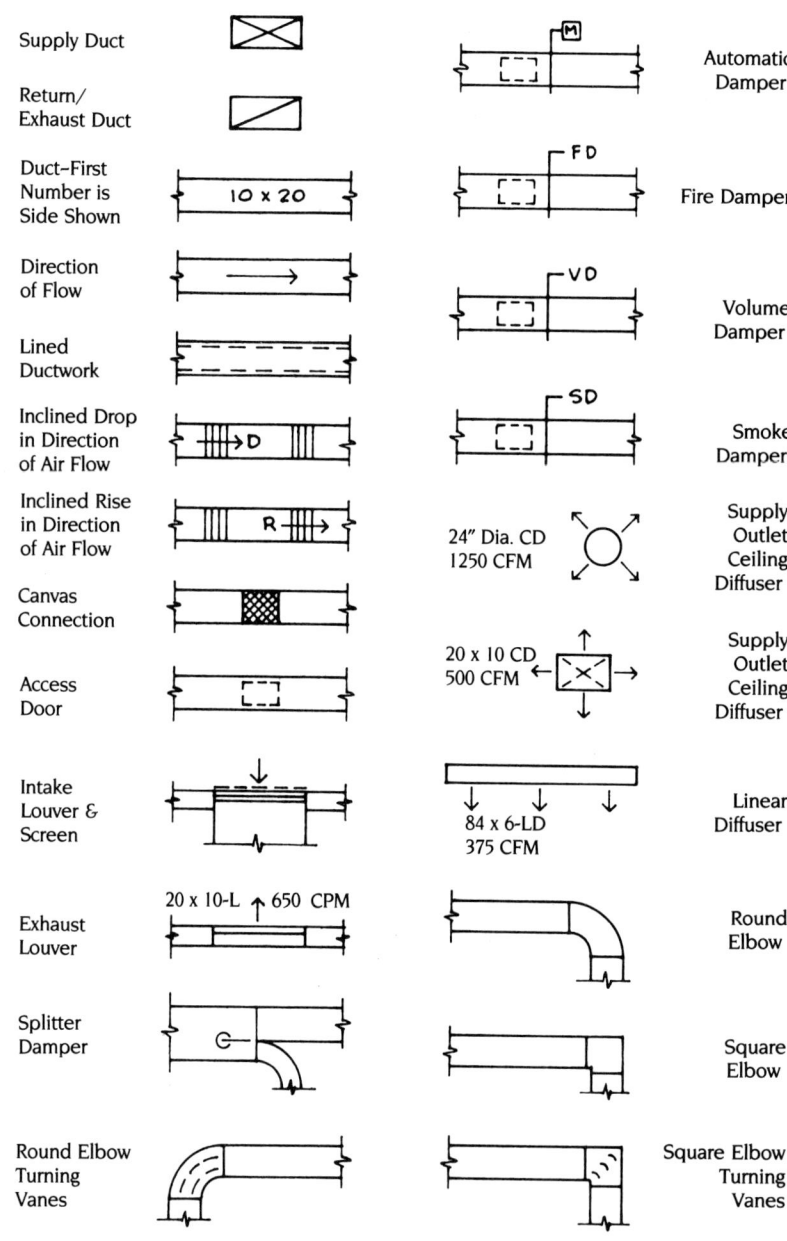

Double Duct Air System

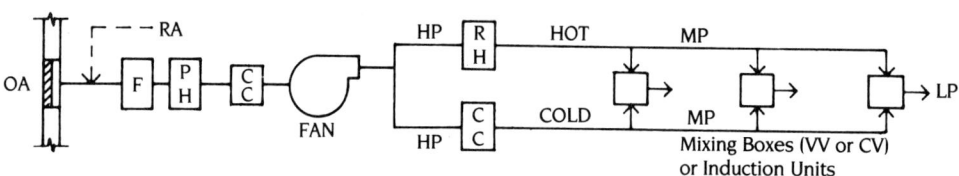

OA = Outside Air
RA = Return Air
F = Filter
PH = Preheat Coil

CC = Cooling Coil
RH = Reheat Coil
HP = High Pressure Duct
MP = Medium Pressure Duct

LP = Low Pressure Duct
VV = Variable Volume
CV = Constant Volume

Appendix F

A

Abbreviations 364–366
Absorption type water chillers 185
Access doors 199
Acknowledgements viii
Acoustic lining 199, 200
Air conditioning 177–198
Air cooled condensers 185
Air filters 191, 196
Air handling central station 193
    equipment takeoff 195
    equipment units 195
    system integrated 204
    units 191
Alarm valves 171
Aluminum ductwork 199
Anchors 215–218
Angle iron 199, 215, 217
Angle valves 140
Appliances plumbing 151
Area drains 157
Assemblies costs how to use 235
Assemblies estimate definition 10
Automatic temperature
    controls 211–213
    takeoff 213
    units 211

B

Backwater valves 157
Ball valves 140
Baseboard radiation 182
Bathtub 156
Bathtub-shower combination 156
Belt drive 195
Bidding analysis 66
    strategies 65–69
Blanket insulation 207
Boiler controls 211
    grates 182
    gross output 177
    horsepower 178
    takeoff 182
    units 181
Boilers 177
    cast iron 177
    electric 177
    fire tube 177
    heating 177
    high efficiency 178
    modified Scotch Marine 180
    packaged 177
    pulse type 178
    sectional 177
    steel 177
    water tube 177
Bonding 41
Brass fittings 131
Breeching 107
Butterfly valves 140

C

C.P.M. diagram 60
Calcium silicate insulation 207
Can washers 158
Carbon dioxide fire suppression 172
Carrier and drain takeoff 162
    units 161
Carriers 157–162
Cash flow 82
    chart 84
Cast iron boilers 177
    flanged fittings 118
    pipe 126
    screwed fittings 117
    screwed flanges 117
    soil pipe fittings 127
Cellular insulation 207
Central station air
    handling unit 193
Centrifical pumps 150
Centrifugal water chillers 185
Check valves 140
Chemical fire suppression 172
    foam and gas takeoff 173
    foam and gas units 173
Chillers 185
City cost index
    how to use 237
Cleanouts 159
Closed cell polyethylene
    insulation 207
Coils 182, 191
Come-a-long 182
Completion analysis 78
Compression fittings 133
Computer estimating 87–108
    reports samples 90–99
    room cooling units 193
Concrete inserts 142
Condensate meters 182
Condenser water 188
Condensers air cooled 185
Conduit 137
Consolidated estimate sample
    form 25
Contingencies 48, 50
Control dampers 212
    diagrams 213
    electric 211
    electronic 211
    pneumatic 211
    points 213
    tubing averages 213
Controls HVAC 211
    boiler 211
    furnace 211
Convectors 182
Converters 181
Cooling hydronic 185
    takeoff 191
Cooling tower 188
    mechanical draft 188
    natural draft 188
Cooling units 190
    computer room 193
Copper fittings 131
    pipe 129
    tubing 129

Cost control 71–86
    definitions 229–232
    indexes how to use 237
    information sources 27–28
    record 75, 76
    types of 30
    life cycle 86
Cranes 182, 191
Crew costs how to use 226
Crews split 204
CSI Masterformat divisions 221
Cubic foot costs how to use 232
Cubic foot estimate definition 10

D

Dampers 191, 199
    control 212
    motorized 204
    schedule 204
Deluge sprinkler system 169
Diffusers 199
Direct costs 30–44
Direct radiation 182
Direct-expansion 185
Draft adjuster 201
Drain and carrier takeoff 162
Drain pans 192
Drains 157–162
    area 157
    floor 157
    storm 157
    swimming pool 157
    units 161
Drinking fountains 152
Dry pipe sprinkler system 166
Duct bracing 199
    coils 182
    heater electric 204
    lining 199
Ductile iron fittings 118
Ductwork 199–205
    aluminum 199
    fiberglass board 199
    galvanized steel 199
    height modifications 205
    insulation 207
    plastic 199
    schedule sample form 24
    scrap allowance 202, 203
    spiral wound 199
    stainless steel 199
    symbols 370
    takeoff 205
    units 199, 203
    weight per lineal foot 203
    weight per square foot 202
    weights 202

E

Efficiency 79
Electric boilers 177
    control 211
    duct heater 204
    wiring 212, 213

Appendix F

Electronic control 211
Energy management systems 211
Equal leg angles 217
Equipment 39–40
Equipment insulation 207
Equivalent pipe footage
    (insulation) 209
Estimate components 3–4
    definitions 7
    preliminaries 15–20
    summary 51–58
    types 7–14
Estimating example
    square foot 245–247
    systems 250
    unit price 287
Estimating guidelines 3–5
    references how to use 237

F

Fabricated piping 171
Fan coil units 182
Fan groupings 196
Fans 195
Fans and ventilator takeoff 196
    units 196
Fiberglass board ductwork 199
    insulation 207
Field fabricated stand 215, 218
Fin tube enclosure 185
    radiation 182
Fire classification 174
    symbols 175
Fire dampers 174
    detectors 173
    extinguishers, portable 174
    protection 163–176
    protection sample estimate
        unit price 303
    protection systems
        sample estimate 256
    standpipes 163
Fire suppression carbon dioxide 172
    chemical 172
    foam 172
    gas 172
    halon 172
Fire tube boilers 177
Firecycle sprinkler systems 166
Firehose cabinet 164
Fittings 116
    brass 131
    cast iron 117
    cast iron flanged 118
    cast iron soil pipe 127
    covers premolded 108
    compression 133
    copper 131
    ductile iron 118
    flared 131
    grip type 125, 173
    malleable iron 118
    plastic 135
    solder type 131
    steel weld 120
    takeoff 118
    units 118
Fixture carriers 157, 160
Fixtures plumbing 151
Flanges 117
    cast iron 117
    copper 131
    insulation 208
    steel weld 120
Flared fittings 131
Flexible duct connector 199, 200
Floor drains 157
    register 201
    sinks 157
Flushing rims 158
Foam fire suppression 172
Foreword ix
Fuel oil heaters 182
Fuel types 177
Furnace controls 211

G

Galvanized steel ductwork 199
Gas fire suppression 172
Gate valves 140
Geometric formulas 356
Globe valves 140
Grates boiler 182
Gravity ventilators 196
Grease interceptors 159
Grilles 199
Grip type fittings 125, 173
Guides 215–218
    takeoff 218
    units 218

H

Halon concentrations 172
    fire suppression 172
Handicapped water coolers 155
Hangers oversized pipe 208
    spacing 147
    pipe 141
Heads non-freeze pendent 171
    sprinkler 171
    wax coated 171
Heat exchangers 181, 182
Heaters fuel oil 182
    unit 182
    water 182
Heating 177–198
Heating and cooling systems
    sample estimate 277
Heating boilers 177
Heating, ventilating, and
    air conditioning 177–198
Height modifications piping 115
Helicopter 182
High efficiency boilers 178
Historical cost index how to use 237
Hoods 196
Humidifiers 182, 204
HVAC controls 211
    sample estimate unit price 315
Hydrants 159
    wall 159
    yard 159
Hydronic cooling 185
    systems 177
    terminal unit takeoff
        procedures 185
    terminal units 182
    terminal units of measure 182

I

Indirect costs 45–50
Indirect radiation 182
Induction boxes 204
Inserts concrete 142
Installed control systems 211
Insulating block 207
    board 207
Insulation 207–209
    calcium silicate 207
    cellular 207
    cement 209
    closed cell polyethylene 000
    ductwork 207
    equivalent pipe footage 209
    fiberglass 207
    flange 208
    jackets 208
    phenolic foam 207
    pipe 207
    polyurethane foam 207
    protection saddles 208
    rigid urethane 207
    rock wool 207
    takeoff 209
    thermal 207
    units 207
Insurance 45
Integrated air handling systems 204
Intelligent building concept 211
Interceptors 157
    grease 159
    oil 159

J

Job progress report 77

K

Kitchen hoods 174, 176, 204

L

Labor 33–39
    cost record 75
Lavatories 153
Letter of transmittal sample form 34
Life cycle costs 86
Louvers 204
Lubricated plug valves 140
Lump sum quotations 185

M

Malleable iron fittings 118
Market analysis 65, 66

Material 30–31
   cost record 76
Measurement conversion
   tables 357, 358
Mechanical draft cooling
   tower 188
   plan reproductions 288–292
Meters condensate 182
Minimum plumbing
   fixture requirements 252
Mixing box 194, 199, 204
Modified Scotch Marine boilers 180
Motorized dampers 204
   valves 212
Multizone unit 201

N
Natural draft cooling tower 188
Non-freeze pendent heads 171

O
Oil interceptors 159
Operating overhead 46–49
Order of magnitude
   estimate definition 8
Outline specification 354–355
Overhead 40–43, 46–49
   costs 231–232
   final markup 346
Oversized pipe hangers 208
Overtime costs 80

P
Packaged boilers 177
   water chillers 185
Painting 218
Percentage complete analysis 78
Perforated channel 215
Phenolic foam insulation 207
Pipe bundling 114
   cast iron 126
   copper 129
   fittings 116
   flanges 117
   hanger and support takeoff 147
   hanger and support units 143
   hanger spacing 147
   hangers 141
   insulation 207
   plastic 134
   preinsulated 137
   schedules 113
   steel 112
   support 141
   weights 113
Piping 111–147
   definitions 111
   fabricated 171
   schedule sample form 23
   symbols 368–369
   takeoff 115
   takeoff example 124
   units 15
Plastic ductwork 199

fittings 135
pipe 134
tubing 134
Plug valves 140
   valves lubricated 140
Plumbing appliances 151
   fixture requirements minimum 252
   fixture symbols 367
   fixtures 151
   sample estimate unit price 293
   systems 151
   systems sample estimate 252
Pneumatic control 211
Polyurethane foam insulation 207
Pork chop sections 181
Portable fire extinguishers 174
   takeoff 175
   units 174
Pre-bid scheduling 59–63
Preaction sprinkler systems 167
Precedence diagram 61
Preinsulated pipe 137
Premolded fitting covers 208
Pricing procedures 27–30
   alternate method 317
Productivity 79
Profit 48
   final markup 349
   to-volume-ratio 67–69
Progress report 77
Project overhead 40–43
   schedule 63, 346
Pulse type boilers 178
Pump piping detail 119
   takeoff 150
   units 149
Pumps 149–150
   contrifical 150
   sump 150

Q
Quantity takeoff guidelines 26
   procedure 21–26

R
Radiation baseboard 182
   direct 182
   fin tube 182
   indirect 182
Radiators 182
Recap sheet 346
Reciprocating water chillers 185
Register elbow 201
Registers 199
Remote water chillers 151
Repair and remolding costs
   how to use 235
Resource analysis 65
Retainage 82
Return air duct 201
   grille 201
Rigid inserts 209
   urethane insulation 207
Risk determination 66, 67
Rock wool insulation 207

Roof curb 194
   mounting frame 194
   top unit 201
Rounding off guidelines 22, 223

S
Sales tax 44
Sectional boilers 177
Self-acting shutters 196
Separators 157
Sheet metal calculator 202
Shower stall 156
Single zone unit 201
Sinks 153
   floor 157
Smoke detectors 173
   pipe 201
Solder quantities 133
   type fittings 131
Sound lining 199
Specification outline 354–355
Spiral wound ductwork 199
Split crews 204
Spot weld pins 209
Spray pond 188
Sprinkler heads 171
Sprinkler system classifications 170
   deluge 169
   dry pipe 166
   firecycle 166
   preaction 167
   takeoff 171
   units 171
   wet pipe 165
Square foot and cubic foot costs
   how to use 232
   estimate definition 10
   estimating example 245
Staging area 191
Stainless steel ductwork 199
Standpipe classifications 163
   takeoff 165
   types 163–164
   units 165
Steel boilers 177
   pipe 112
   tables 216
   weld fittings 120
   weld flanges 120
Stokers 182
Storm drains 157
Structural shapes 215, 216
Subcontractor bids 346
Subcontractors 40
Sump pumps 150
Supervised control systems 211
Supply and return faces 203
Supply plenum 201
Supports 215–218
   pipe 141
   takeoff 218
   units 218
Swimming pool
   drains 157
Systems estimate definition 10
   estimating example 250

Appendix F

T

Table of contents .... v–vii
Taxes 45
Telephone quotation sample form 29
Thermal insulation 207
Thermostats 211
Timesheet daily 73
    weekly 74
Trap primers 158
Tubing copper 129
    plastic 134
Turning veins 200
Types of fuel 177

U

Under floor distribution 193
Unequal leg angles 217
Unit heaters 182
Unit price costs how to use 223, 225
    estimate definition 11
    estimate summary example 346
    estimating example 287
Units definition of 5
Urinals 152

V

Valve enclosures 185
    water regulating 188
Valves 138
    alarm 171
    angle 140
    backwater 157
    ball 140
    butterfly 140
    check 140
    gate 140
    globe 140
    lubricated plug 140
    motorized 212
    plug 140
    takeoff 141
    units 141
Vapor barrier 207, 208
Vending machines 151
Ventilating 177–198
Ventilators 195
    gravity 196

W

Wall hydrants 159
Wash fountains 153
Water tube boilers 177
Water chillers absorption type 185
    centrifugal 185
    packaged 185
    reciprocating 185
    remote 151
Water closets 152
Water cooled chillers 185
Water coolers 155
    handicapped 155
Water heaters 154, 182
Water regulating valve 188
Wax coated heads 171
Weight per lineal foot ductwork 203
Weight per square foot ductwork 202
Welding output 123
Wetpipe sprinkler system 165
Wire mesh 209
Wiring electric 212, 213

Y

Yard hydrants 159